Optical Wireless Communication Theory and Technology

Series Editor

Xizheng Ke, School of Automation and Information Engineering, Xi'an University of Technology, Xi'an, Shaanxi, China

The book series Optical Wireless Communication Theory and Technology aims to introduce the key technologies and applications adopted in optical wireless communication to researchers of communication engineering, optical engineering and other related majors. The individual book volumes in the series are thematic. The goal of each volume is to give readers a comprehensive overview of how the theory and technology in a certain optical wireless communication area can be known. As a collection, the series provides valuable resources to a wide audience in academia, the communication engineering research community and anyone else who are looking to expand their knowledge of optical communication.

Xizheng Ke

Partially Coherent Optical Transmission Theory in Optical Wireless Communication

Science Press
Beijing

Springer

Xizheng Ke
School of Automation and Information
Engineering
Xi'an University of Technology
Xi'an, Shaanxi, China

ISSN 2731-5967 ISSN 2731-5975 (electronic)
Optical Wireless Communication Theory and Technology
ISBN 978-981-97-7157-8 ISBN 978-981-97-7158-5 (eBook)
https://doi.org/10.1007/978-981-97-7158-5

Jointly published with Science Press
The print edition is not for sale in China mainland. Customers from China mainland please order the print book from: Science Press.
ISBN of the Co-Publisher's edition: 978-7-03-050358-9

Translation from the Chinese Simplified language edition: "无线光通信中的部分相干光传输理论" by Xizheng Ke, © Science Press 2016. Published by Science Press. All Rights Reserved.

This Springer imprint is published by the registered company Springer Nature Singapore Pte Ltd.
The registered company address is: 152 Beach Road, #21-01/04 Gateway East, Singapore 189721, Singapore

Contents

Chapter 1
Introduction

1.1 Wireless Optical Communication and Atmospheric Turbulence

Optical communication can be traced back to the Western Zhou Dynasty more than 700 years ago when the beacon used to transmit military information was a primitive application of wireless optical communication technology. Bell invented the "optical telephone" in the 1880s, which marked the beginning of the modern wireless optical communication era [1]. Owing to the influence of the atmosphere and environmental factors, coupled with the poor coherence of initial light sources, the modulation/demodulation of signals was very difficult, and the further development of wireless optical communication encountered several obstacles. During World War II, wireless optical communication technology was further developed and evolved into infrared telephony [2]. The advent of the ruby laser in 1960 provided a new opportunity for the development of wireless optical communication technology [3]. Given that the output of a laser exhibits high monochromaticity, coherence, directionality, and power, the transmission distance is long, and modulation/demodulation is possible, which is significantly advantageous compared to the early incoherent light sources such as sunlight and incandescent lamps.

Laser communication can be divided into two categories, wired laser communication, and wireless laser communication, according to the utilized transmission channels. The former refers to the use of an optical fiber and other wired channels as the carrier of information, as described in the research article "Dielectric Fiber Surface Waveguide for Optical Frequency" published by C. K. Kun et al. proposed the basic principle of using quartz glass to fabricate optical fibers for application to communication. An important communication technology was developed, which has had a far-reaching impact on information access. Wireless laser communication, namely Free Space Optical communication (FSO), is a technology that uses laser beams as carriers to transmit information such as data, voice, and images [4]. Given that an artificial transmission medium is not required, a surge in global research has

X. Ke, *Partially Coherent Optical Transmission Theory in Optical Wireless Communication*, Optical Wireless Communication Theory and Technology,
https://doi.org/10.1007/978-981-97-7158-5_1

been triggered, and western developed countries are investing considerable resources in research on wireless laser communication. However, given that atmospheric turbulence induces coherence transformation, intensity attenuation, beam divergence, and spot jitter during long-distance laser transmission, coupled with the limitations of optoelectronic devices, research on wireless laser communication was almost stagnant for a long period. However, in recent years, with the continuous development of optoelectronic technology and the need for access to new radio spectra, research on wireless optical communication technology is quietly being revived.

The technology wherein laser beams are transmitted in the near-surface atmosphere for communication is called atmospheric laser communication. Atmospheric laser communication technology has received extensive attention owing to its advantages which include a large channel capacity, good confidentiality, and strong anti-interference ability. The atmospheric laser communication system is mainly composed of a source/sink, channel, modulator/demodulator, light source (laser), and optical transceiver antenna. Atmospheric turbulence is a key factor that restricts the practical application of wireless laser communication systems and affects communication along atmospheric channels. Therefore, an in-depth study of this phenomenon is important to the development of wireless laser communication.

Laminar and turbulent flows exist in fluid movement. Laminar flow is a regular movement, whereas turbulent flow is an irregular movement. Atmospheric turbulence is an irregular random motion in the atmosphere that can be categorized as turbulent flow. Physical properties such as pressure, velocity, and temperature, fluctuate randomly at each point in a turbulent flow. Atmospheric turbulence often occurs in the boundary layer at the bottom of the atmosphere, inside the body of a convective cloud, and in the westerly jet stream in the upper troposphere. The occurrence of atmospheric turbulence requires certain dynamics (such as wind shear in the air layer) and thermodynamic conditions (the air layer must be unstable), among which the most favorable condition is that the air temperature in the upper layer must be less than that of the lower layer. Atmospheric turbulent motion is formed by the continuous distribution and superposition of vortices at various scales. Large vortex scales can reach hundreds of meters, with a minimum scale of approximately 1 mm. Turbulent flow is the outer scale (denoted as L_0), which is equivalent to the height of the airflow from the ground, usually in the range of several meters to hundreds of meters. When the speed of the turbulent flow increases, the large turbulent vortices exhibit instability and are split into smaller vortices, thereby transferring energy to them in the process. As a result, the Reynolds number decreases accordingly. The energy of the initial large turbulent vortex originates from solar radiation and various meteorological factors. Although there is such a cascade relationship, due to the viscosity of the fluid and other factors, the splitting of the turbulent vortex will not continue forever. When the energy of the turbulent vortex is equal to its kinetic energy, splitting ceases to occur. The smallest scale of the turbulent vortex is the inner scale, which is generally on the order of several millimeters.

Owing to turbulent motion in the atmosphere, all the properties related to flowing gas, such as temperature, refractive index, and aerosol particle distribution, are subject to turbulent mixing. For wireless optical communication, researchers are particularly concerned about the refractive index in turbulent fields. Since the atmospheric refractive index is a function of temperature, the properties of the refractive index field are closely related to those of the temperature field. The turbulence caused by temperature fluctuation is called optical turbulence, and the parameter used to quantify the intensity of this refractive index fluctuation is called the atmospheric refractive index structure constant, which is defined as follows [5] in the case of uniform isotropic turbulence:

$$C_n^2 =< [n(\vec{x}) - n(\vec{x}+\vec{r})]^2 > r^{-2/3}, l_0 \ll r \ll L_0 \tag{1.1}$$

There, n is the atmospheric refractive index, $\vec{x}$ and $\vec{r}$ is the position vector, r is the mode of the vector $\vec{r}$, l_0 and L_0 is the inner scale and outer scale of turbulence respectively. C_n^2 is not a constant in the real sense, but related to the atmospheric conditions and the altitude.

The fundamental cause of atmospheric turbulence is that the atmosphere is affected by solar radiation and various meteorological factors (temperature, humidity, and air pressure changes), resulting in random changes in atmospheric wind speed. The cumulative effects of these changes cause random fluctuations in the atmospheric refractive index. When the beam propagates through atmospheric turbulence, it experiences different degrees of light intensity peak reduction, beam expansion, and angular expansion. The extent and effect of atmospheric turbulence on the beam depend on the relative magnitude between the beam diameter and the turbulence scale. When the turbulence scale is much larger than the beam diameter, the main effect of turbulence causes random deflection of the beam. In addition, the transmission direction of the beam and the projection on the receiving surface are random (i.e., the common beam drift phenomenon). When the turbulence scale is the same as that of the beam diameter, random deflection of the beam cross-section occurs, resulting in fluctuations in the angle of arrival and random jitter on the receiving surface. Each of them will diffract a small part of the beam; thus, the intensity and phase of the beam will be randomly distributed in space and time. In addition, the coherence will be degraded, the beam area will be enlarged, the light intensity will be attenuated, and the beam size will be expanded on the receiving surface. If these turbulent effects occur simultaneously, the light transmission efficiency is greatly reduced, the light spot at the receiving end is distorted, and the beam quality is degraded. For example, for a powerful laser weapon system, owing to the influence of atmospheric turbulence, the expansion of the target spot will reduce the tracking accuracy. Moreover, when the laser power is sufficiently high, the phenomenon of nonlinear thermal halo will occur [6–9]. When the light intensity flickering effect occurs, the target cannot be reliably identified owing to the flickering light intensity, resulting in the inoperability of a lidar system owing to the high probability of false alarms. The high coherence of lasers is disadvantageous in the heat treatment of laser material surfaces, atmospheric laser optical communication, laser coating, and target surface irradiation in inertial confinement nuclear fusion. The higher the coherence of a laser's output, the greater the influence of atmospheric disturbance. Thus, speckle phenomenon readily occurs resulting in an uneven light intensity distribution.

Conversely, partially coherent light with a low coherence can effectively suppress these undesired effects. Strictly coherent light or incoherent light cannot be practically achieved; they are simply ideal mathematical models. In the case of a laser, the sharpest spectral line also has a certain spectral line width, which cannot be a point light source. Therefore, the output is always partially coherent. However, when a laser beam propagates through atmospheric turbulence, if the transmission distance is long and the turbulent fluctuation is intense, its coherence will be adversely affected and the beam will be transformed into partially coherent light.

In summary, owing to the random fluctuation of the atmospheric refractive index when a laser's output propagates through the atmosphere, the beam experiences turbulence effects such as expansion, drift, fluctuation of the arrival angle, and light intensity scintillation, which reduces the beam quality. Therefore, it is necessary to study the transmission of a laser beam in atmospheric turbulence to develop techniques and solutions that can suppress the effect of this phenomenon.

1.2 Research Progress on Atmospheric Turbulence Effects

The influence of atmospheric turbulence on optical transmission cannot be ignored, and the various turbulent effects caused by fluctuations in the atmospheric refractive index must be considered in wireless optical communication. Research on the transport of light in atmospheric turbulence can be divided into the following three stages.

(1) 1940s: Initial theoretical methods were established based on experiments, such as geometric optics approximation [10–14], Rytov approximation [15–17], and several other techniques. In the late 1950s, the former Soviet Union scholar Tatarskii introduced the turbulence theory developed by Kolmogorov and Obukhov, and used the Rytov approximation to express the electric field of a propagating beam in exponential form, expanded it into a series, and finally simplified the solution amplitude by obtaining the exponential series solution. The approximation method has also become an indispensable classical theory for studying the characteristics of light transmission under weak turbulence.

(2) In the 1960s and the 1970s, researchers explored and used new methods to conduct more experimental work on different wave sources, such as acoustic waves, radio waves, and lasers, and discovered many new experimental phenomena, such as the saturation effect of light wave fluctuations and inflation, fall phenomenon, etc. To better explain these phenomena, strong fluctuation theories such as the Markov approximation [18–22], heuristic theory [23, 24], Feynman path integral method [25] and phase screen theory [26] have been proposed. The Markov approximation method is based on the Rytov approximation method. The optical transmission process is regarded as a Markov process, and the equations for each order moment were derived. This method

is appropriate for the transmission of light in strong turbulence, but is not suitable for describing propagation through moderate turbulence. These phenomena cannot be explained using classical perturbation theory, which has motivated the development of new theories.

(3) After the 1980s, research in the field began to focus on solving higher-order moments. Wilbur et al. used staircase approximation to derive the second moment of wave propagation in random media [22]. Filinov calculated the flicker exponent and second and fourth moments of a field using the Monte Carlo method [23]. Wu discussed the solutions of the second-order moment using an iterative method of integral equations under a more general scattering mechanism [24].

The second-order moment cannot be used to accurately determine the transmission characteristics of light waves. However, higher-order moments can provide information about the tail of the probability density function, and even- and odd-order higher-order moments facilitate the determination of the distribution width and the deviation based on symmetry information, respectively. Therefore, to study the transmission of light waves in the atmosphere with greater accuracy, it is necessary to consider the information at the end of the probability density function, which requires the study of higher-order moments. Initially, many experts discussed the general form of the solution of the higher-order moment equation and developed different methods for its solution. Lee et al. obtained Gaussian solutions for higher-order symmetric moments [25]. Shigeo et al. studied the transmission characteristics of light beams from both theoretical and experimental perspectives, and a tenth-order moment was eventually obtained [26]. These works focused on the study of the statistical properties of the intensity of light beams propagating in a turbulent atmosphere, but analytical expressions were not established for the corresponding higher-order moments. Wu Jian et al. proposed the concept of moment space, solved the forward multiple scattering order moment equation, and derived expressions for the first-order moment, second-order moment, and fourth-order symmetric moment. The results were then applied to the investigation of ionospheric scintillation. In this study, the expression of the flicker index was obtained, and the correlation function of the amplitude flicker was discussed [27]. On this basis, Li Guizhen et al. deduced the general expression of the second-order symmetric moment using perturbation theory and obtained an analytical solution of the two-frequency and two-point mutual interference functions represented by the second-order moment [28]. Li et al. [21] analyzed the high-order intensity moments of a beam as it propagated through atmospheric turbulence, and derived expressions for the first-order, second-order, third-order, and fourth-order intensity moments. Using a Gaussian beam as an example, the propagation law of the beam flatness parameter, K, in atmospheric turbulence was examined [27].

Most of the research on turbulence effects is on beam expansion and the scintillation characteristics of the intensity of light [27–29]. In 1970, Whitman [30] et al. derived the expansion formula for laser transmission in random media based on the scalar wave equation and performed a detailed analysis using the Kolmogorov spectrum. The results showed that the longer the transmission distance, the larger the expansion radius. In 1972, Poitier et al. [31] applied the Born approximation method to discuss the broadening effect of focused beams in a turbulent flow. In 1975, Fante [32] derived a long-term beam expansion formula using the modified von Karman spectrum and obtained a short-term beam expansion formula using the mutual coherence function. In 1999, Fan [33] used the Kolmogorov spectrum and Mellin transform to obtain the light intensity distribution and beam expansion of a Gaussian collimated beam, and obtained the average light intensity and long-term beam radius using the von Karman spectrum based on the near-surface atmosphere measured in the Kunming area. The height distributions of the turbulent intensity and beam-spread conditions at various target heights were calculated. In 1999, Andrews et al. [34–38] studied the influence of the inner and outer scales of turbulence on the long-term beam radius and used numerical simulation to perform long-term beam calibration, and Wei et al. [39] determined the spread radius of the laser beam in the oblique transmission path In 2002, Guo et al. [40, 41] used the modified Rytov method to extend the classical Rytov method to moderate and strong turbulent regions, based on the turbulent atmosphere proposed by the International Telecommunications Union-Radio Communication Sector (ITU-R), which generalizes horizontal transmission to oblique transmission. The variation law of the scintillation index of atmospheric turbulence zero inner and outer scale plane waves and spherical waves were then obtained using the modified Rytov variance. In 2004, Wu [42] and Wei [43] examined the scintillation index characteristics of oblique Gaussian beams. In 2007, Zhang et al. [44] investigated the relationship between the short-term beam expansion factor and the Gaussian beam equivalent radius as a function of the turbulent outer scale, transmission distance, initial beam diameter, and light wavelength, using the modified von Karman turbulence spectrum. Owing to the influence of atmospheric turbulence, a laser spot was "broken" into several fragments according to the transmission characteristics of the laser in the near-surface layer. Zhang et al. [45] calculated the propagation characteristics of the emitted beam with different shapes and beam qualities under different turbulence intensities in order to study the influence of spot debris on beam propagation, The characteristics of fragmentation spots were statistically analyzed, and the results showed that as the beam quality deteriorated and the turbulence intensity increased, the beam spot expanded, the number of fragmentation spots increased, and the occupancy rate increased. But the radius of the speckles changed little, whereas the radius of the fragmentation spot, which can be approximated as the radius of the beam containing 63.2% of the total energy for vacuum diffraction, does not change significantly. Andrews et al. [17] derived flicker expressions for the light intensity of plane waves, spherical waves, Gaussian beams, etc., using the Kolmogorov spectrum. In 2012, Zhang et al. [46] used a partially coherent flat-top beam as an example and derived an analytical expression for the root-mean-square beamwidth for oblique propagation in atmospheric turbulence. In 2013, Duan et al. [47] used Hermitian-Gaussian beams to study beam expansion in sloping-propagating atmospheric turbulence.

It has been observed that the initial beam is generally completely coherent or completely incoherent, and partially coherent states have not been extensively investigated. Provided that a saturable gain medium is present and no mode selection measures are added to the laser, many transverse modes oscillate simultaneously in the laser, and the optical fields of different modes are incoherent. Therefore, the multi-mode optical resonator cavity and the output beam are always partially coherent. Owing to the effect of atmospheric turbulence, fully coherent light is limited in most practical applications, whereas a Gaussian-Schell model (GSM) beam, which is a typical partially coherent light, has a relatively uniform light intensity, and can reduce certain turbulence-induced features such as speckle. Therefore, when addressing the problem of partially coherent light, a partially coherent Gauss-Schell beam is usually used, or incoherent Hermitian-Gaussian beams or Laguerre-Gaussian beams are superimposed to form partially coherent beams. Research on the transmission characteristics of partially coherent light in atmospheric turbulence has attracted the attention of researchers worldwide.

1.3 Atmospheric Turbulence and Its Effects

1.3.1 The Formation of Turbulent Flow

The atmosphere is an important random and non-uniform medium that is critical to our survival. Random changes in atmospheric temperature and velocity fields are caused by processes such as wind shear owing to air drag on the Earth's surface, differences in solar radiation heating at different locations, and thermal convection caused by surface thermal radiation. Random changes in atmospheric temperature, in turn, lead to changes in wind speed, resulting in random changes in atmospheric density and pressure, which lead to atmospheric turbulence. The existence of atmospheric turbulence significantly enhances the exchange of momentum, heat, and water vapor in the atmosphere, which is far greater than the exchange associated with molecular motion. Another important phenomenon associated with turbulence is random fluctuations in the atmospheric refractive index. which adversely affects the propagation of optical signals. Its fluctuation characteristics are mainly related to atmospheric temperature, humidity, and wind shear instability [48].

The inhomogeneity of the atmospheric refractive index is called the turbulent "vortex." Under normal circumstances, vortex elements generated by atmospheric turbulence continue to be generated and disappear with the movement of the wind. Based on these changes, there are usually many airflow vortices with different temperatures, pressures, flow velocities, etc., in a turbulent state. These vortices continue to move and change, and the large vortices continue to shrink with the flow until they disappear [49]. They are generated and subsequently reduce in size until they disappear, and this process is repeated continuously. In addition, their motions are superimposed on each other to form random turbulent motion, which is referred to as atmospheric turbulence.

1.3.2 Reynolds Number

In the field of atmospheric optics, turbulence is mainly caused by random changes in the refractive index owing to random changes in the local temperature and pressure in the atmosphere. This phenomenon typically occurs at the junction of airflow layers moving at different speeds. Reynolds conducted the first experiment on turbulence in 1883. Water was poured into a round glass tube at one end, which was kept still. To observe the flow state of the water in the glass tube, a dyed liquid was used. The experimental results showed that if the flow rate of the water in the tube was sufficiently low, the colored part of the liquid maintained a straight line from the injection point to the outlet of the glass tube, and it could be observed to flow downstream, without dispersion. However, when the flow rate of the water in the glass tube increased and exceeded a threshold value, the colored liquid separated quickly and mixed with the surrounding water without the dye. The color of the water then faded, and there was no clear distinction between the colored and uncolored streams. The state of the colored water flow after mixing with the surrounding water was very irregular, its movement became chaotic and irregular, and randomness was enhanced. This fluid movement is called "turbulent flow." For a thin flow prior to mixing, the streamline formed by the dyed liquid was smooth and straight, which is called "laminar flow." Reynolds predicted that the flow of a liquid would always change from laminar to turbulent flow at the same Reynolds number. Related experiments also revealed that random turbulent flow can be produced by regular laminar flow as the flow velocity increases [50].

In related theories, atmospheric turbulent motion is regarded as the superposition of the continuous distribution of vortices of various scales, and the Reynolds number is used to describe the dynamic properties of turbulent flow, which is expressed as the ratio of kinetic energy to dissipation energy [51], as follows:

$$\mathrm{Re} = \frac{\rho \upsilon L}{\mu} \tag{1.2}$$

Experiment results have shown that there is a critical Reynolds number when $\mathrm{Re} < \mathrm{Re}_c$, for which the atmospheric flow is laminar; when $\mathrm{Re} > \mathrm{Re}_c$, the atmospheric flow is turbulent, as shown in Fig. 1.1 [52].

The Reynolds number for turbulent flow can also be explained by the interrelationship between inertial and frictional forces [53]. When a friction force exceeds the inertial force, Re is reduced, the effect of the internal friction force increases, and possible disturbances are converted into heat energy. Thus, the fluid is relatively stable and laminar flow occurs. When Re is large, the turbulent flow is more affected by the inertial force, and the energy of the turbulent flow is continuously transferred to the disturbance motion, thereby forming atmospheric turbulence. Turbulent motion has no fixed shape and is disordered. When turbulent flow moves, its energy is gradually transferred from the large-scale turbulence to the small-scale turbulent vortex, and energy dissipation does not occur in the process.

Fig. 1.1 Schematic diagram of laminar-turbulent flow [52]

$$\frac{dE}{dt} = W_k - \varepsilon \tag{1.3}$$

$$W_k \sim \frac{V^2}{2\tau} \sim \frac{V^3}{l} \tag{1.4}$$

$$\varepsilon \sim \frac{v}{2}\left(\frac{\partial V_i}{\partial x_i} + \frac{\partial V_j}{\partial x_j}\right)^2 \sim v\left(\frac{V}{l}\right)^2 \tag{1.5}$$

Equation (1.5) represents the balance equation of turbulent energy. When $W_k < \varepsilon$, $\frac{dE}{dt} < 0$ the fluctuation decreases and tends to disappear, atmospheric motion will not form turbulence; When $W_k = \varepsilon$, the energy E is constant, the atmospheric motion forms a steady state turbulence; When $W_k > \varepsilon$, $\frac{dE}{dt} > 0$, when the disturbance energy increases, the motion will be unstable.

For turbulence in the boundary layer, the Reynolds number is very large, usually in the range of 10^6–10^7. When $\mathrm{Re} \gg \mathrm{Re}_c$, the motion of the fluid was completely random and was no longer affected by the initial conditions. At the same time, $W_k \gg \varepsilon$, almost all kinetic energy is transferred from large-scale turbulence to small-scale turbulence.

1.3.3 Inner and Outer Scales of Turbulence

When the fluid velocity is greater than a threshold value, laminar motion may be transformed into turbulent flow. At the beginning of this transformation, a giant vortex with the characteristic scale, L_0, of the fluid first appears, The size of L_0 depends on the fluid geometry. For atmospheric turbulence, L_0 is equivalent to the height of the airflow from the ground, usually tens to hundreds of meters, and this parameter is the outer scale of the turbulence. Vortex scales larger than L_0 are generally anisotropic, whereas those smaller than L_0 are generally isotropic.

A vortex with an outer scale L_0 has large fluctuation energy, which constitutes the initial turbulent energy. To study the flow characteristics of a vortex, the inner

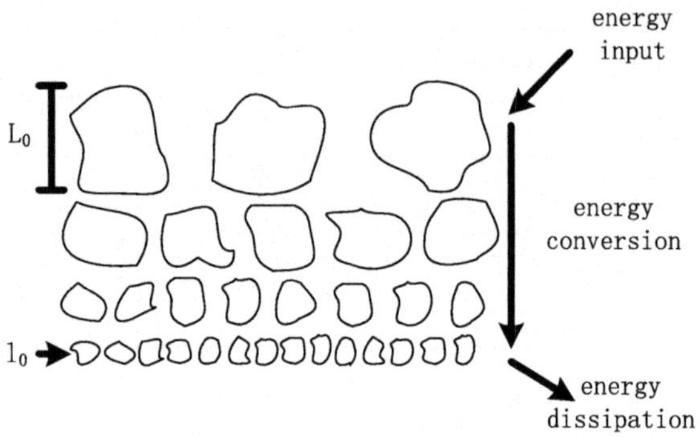

Fig. 1.2 Schematic diagram of turbulent energy conversion [55]

Reynolds number is introduced, which is the ratio of the energy conversion rate and the energy dissipation rate, which is expressed as follows [54]:

$$R_{e1} = \frac{lv_1}{\upsilon} \tag{1.6}$$

where, l is the characteristic scale of vortex flow, v_1 is the fluctuation speed equivalent to the characteristic scale.

The inner Reynolds number of this giant vortex increases with the Reynolds number of the fluid flow. When R_{el} is greater than a critical value, the vortex exhibits instability, and splits into smaller vortices. Energy is transferred to these small vortices in the process. This cascade process continues, the characteristic scale of the vortex becomes increasingly smaller, and the corresponding inner Reynolds number also becomes smaller. Unless the inner Reynolds number decreases to a certain value, the viscosity of the fluid is the primary factor that affects motion, and kinetic energy is converted into heat energy, and cannot form smaller vortices. Thus, the characteristic scale of the smallest vortex is turbulent. The internal scale is generally on the order of millimeters. Figure 1.2 shows a schematic diagram of turbulent energy conversion.

When the turbulence is fully developed, the vortex gradually disappears, and new vortices are generated simultaneously. Therefore, there is a continuous spectrum of vortices with scales in the range of $l_0 \leq l \leq L_0$ in the flow. These motions are superimposed on each other to form a complex turbulent structure.

1.3.4 Characteristics of Turbulent Flow

Atmospheric turbulence is the result of chaotic motion with the following properties:

(1) The turbulent motion exhibits irregular random characteristics

Atmospheric turbulence is a type of motion that is produced by external forces. As the external force increases, the motion state of the fluid changes from laminar to turbulent flow, and the motion gradually loses its stability and becomes irregular and disorderly nonlinear motion.

(2) The turbulence parameters exhibit statistical regularity characteristics

Although turbulent motion is irregular, its motion parameters on adjacent spatial points have certain correlation characteristics. Therefore, statistical laws, such as the statistical averaging method, can be used to estimate and predict turbulence.

(3) Sensitive dependence of turbulent flow on initial conditions

Professor Lorenz first deduced that the atmosphere is sensitive to initial conditions [56]. Subsequently, Berry used accurate numerical calculation results as proof of Professor Lorenz's inference and concluded that atmospheric turbulence is also sensitive to initial conditions.

1.3.5 Atmospheric Structure Constants

The atmospheric refractive index structure constant C_n^2 is one of the fundamental parameters in atmospheric optics, and researchers have shown great interest in the properties of this parameter for decades by summarizing a variety of measurement methods and developing new theories to account for the results. In the fields of satellite-ground communication, laser guidance, remote sensing mapping, etc., owing to the inhomogeneous atmospheric structure constant of oblique transmission paths, its magnitude changes with a change in height, whereas the atmospheric structure constant is generally associated with horizontal transmission. Typical values can be obtained; however, the underlying processes are more complicated compared to the case of horizontal transmission. The atmospheric structure constant must be derived to study the transmission characteristics of light beams under the influence of atmospheric turbulence.

Atmospheric turbulence can be divided into two categories. One type is the near-surface atmosphere, wherein the turbulent state is greatly affected by ground conditions, called "boundary layer" turbulence. The other type is associated with the atmosphere above the ground and is largely unaffected by the ground, called "free atmosphere" turbulence. The boundary between the two layers is affected by many factors, such as the geographical environment, macro-meteorological conditions, and ground conditions. Typically, the thickness of the "boundary layer" is on the order of tens to hundreds of meters.

Boundary layer turbulence is mainly affected by surface temperature, humidity, and pressure. Using dimensional and similarity theory, Wyngaard et al. estimated the atmospheric structure constants at different heights based on the atmospheric structure constants at a reference height and the height of the convective boundary layer. It was demonstrated that the atmospheric structure constants in the boundary layer are power-dependent with height. Since the Wyngaard model utilizes empirical equations based on experimental results in its theoretical derivation, it is a semi-empirical

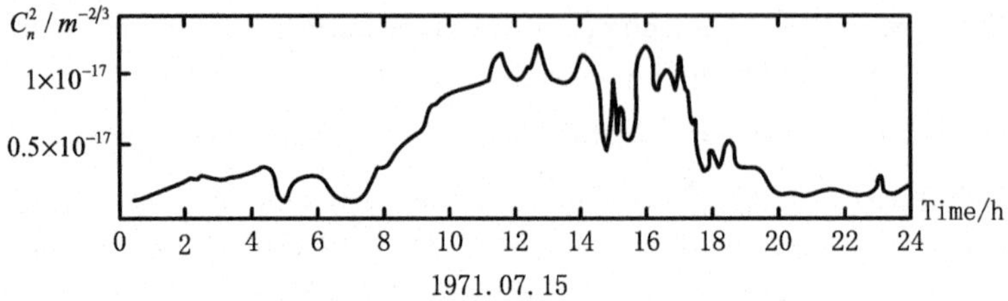

Fig. 1.3 24 h characteristics for a sunny day [57]

theory. However, the result based on the − 3/4 power relationship can accurately describe the variation law of the atmospheric structure constant for distances of 100 m above the ground. Subsequently, Neff et al. used sodar to study the effect of altitude, and also verified the − 3/4 power relationship of Wyngaard's theory. Subsequently, Kukharets and Tsvang proposed an exponential model to describe the structure constants of the atmospheric refractive index.

The atmospheric refractive index structure constant C_n^2, in addition to varying with altitude, fluctuates significantly within a day [57]. Figure 1.3 is an example of a day and night 24 h characteristic at a distance of 2 m from the ground.

The experimental data in Fig. 1.3 were obtained from a leeward platform on the Rocky Mountains in Boulder, Colorado [57]. The values are relatively small at night and dawn, however, C_n^2 increases rapidly as the sun rises and decreases significantly after the sun sets. A typical value is in the range of $10^{-17}\text{m}^{-2/3} < C_n^2 < 10^{-12}\text{m}^{-2/3}$.. From 14:00 to 16:00 (Fig. 1.3), the steep decrease of C_n^2 was caused by clouds that temporarily blocked the sun. In addition, different land surfaces influence the size of atmospheric structures. In general, bare surfaces are more turbulent than vegetated surfaces; therefore, turbulence is often high in desert areas.

The structure constant of free atmospheric turbulence is the most important part of slope transport. To simplify the analysis of the problem, the negative power exponential model is typically used, or the B-spline interpolation method is employed to construct a model according to the value of the atmospheric structure constant at typical heights. Ochs measured the variation of atmospheric structure constants as a function of height h below 3.2 km; Hufnagel et al. derived an empirical formula C_n^2 that is applicable in the range of 3–24 km based on the measured data.

$$C_n^2 = [2.2 \times 10^{-53} h^{10} (W/27)^2 e^{-h/1000} + 10^{-16} e^{-h/1500}] \cdot e^{J(h,t)} \tag{1.7}$$

where J is a uniform Gaussian random variable with zero mean, and its correlation function is given by:

$$\langle J(h+h_1, t+\tau) \rangle = A\left(\frac{h_1}{100}\right) e^{-\tau/5} + A\left(\frac{h_1}{100}\right) e^{\tau/80} \tag{1.8}$$

where,

$$A\left(\frac{h_1}{100}\right) = \begin{cases} 1 - h/100 & h < 100 \\ 0 & h \geq 100 \end{cases} \tag{1.9}$$

The function W represents the root mean square velocity at altitudes between 5 and 20 km, and is expressed as

$$W = \left[\frac{1}{15}\int_5^{20} \upsilon^2(h)\mathrm{d}h\right]^{1/2} \tag{1.10}$$

$\upsilon(h)$ is the wind speed at this height. The typical data W proposed by Hufnagel [58] is that the mean value is 27 m/s and the standard deviation is 9 m/s. In 2001, the ITU-R published a model of atmospheric structure that varied with altitude. The development of this model facilitates the theoretical analysis of the transmission characteristics of light waves for oblique-path transmission [58]. Figure 1.4 shows the variation of the atmospheric structure constant with altitude, that was published by ITU-R in 2001. It is evident that the atmospheric structure constant gradually decreases with the increase in height, then decreases rapidly after a period of increase at 10 km. This is caused by the absorption of the atmosphere, the atmospheric structure constant at the surface, and wind speed at high altitudes. At 1–4 km, the atmospheric structure constant was not affected by wind speed and $C_n^2(0)$ at the surface. However, at altitudes greater than 10 km, it was mainly affected by wind speed.

$$C_n^2(h) = 8.148 \times 10^{-56} v_{RMS}^2 h^{10} e^{-h/1000} + 2.7 \times 10^{-16} e^{-h/1500} + C_0 e^{-h/100} \tag{1.11}$$

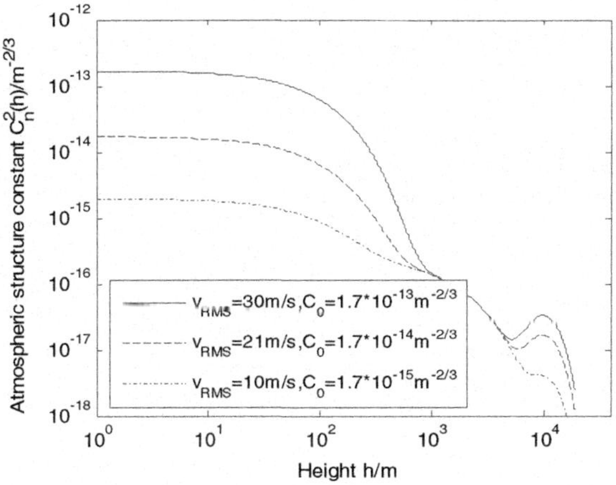

Fig. 1.4 Variation of atmospheric structure constant with height h

1.4 Atmospheric Turbulent Refractive Index Fluctuation Spectrum Model

When light waves travel in the atmosphere, they are affected by the refractive index of the air. Although the refractive index of the atmosphere is not significantly different from that of a vacuum, its cumulative effect in the atmosphere is unpredictable. Therefore, the description of the random fluctuation state of the refractive index atmospheric turbulent is the primary problem that must be solved.

(1) Input area

In this region, $k < 2\pi / L_0$, and energy is inputted to the turbulent flow owing to wind shear and temperature gradients. The shape of the spectrum in this region depends on how a particular turbulence occurs, and the characteristic physical quantities of random media are generally nonuniform and anisotropic. Given that the turbulence spectrum is related to the generation of atmospheric turbulence, there is currently no general equation that describes the turbulent characteristics of this region.

(2) Inertial zone

In this region, $k > 2\pi / L_0$, the Reynolds number is large, and the kinetic energy of the turbulent vortex exceeds the viscous dissipation. Thus, the motion process is unstable, and atmospheric turbulence occurs. The shape of the power spectrum is determined by the physical laws that restrict the breaking of large turbulent eddies into small eddies, and the turbulent flow in this region is isotropic.

In the inertial region, the traditional Gaussian and exponential spectra cannot fully describe beam propagation and scattering phenomena. Kolmogorov proposed the Kolmogorov spectral model for turbulence, as shown in Fig. 1.5.

$$\phi_n(\kappa) = 0.033 C_n^2 \kappa^{-11/3} \quad (\kappa_0 \ll \kappa \ll \kappa_m) \tag{1.12}$$

The disadvantage of this spectral model is that when the wave number is $\kappa \to 0$, the $\kappa^{-11/3} \to \infty$, $\phi_n(\kappa) \to \infty$ there will be unreasonable results in this formula.

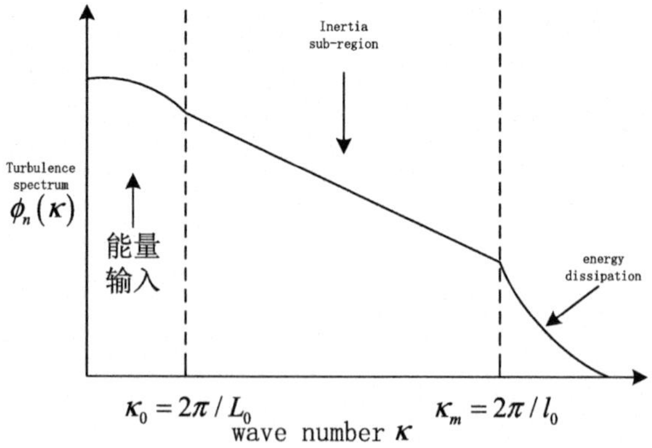

Fig. 1.5 Kolmogorov power spectrum of turbulence

(3) Dissipative areas

When κ increases to the critical value of $\kappa_m = 2\pi / l_0$, i.e., $\kappa > 2\pi / l_0$, the shape of the power spectrum $\phi_n(\kappa)$ changes. In this region, the Reynolds number is small and the viscous dissipation energy of the turbulent vortex is greater than the kinetic energy. Therefore, the spectrum is extremely small. The dissipative region is described by the Tatarskii spectral model [58].

$$\phi_n(\kappa) = 0.033C_n^2\kappa^{-11/3}\exp\left(-\kappa^2 / \kappa_m^2\right) \qquad (\kappa > \kappa_m) \tag{1.13}$$

The Tatarskii spectral model considers the internal scale of turbulent flow, and there is a Gaussian attenuation factor in the region with a large wave number. Therefore, unreasonable results for the Kolmogorov spectrum can be addressed. These two spectral models are still not widely used because of the presence of nonpositive poles at the origin. In reality, the atmosphere around Earth is limited. Therefore, when $\kappa \to 0$, $\phi_n(\kappa)$ is impossible to approach.

In 1948, the von Karmon spectrum was proposed to address this problem. The power spectrum of the dissipative region can be approximated using the von Karmon turbulence spectrum model [59] as follows:

$$\phi_n(\kappa) = 0.033C_n^2\left(\kappa^2 + \kappa_0^2\right)^{-11/6} \tag{1.14}$$

Subsequently, the internal and external scales of turbulence were introduced into the spectral model, and a modified von Karmon spectrum was proposed as follows:

$$\phi_n(\kappa) = 0.033C_n^2\kappa^{-11/3}\left(\kappa^2 + \kappa_0^2\right)^{-11/6}\exp\left(-\kappa^2 / \kappa_m^2\right) \tag{1.15}$$

The two spectral models based on the Tatarskii spectrum and von Karmon spectrum have a common limitation in that both do not consider the abrupt factors in the high wavenumber region that significantly affect light transmission. Subsequently, Andrews and Hill obtained a more accurate numerical model based on experiments [60].

$$\Phi_n(k) = 0.033C_n^2\kappa^{-11/3} \times \left\{\exp\left(-1.2\kappa^2 l_0^2\right) + 1.45\exp\left[-0.97(\ln \kappa l_0 - 0.452)^2\right]\right\} \tag{1.16}$$

The form of this model is relatively complex, and is not ideal for theoretical research. Therefore, Andrews et al. proposed a modified Hill spectral model using data interpolation and other methods [61].

$$\Phi_n(k) = 0.033C_n^2(h)(1 + a_1(\kappa/\kappa_l) - a_2(\kappa/\kappa_l)^{7/6})\frac{\exp(-\kappa^2/\kappa_l^2)}{(\kappa_0^2 + \kappa^2)^{11/6}} \tag{1.17}$$

where. $\kappa_l = 3.3/l_0$, $\kappa_0 = 2\pi/L_0$, $a_1 = 1.802$, $a_2 = 0.254$, It can be seen from the above formula that the spectral model can be simplified to von karmon spectral model after substituting $a_1 = a_2 = 0$ and $\kappa_l = \kappa_m$. When $\kappa_0 = l_0 = 0$, it can degenerate to Kolmogorov spectral model.

The Tatarskii spectral model was used in the early stages of the atmospheric turbulence research. However, because the form of the dissipation region is inconsistent with real-world situations, Hill proposed a spectral model that is suitable for application to turbulent dissipation regions. However, it has not been generally recognized in the field. Tatarskii and von Karmon's spectral models are usually used in theoretical research.

1.4.1 Beam Expansion

As shown in Fig. 1.6, let F_0 be the wavefront curvature radius at the emission plane be given by, where $F_0 = \infty$, $F_0 > 0$ and $F_0 < 0$ correspond to collimated, focused, and diverging beams, respectively. In experiments, the beam is typically collimated or focused (Fig. 1.7).

Fig. 1.6 Schematic diagram of different beams

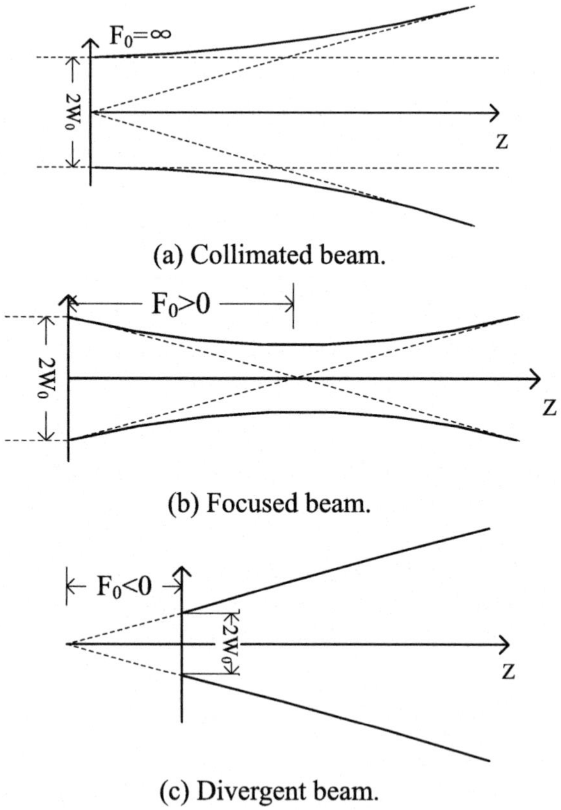

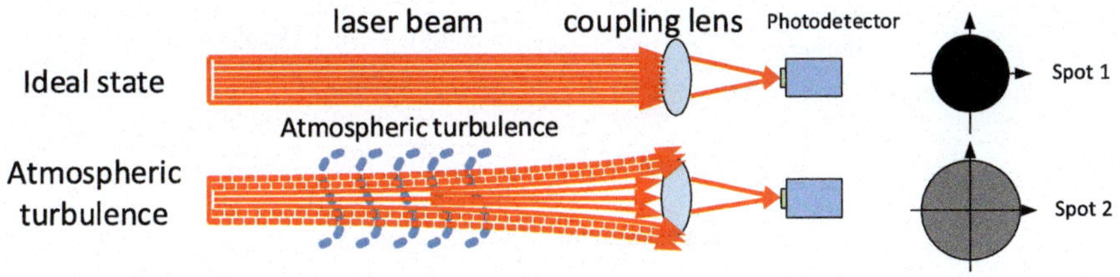

Fig. 1.7 Influence of beam expansion on atmospheric laser communication system

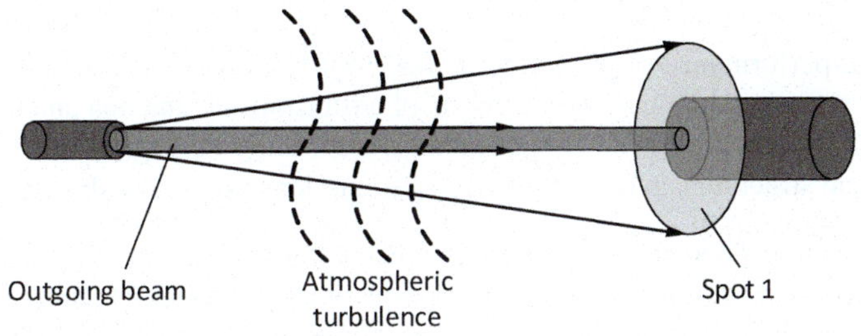

Fig. 1.8 Schematic diagram of beam expansion

In a wireless laser communication system, after the beam is transmitted through the atmospheric channel, the output is affected by light intensity flicker, beam expansion, arrival angle fluctuation, and spot drift. These turbulent factors adversely affect the beam quality, reduce the energy coupling efficiency of the photodetector's surface, increase the bit error rate of the communication system, and lead to deterioration of the performance of the laser communication system. When the laser beam is transmitted in an ideal state, it is focused on the surface of the photodetector through the coupling lens, then photoelectrically converted to obtain an electrical signal. Under ideal conditions, there is almost no loss of beam energy during transmission, and an ideal spot is obtained on the receiving surface, as denoted by spot 1 in the figure. For propagation through atmospheric turbulence under the influence of fluctuation of the atmospheric refractive index, the beam expands, and the energy of the beam cross-section decreases. Therefore, part of the beam cannot pass through the coupling lens to reach the photodetector, such as the portion of the beam represented by the dotted line. A light spot is detected on the receiving surface, for which the spot radius increases and the beam energy is reduced owing to atmospheric turbulence (Fig. 1.8).

1.4.2 Beam Drift

Owing to the influence of atmospheric turbulence, the light spot on the receiving plane of the wireless optical communication system expands and drifts, which seriously

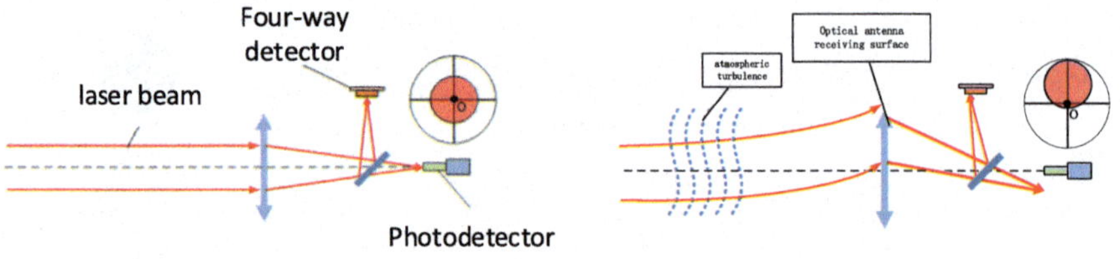

(a) Ignoring atmospheric turbulence (b) Atmospheric turbulence

Fig. 1.9 Influence of spot drift on wireless optical communication system

affects the performance of photoelectric detection. As shown in Fig. 1.9, the wireless laser communication system consists of a transmitting antenna and a receiving antenna. Transmitting and receiving antennas are usually Cassegrain telescopes, and their optical structure can be simplified as a single-lens system, as shown in Fig. 1.9.

(1) When atmospheric turbulence is ignored, the transmitting beam is transmitted to the receiving surface of the receiving optical antenna, and its main optical axis coincides with the viewing axis of the receiving antenna. The beam is focused on the receiving end of the communication detector through the receiving optical system. A beam-splitting prism was placed in the optical path inside the optical system, which divided the receiving optical path into two components: one beam was used for communication detection, and the other beam was used for spot position detection. A four-quadrant detector was used as the detection sensor. Using the position of the light spot on the receiving surface of the four-quadrant detector, the center position of the light spot on the detector can be obtained, and the center of the light spot coincides with the center of the detector.
(2) When considering atmospheric turbulence, the beam drifts owing to its influence during transmission, resulting in the random movement of the beam in the direction perpendicular to the optical axis, causing deviating relative to the receiving surface. Thus, the optical axis of the transmitted beam is no longer coincident with the boresight axis of the receiving antenna, and the beam propagates at an angle. This causes its focus to shift, and the shifted focus is no longer incident on the receiving surface of the communication detector. Therefore, the beam power received by the communication detector is significantly reduced, which adversely affects the signal detection of the communication system. The four-quadrant detection can detect the offset of the centroid of the light spot relative to its center. As the turbulence becomes more extensive, the transmitted beam may completely drift out of the receiving surface of the receiving antenna. In this case, the receiving antenna cannot detect the transmitted beams, which interrupts the communication link.

Beam spreading refers to a change in the radius or area of the spot on the receiving plane owing to turbulent eddies. When light waves propagate as small-scale vortices, diffraction effects occur, resulting in beam expansion. Conversely, when light waves propagate as large-scale vortices, refraction occurs, resulting in beam drift. As shown

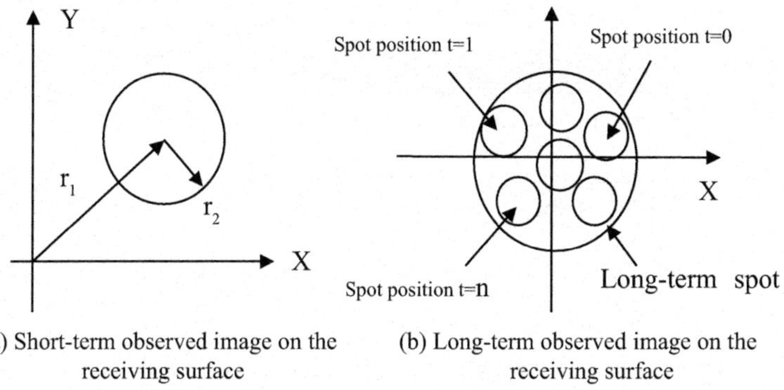

Fig. 1.10 Schematic diagram of the light spot observed on the receiving plane

in Fig. 1.10, for a short observation time at the receiving plane, the distance that a broad light spot with a radius of 1 deviates from the origin is the beam drift, and if the time is sufficiently long, it can be seen that a random walk of the light spot produces a large spot with a mean square radius; the former is called short-term beam expansion and the latter is called long-term beam expansion [62, 63].

Owing to the atmospheric transmission environment and the detection system device, the received light spot jitters and drifts. To effectively receive and detect optical signals, the receiving device must be able to track the drift of the light spot in real-time and automatically adjust the position of the receiving device to align to the incident light spot.

1.4.3 Light Intensity Flicker

As shown in Fig. 1.11, when the laser is transmitted through atmospheric turbulence and the beam diameter is much larger than the turbulent scale, the beam cross-section contains multiple turbulent eddies, and each vortex is a part of the beam that is scattered and diffracted independently, causing continuous spatial and temporal changes in the optical intensity at the detector plane. Specifically, the intensity of the light beam is suddenly large and small, and the light spot flickers. This change is called light intensity flicker (light intensity fluctuation). The atmospheric channel deflects the laser beam in a manner related to the time required to propagate through the turbulent gas. If the intensity of light is measured at a certain distance from the light source in a turbulent atmosphere, there are random fluctuations of the light intensity I around the mean value $\langle I \rangle$, with time, which is observed as light intensity flickering. The most direct impact of the light intensity flicker effect on the communication system is an increase of the bit error rate of the system. This is because atmospheric turbulence causes an increase in the range of light intensity variation, which causes the detected power to be too large. This causes saturation of the detector at certain

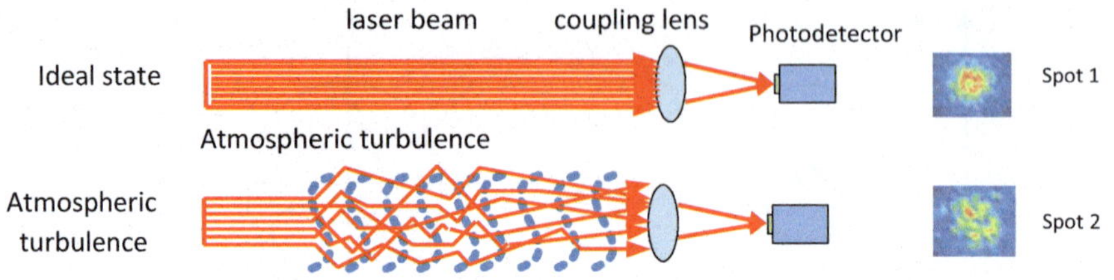

Fig. 1.11 Influence of light intensity flicker on atmospheric laser communication system

moments, whereas the detected power is lower than the threshold detector sensitivity at certain moments, resulting in an increase in the noise of the detection system. The size of the scintillation is closely related to the strength of the turbulence and has an obvious diurnal variation rule, similar to the turbulence; it is large during the day, small at night, and has a minimum value before and after sunrise and sunset, respectively. Given that the light intensity is a physical quantity that can be directly detected by the human eye, it is involved in practical applications. Therefore, the study of light intensity flicker is an area of interest that has received wide attention in the fields of astronomical observation and laser communication.

1.4.4 Arrival Angle Fluctuation

There are several turbulent effects caused by atmospheric turbulence, of which, the angle of arrival fluctuation is a major concern. A laser beam transmitted in a homogeneous medium has a uniform wavefront. In a turbulent atmosphere, owing to the fluctuation of the atmospheric refractive index in different regions of the beam cross-section, different parts of the beam wavefront will experience different phase changes, leading to random changes. The isophase surfaces of the undulating shapes are shown in Fig. 1.12. This phase deformation leads to a continuous change in the arrival angle of the wavefront of the beam, which is known as the arrival angle fluctuation effect. The specific performance is the random inclination of each point on the wavefront relative to the receiving plane when the beam propagates in the atmosphere, particularly when the turbulence scale is approximately equal to the beam diameter. The most direct effect of the angle of arrival fluctuation on the communication system is jitter of the image point in the optical antenna (telescope). If the telescope is used for observation, it will not be possible to focus the image point at the same position on the focal plane; this phenomenon is called image point jitter, which makes it difficult for the receiving end to receive the beam. Therefore, to accurately detect and receive optical signals, it is necessary to theoretically study and analyze the effect of the angle of arrival fluctuation, thereby establishing a robust foundation for the practical application of atmospheric laser communication technology.

When a beam is transmitted in a turbulent atmosphere, the random fluctuation of the transmission direction causes the beam to deviate from the expected position and

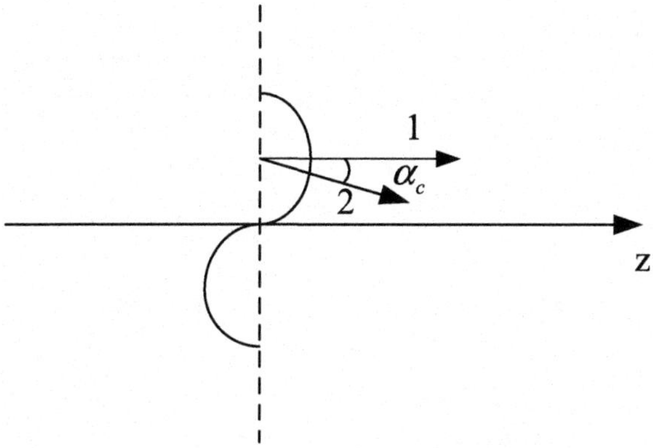

Fig. 1.12 Schematic diagram of 1—uniform phase front normal; 2—equal phase normal

spot drift occurs. The degree of drift is typically measured using the drift angle or drift amplitude. In addition to the drift, the arrival angle of the beam on the receiving plane also fluctuates randomly owing to the turbulence. The angle at which the center of the light spot deviates from its average position is called the drift angle and the divergence angle when the beam is incident on the receiving surface is called the angle of arrival.

Atmospheric turbulence directly affects the performance of wireless optical communication. As shown in Fig. 1.13, this phenomenon may cause laser power attenuation, pulse broadening, intensity fluctuation, beam drift, angle-of-arrival fluctuation, and beam expansion. This results in a decrease in the signal-to-noise ratio of the communication system, an increase in the fluctuation variance, and an increase in the bit error rate. Compared to fully coherent light, partially coherent light can suppress spot drift, intensity flicker, and angle-of-arrival fluctuations. In this work, the mechanism by which partially coherent light suppresses atmospheric turbulence effects is analyzed.

1.5 Partially Coherent Light

The fully coherent nature of a laser beam is not a necessary condition for producing good directivity, and a partially spatially coherent light source can produce the same far-field intensity distribution as a laser. Beams with good directivity but only partial spatial coherence are called Gauss-Schell model beams, and they can be used as physical models to describe multi-mode lasers under certain conditions.

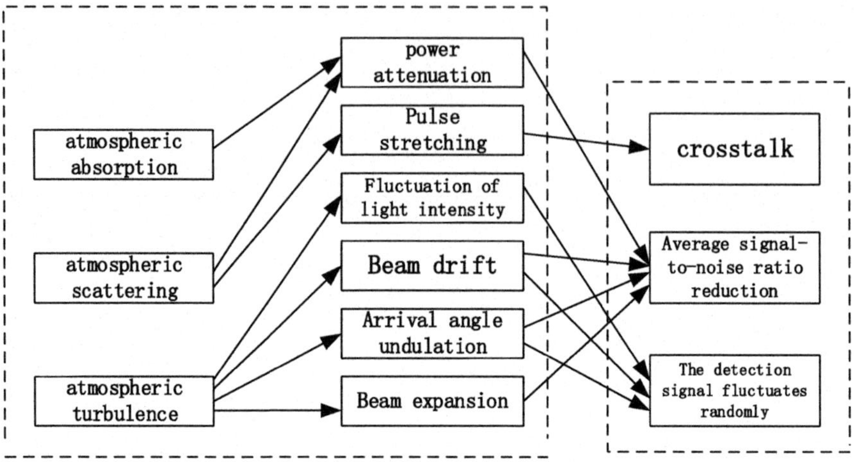

Fig. 1.13 Atmospheric turbulence and local angle of arrival impact on optical communication

1.5.1 Definition of Partially Coherent Light

The coherence of a beam includes both spatial and temporal coherences. In general, a partially coherent light source can be obtained by superimposing a random amplitude and phase distribution over time on a fully coherent light field. The added random amplitude and random phase distribution can be regarded as a random phase screen that represents the temporal and spatial coherence of the light field. Assuming that the light field distribution of the initial fully coherent light source is $U_0(x, y, 0)$, the time-varying amplitude and phase distributions are expressed as $t_A(x, y; t) = \exp[i\xi(x, y; t)]$, and the light field distribution of the partially coherent light at the emission can be expressed as follows: [64]

$$\begin{aligned} U\prime(x, y, 0; t) &= U_0(x, y, 0)t_A(x, y; t) \\ &= U_0(x, y, 0)\exp[i\xi(x, y; t)] \end{aligned} \tag{1.18}$$

where, $\xi(x, y; t)$ represents the partial coherence of the light field of the light source.

1.5.2 Space–Time Domain

Suppose $\rho = (x, y, z)$ represents the spatial position vector where z is the transmission distance of the beam, $\vec{\rho}_1$ and $\vec{\rho}_2$ are any two points in the light field, $V(\vec{\rho}_1, t + \tau)$ and $V^*(\vec{\rho}_2, t)$ represent the light field spatial points of partially coherent light, respectively, and $\vec{\rho}_1$ and $\vec{\rho}_2$ are complex amplitudes at times t and $t + \tau$, respectively. In the space–time domain, these two light field points can be described by the mutual coherence function $\Gamma(\vec{\rho}_1, \vec{\rho}_2, \tau)$, the mutual interference function is defined as [65]:

$$\Gamma(\vec{\rho}_1, \vec{\rho}_2, \tau) = \langle V(\vec{\rho}_1, t+\tau)V^*(\vec{\rho}_2, t)\rangle \tag{1.19}$$

In the equation, $\langle \cdot \rangle$ represents the ensemble mean.

Assuming that the radiation field satisfies the ergodicity of various states, the ensemble average of the light field of the beam can be obtained by averaging the calculation times.

$$\langle V(\vec{\rho}_1, t+\tau)V^*(\vec{\rho}_2, t)\rangle = \lim_{T\to\infty} \frac{1}{2T} \int_{-T}^{T} V(\vec{\rho}_1, t+\tau)V^*(\vec{\rho}_2, t)\mathrm{dt} \tag{1.20}$$

where T is the measurement time.

Let $\vec{\rho}_1 = \vec{\rho}_2 = \rho, \tau = 0$ in Eq. (1.31). The average light intensity at the point $\vec{\rho} = (x, y, z)$ can be represented as follows:

$$I(\vec{\rho}) = \langle V(\vec{\rho}, t)V^*(\vec{\rho}, t)\rangle = \Gamma(\vec{\rho}, \vec{\rho}, 0) \tag{1.21}$$

The spatial coherence of the light field can be described by the degree of complex coherence $\gamma(\vec{\rho}_1, \vec{\rho}_2, \tau)$, called the normalized mutual coherence function, which is expressed as follows:

$$\gamma(\vec{\rho}_1, \vec{\rho}_2, \tau) = \frac{\Gamma(\vec{\rho}_1, \vec{\rho}_2, \tau)}{\sqrt{\Gamma(\vec{\rho}_1, \vec{\rho}_1, 0)\Gamma(\vec{\rho}_2, \vec{\rho}_2, 0)}} = \frac{\Gamma(\vec{\rho}_1, \vec{\rho}_2, \tau)}{\sqrt{I(\vec{\rho}_1)I(\vec{\rho}_2)}} \tag{1.22}$$

The light field can be divided into three categories according to the value of $|\gamma(\vec{\rho}_1, \vec{\rho}_2, \tau)|$:

$$\begin{cases} |\gamma(\vec{\rho}_1, \vec{\rho}_2, \tau)| = 0 & \text{completely incoherent light} \\ 0 < |\gamma(\vec{\rho}_1, \vec{\rho}_2, \tau)| < 1 & \text{partially coherent light} \\ |\gamma(\vec{\rho}_1, \vec{\rho}_2, \tau)| = 1 & \text{perfectly coherent light} \end{cases}$$

Thus, the normalized mutual coherence function modulus $|\gamma(\vec{\rho}_1, \vec{\rho}_2, \tau)|$ assumes a value from 0 to 1, and the larger the value, the higher the visibility of the interference fringes and the greater the coherence of the light field.

The temporal coherence of the light field is denoted by $\gamma(\vec{\rho}, \vec{\rho}, \tau)$, also known as the self-coherence function, which is expressed as

$$\Gamma(\tau) = \Gamma(\vec{\rho}, \vec{\rho}, \tau) = \langle V(\vec{\rho}_1, t+\tau)V^*(\vec{\rho}_2, t)\rangle \tag{1.23}$$

Simultaneously, the average light intensity $I(\vec{\rho})$ can also be represented by the self-coherence function given by

$$I(\vec{\rho}) = \langle V(\vec{\rho}, t)V^*(\vec{\rho}, t)\rangle = \Gamma(\vec{\rho}, \vec{\rho}, 0) = \Gamma(0) \tag{1.24}$$

The normalized self-coherence function, also known as the complex self-coherence degree, is expressed as:

$$\gamma(\tau) = \frac{\Gamma(\tau)}{\Gamma(0)} \tag{1.25}$$

where $\gamma(0) = 1, 0 \leq \gamma(\tau) \leq 1$.

1.5.3 Space-Frequency Domain

In the space-frequency domain, the cross-spectral density function is used to describe the coherence of the light field, and is expressed as

$$W(\vec{\rho}_1, \vec{\rho}_2, \omega) = \langle \hat{V}(\vec{\rho}_1, \omega)\hat{V}^*(\vec{\rho}_2, \omega)\rangle \tag{1.26}$$

where $V(\vec{\rho}_1, \omega)$ and $\hat{V}(\vec{\rho}_1, \omega)$ are the field function and its Fourier transform, respectively:

$$\hat{V}\left(\vec{\rho}_j, \omega\right) = \int V\left(\vec{\rho}_j, t\right) \exp(i\omega t)d \quad j = 1, 2 \tag{1.27}$$

Therefore, the cross-spectral density function and mutual coherence function are related as follows:

$$W(\vec{\rho}, \vec{\rho}, \omega) = \int \Gamma(\vec{\rho}_1, \vec{\rho}_2, \tau) \exp(i\omega t)d\tau \tag{1.28}$$

$$\Gamma(\vec{\rho}_1, \vec{\rho}_2, \tau) = \frac{1}{2\pi}\int W(\vec{\rho}, \vec{\rho}, \omega) \exp(-i\omega t)d\omega \tag{1.29}$$

Let $\vec{\rho}_1 = \vec{\rho}_2 = \rho$, at the spatial point p, the average light intensity at frequency ω is:

$$I(\vec{\rho}, \omega) = W(\vec{\rho}, \vec{\rho}, \omega) \tag{1.30}$$

In the space-frequency domain, the temporal coherence is represented by the spectral density function S(ω), which is defined as follows:

$$S(\omega) = W(\vec{\rho}, \vec{\rho}, \omega) \tag{1.31}$$

The spectral density function, S(ω), changes even when partially coherent light is transmitted in free space, which is called the Wolf effect [66].

By normalizing the cross-spectral density, the degree of complex spatial coherence, also known as the degree of spectral coherence, can be obtained as follows:

$$\mu(\vec{\rho}_1, \vec{\rho}_2, \omega) = \frac{W(\vec{\rho}_1, \vec{\rho}_2, \omega)}{\sqrt{W(\vec{\rho}_1, \vec{\rho}_1, \omega)W(\vec{\rho}_2, \vec{\rho}_2, \omega)}} = \frac{W(\vec{\rho}_1, \vec{\rho}_2, \omega)}{\sqrt{S(\vec{\rho}_1, \omega)S(\vec{\rho}_2, \omega)}} \tag{1.32}$$

1.6 Research Methods for Partially Coherent Light

1.6.1 Generalized Huygens-Fresnel Principle

Figure 1.14 is a geometric schematic diagram of the Huygens-Fresnel principle. The left side of the figure represents the plane where the light source is located and the right side represents the receiving plane. The field distribution $B(x, y, z)$ at any point $E(x, y, z)$ on the receiving surface is formed by the superposition of the waves emitted by all the spherical wave source points of the light source O, namely, [67]

$$E(x, y, z) = \frac{1}{i\lambda}\iint E_0(x_0, y_0, 0)\frac{e^{ikr}}{r}\frac{(1+\cos\theta)}{2}\mathrm{d}x_0\mathrm{d}y_0 \tag{1.33}$$

When light waves travel in a vacuum, using the paraxial approximation, we obtain,

$$E(x, y, z) = \frac{e^{ikr}}{i\lambda z}\iint E_0(x_0, y_0, 0)\exp\left\{\frac{ik}{2z}\left[(x-x_0)^2 + (y-y_0)^2\right]\right\}\mathrm{d}x_0\mathrm{d}y_0 \tag{1.34}$$

When the beam travels through turbulent atmosphere and the light in the source plane travels from $(x_0, y_0, 0)$ to (x, y, z), the field change can be represented as a negative phase factor:

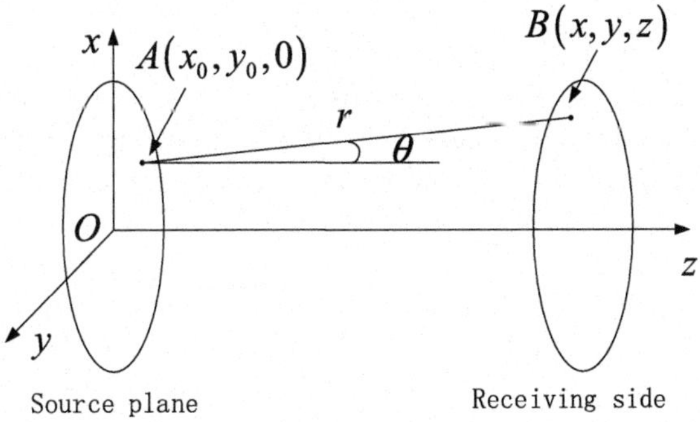

Fig. 1.14 Geometric schematic diagram of Huygens-Fresnel principle

$$F_1 = \exp(\chi + iS_1) = \exp[\psi(\rho', r)] \tag{1.35}$$

Equation (1.29) can be written as follows:

$$E(x, y, z) = \frac{e^{ikr}}{i\lambda z} \iint E_0(x_0, y_0, 0) \exp\left\{\frac{ik}{2z}\left[(x - x_0)^2 + (y - y_0)^2\right]\right\} \times \exp[\psi(\rho', r)] \mathrm{d}x_0 \mathrm{d}y_0 \tag{1.36}$$

The generalized Huygens-Fresnel principle does not only consider the phase fluctuations caused by atmospheric turbulence, but also takes into account the phase fluctuations caused by other factors, such as aerosol particle scattering and mechanical jitter in the emission system.

1.6.2 Cross Spectral Density Function

The two-point cross-spectral density function of a partially coherent light source is given by [68]

$$W_0(\boldsymbol{\rho}_1, \boldsymbol{\rho}_2) = \langle U'^*(\boldsymbol{\rho}_1) U'(\boldsymbol{\rho}_2) \rangle \tag{1.37}$$

where $\boldsymbol{\rho}_1, \boldsymbol{\rho}_2$ represents a two-dimensional vector at the light source plane.

The cross spectral density of a partially coherent Schell beam is given by:

$$W_0(\boldsymbol{\rho}_1, \boldsymbol{\rho}_2) = \sqrt{I_0(\boldsymbol{\rho}_1)} \times \sqrt{I_0(\boldsymbol{\rho}_2)} \mu_0(\boldsymbol{\rho}_2 - \boldsymbol{\rho}_1) \tag{1.38}$$

where $I_0(\boldsymbol{\rho})$ is the average light intensity, and $\mu_0(\rho_2 - \rho_1)$ represents the coherence of the beam at z = 0.

When $I_0(\rho) = A \exp\left(-\frac{|\rho|^2}{2\sigma_s^2}\right)$, $\mu_0(\rho) = \exp\left(-\frac{|\rho_1 - \rho_2|^2}{2\sigma_g^2}\right)$ is a partially coherent Gauss-Schell beam, for which the cross spectral density at z = 0 is given by,

$$W^{(0)}(\boldsymbol{\rho}_{s1}, \boldsymbol{\rho}_{s2}, 0) = A \exp\left(-\frac{|\boldsymbol{\rho}_{s1}|^2 + |\boldsymbol{\rho}_{s2}|^2}{4\sigma_s^2}\right) \times \exp\left(-\frac{|\boldsymbol{\rho}_{s1} - \boldsymbol{\rho}_{s2}|^2}{2\sigma_g^2}\right) \tag{1.39}$$

where $\boldsymbol{\rho}_{s1}, \boldsymbol{\rho}_{s2}$ are the coordinate vectors of two points in the source plane, and the parameters A, σ_s and σ_g represent the light intensity, beam waist width and coherence length of the light source, respectively.

1.7 Analytical Methods for Light Transport in Atmospheric Turbulence

The statistical properties of turbulent flow and the theory of wave propagation in random media are very complicated, and coupled with challenging experimental conditions, the study of light wave propagation is difficult. At present, the main methods used by researchers to study the transmission of light waves in random media are as follows: (1) perturbation approximation is used to investigate the permittivity and radiation field of random media, which involves solving the wave equation in random media and obtaining the radiation field distribution. Given that this method facilitates the direct determination of the light field, the utilization of approximate conditions limits the approach to the transmission problem under weak turbulence; (2) assumptions are made regarding the statistical properties of the dielectric constant of the random media and the statistical moment equation of the radiation field is determined and solved. Although these methods consider multiple scattering, they can only be used for large-scale inhomogeneous random media because of the assumption that only forward scattering exists. (3) Considering multiple scattering, a rigorous wave equation is constructed, and a formal solution of the radiation field is obtained. However, these methods are difficult to solve. Typical Rytov and Markov approximations are explained and analyzed in the following section.

1.7.1 Rytov Approximation

In atmospheric turbulence, it is extremely difficult to describe the constantly changing relative permittivity. Therefore, statistical methods are used to describe the changes in this medium; specifically, random functions of space and time are used to describe the dielectric constant as follows:

$$\varepsilon_r = \varepsilon_r(R, t) = n^2(R, t) \tag{1.40}$$

Assuming that ε_r only varies spatially, i.e., the light wave maintains a single frequency regardless of the frequency shift or spectrum, then we have:

$$\varepsilon_r = \varepsilon_r(R) = n^2(R) \tag{1.41}$$

Using Maxwell's equation, the refractive index n can be written as:

$$\nabla^2 E(R) + K^2 n^2 E(R) + 2\nabla(\frac{\nabla n}{n} \cdot E(R)) = 0 \tag{1.42}$$

Ignoring the depolarization effect, the third term in the equation can be omitted. Considering $U(R)$, which is the electromagnetic field transmitted in the direction perpendicular to the z-axis, the wave equation for electromagnetic wave transmission in random media is given by,

$$\nabla^2 U(R) + K^2 n^2 U(R) = 0 \tag{1.43}$$

In weak turbulence, the electromagnetic field can be expressed as:

$$U(R) = U(r, l) = U_0(r, l) \exp[\psi(r, L)] \tag{1.44}$$

where $U_0(r, L)$ is the scattering field without turbulence and $\psi(r, L)$ is the complex random phase perturbation caused by turbulence. The complex phase perturbation is unwrapped and expressed as

$$\psi(r, L) = \psi_1(r, L) + \psi_2(r, L) \tag{1.45}$$

where $\psi_1(r, L)$ and $\psi_2(r, L)$ are first-order and second-order complex-phase perturbation approximations, respectively.

We define the three fundamental statistical moments as follows:

$$E_1(r, r) =< \psi_2(r, L) > +\frac{1}{2} < \psi_1(r, L) >= -2\pi^2 k^2 L \int_0^\infty \kappa \Phi_n(k) d\kappa \tag{1.46}$$

$$E_2(r_1, r_2) =< \psi_1(r_1, L)\psi_1^*(r_2, L) >= 4\pi^2 k^2 L \int_0^1 \int_0^\infty \kappa \Phi_n(\kappa) \exp\left[-\frac{\Lambda L \kappa \xi^2}{k}\right] \times J_0\left[\kappa \left|[1 - \overline{\Theta}\zeta]p - 2i\Lambda r\right|\right] d\zeta d\kappa \tag{1.47}$$

$$E_3(r_1, r_2) =< \psi_1(r_1, L)\psi_1^*(r_2, L) >= -4\pi^2 kL \int_0^1 \int_0^\infty \kappa \Phi_n(\kappa) \exp\left[-\frac{\Lambda L \kappa \xi^2}{k}\right] \times \exp\left\{-\frac{iL\kappa^2 \zeta}{k}[1 - \overline{\Theta}\zeta]\right\} J_0[\kappa\rho(1 - \overline{\Theta}\zeta - i\Lambda\zeta)] d\zeta d\kappa \tag{1.48}$$

To determine the wavelength and transmission distance, the plane wave variance is usually used to describe the transmission conditions of light waves.

$$\sigma_R^2 = 1.23 C_n^2 k^{7/6} L^{11/6} \tag{1.49}$$

As the *Rytov* variance increases, the scintillation index approaches 1 and does not increase further because multiple scattering gradually increases and becomes non-negligible. Therefore, the traditional *Rytov* approximation cannot be applied to the highly turbulent region. Accordingly, researchers such as Andrews have developed a modified *Rytov* approximation based on specific assumptions i.e., assumptions: (1)

The turbulent atmosphere is statistically inhomogeneous. (2) Given that the small-scale turbulent fluctuations are modulated by large-scale turbulent fluctuations, light intensity scintillation occurs at the receiving end; i.e., the light intensity scintillation is the product of the small-scale turbulent fluctuations and large-scale turbulent fluctuations. (3) The effects of large-scale and small-scale turbulence on the intensity are statistically independent of each other. (4) After spatial frequency filtering is introduced, an approximation can be used to study the scintillation problem of light waves in highly turbulent regions. (5) The study of large-scale turbulent light intensity fluctuations can be based on geometrical optics approximations.

Based on the aforementioned assumptions, the modified approach under weak turbulence can be used in a strong-turbulence region. Therefore, the variance of the light wave intensity can be expressed as follows:

$$\langle I^2 \rangle = \langle x^2 \rangle \langle y^2 \rangle = (1+\sigma_x^2)(1+\sigma_y^2) \tag{1.50}$$

where σ_x^2 is the normalized variance of x, and σ_y^2 is the normalized variance of y. For convenience, we take $\langle I \rangle$ as 1, and the expression for the variance of the light intensity fluctuation is given as,

$$\sigma_1^2 = (1+\sigma_x^2)(1+\sigma_y^2) - 1 = \sigma_x^2 + \sigma_y^2 + \sigma_x^2\sigma_y^2 \tag{1.51}$$

In the Rytov approximation, which is used in the weakly turbulent region, the Rytov solution corresponding to Eq. (2.14) at distance L is only applicable to the single scattering region. Therefore, the Rytov approximation of the multiple scattering region must be corrected. The revised formula is given by,

$$U(r,l) = U_0(r,l)\exp[\psi_x(r,L) + \psi_y(r,L)] \tag{1.52}$$

where $\psi_x(r,L)$ and $\psi_y(r,L)$ are independent and statistically uncorrelated complex random phases that correspond to large-scale and small-scale turbulence, respectively. The addition of the exponential term to the equation can be regarded as the small-scale turbulent fluctuation that is modulated by the large-scale turbulent fluctuation. Thus, the atmospheric turbulence is defined as a spatial filter function as follows:

$$\Phi_n(\kappa) = 0.033 C_n^2 G(\kappa, l_0) \tag{1.53}$$

in

$$\begin{aligned} G(\kappa, l_0) &= G_x(\kappa, l_0) + G_y(\kappa) \\ &= f(\kappa, l_o)\exp\left(-\frac{\kappa^2}{\kappa_x^2}\right) + \frac{\kappa^{11/3}}{(\kappa^2+\kappa_y^2)^{11/6}} \end{aligned} \tag{1.54}$$

k_x is the spatial frequency of large-scale turbulence and k_y is small-scale turbulence. The filter function $G(k, l_0)$ is a combination of low-pass filter function $G_x(k, l_0)$ and high pass filter function $G_y(k, l_0)$.

1.7.2 Markov Approximation

As the atmospheric turbulence and beam travel distance increase, the Rytov approximation no longer reasonable describes the observed experimental phenomenon. Thus, more refined models are required, such as the Markov approximation. It is assumed that the light-wave transmission process can be regarded as a Markov process, based on which various order moment equations can be derived. This theory is called the Markov approximation. Many researchers have obtained second-order moment equations using different methods, but only the Markov approximation facilitates the derivation of higher-order moment equations. These equations are the basis of the solution to the light-wave transmission problem.

The Markov approximation is used to solve the problem of wave transmission under highly turbulent conditions, i.e., the light field is regarded as a Markov process, the parabolic equation that satisfies this process is obtained, and the various order moments of the light wave field are derived. The parabolic equation is given as follows.

Assuming that the beam propagates along the z-axis and the phase changes, the scalar field can be written as follows:

$$u = U(r)e^{ikz} \tag{1.55}$$

$U(r)$ is a slowly varying function of z, and the parabolic equation it satisfies is given by:

$$2ik\frac{\partial U(r)}{\partial z} + \nabla^2 U(r) + k^2 n^2(r) U(r) = 0 \tag{1.56}$$

$$2ik\frac{\partial U}{\partial z} + \nabla_T^2 U + k^2 n_1 U = 0 \tag{1.57}$$

In the equation, $\nabla_x^2 = \partial^2/\partial x^2 + \partial^2/\partial y^2$, the above Eq. (1.67) is also called the quasi-optical approximation equation. The following assumptions are based on this equation; assumptions: (1) The parabolic equation is utilized, (2) the field is a Gaussian random field. Thus, the feature quantity can be represented using a correlation function, namely:

$$< n_1(z, \vec{\rho}) n_1(z', \vec{\rho}') >= \delta(z - z') A(\vec{\rho'} - \vec{\rho}') \tag{1.58}$$

Finally, the relationship between the correlation function $A(\vec{\rho}' - \vec{\rho}')$ and the fluctuation spectrum of the refractive index $\Phi_n(\vec{\kappa})$ is as follows:

$$\Phi_n(\vec{\kappa}) = \frac{1}{(2\pi)^3} \int\int A(\vec{\rho} - \vec{\rho}') \exp[-i\vec{\kappa} \cdot (\vec{\rho} - \vec{\rho}')] d(\vec{\rho} - \vec{\rho}') \quad (1.59)$$

Arbitrary matrix equations were derived by Gurvich and Tatarski [69]. Let a random variable be expressed as.

$$\gamma(z, \rho_1, \rho_2, \ldots \rho_n; \rho_1', \ldots \rho_m') = u(z, \rho_1) u(z, \rho_2) \ldots u^*(z, \rho_m') \quad (1.60)$$

The moment equations of each order are represented as:

$$2ik \frac{\partial \Gamma_{mn}}{\partial z} + (\nabla_{T1}^2 + \cdots + \nabla_{Tn}^2 - \nabla_{T1}'^2 \cdots - \nabla_{Tm}'^2) \Gamma_{nm} + \frac{ik^3}{4} F_{nm} \Gamma_{nm} = 0 \quad (1.61)$$

in the following equation,

$$F_{nm} = \sum_{i=1}^{n} \pm \sum_{k=1}^{m} A_n(|\rho_i - \rho_k|) - 2 \sum_{i=1}^{n} \sum_{j=1}^{m} A_n\left(\left|\rho_i - \rho_j'\right|\right) + \sum_{j=1}^{m} \sum_{k=1}^{m} A_n\left(\left|\rho_j' - \rho_i'\right|\right) \quad (1.62)$$

The conditions for the establishment of the parabolic equation and the Markov approximation are $kl_0 \gg 1$ and $kC_n^2 L_0^{5/3} \ll 1$.

1.8 Generation of Partially Coherent Gaussian-Schell Model Beams

Laser speckle phenomenon often occurs in highly coherent laser beams [70, 71]. In high-power-driven inertial confinement fusion, which requires high uniformity of the irradiation energy on the target surface, interference speckle caused by the high coherence of the laser will appear. Partially coherent light can effectively overcome these shortcomings, resulting in inhomogeneous light intensity distribution [72].

1.8.1 *Coherence of Fundamental Mode Gaussian Beam and Gaussian-Schell Model Beam*

In 1978, Collett and Wolf proposed a typical partially coherent source (Collett-Wolf source), also called a Gaussian-Schell model beam, which has a Gaussian intensity distribution and spatial coherence distribution [73]. In 1979, P. De Santis and F. Gori et al. used the light emitted by a laser to irradiate a rotating frosted glass to generate false thermal light. A Collett-Wolf source was then then produced using a Gaussian filter, i.e., a Gaussian-Schell model beam [74]. In 1981, Deschamps et al. irradiated a rotating scatterer using a laser. After propagation through the scatterer, the coherence of the laser light was destroyed. Partially coherent light was generated after reflection when the beam propagated through an interferometer [75]. In 1991, Tervonen and Friberg used the acousto-optic hologram proposed by J. Appl to convert a Gaussian beam into a partially coherent arbitrary anisotropic Gaussian-Schell (AGASM) beam [76]. In 2004, Shirai et al. used a spatial light modulator placed in the two arms of a Mach–Zehnder interferometer to modulate two coherent polarized plane waves to generate a partially coherent Gaussian-Schell model beam [77]. The use of liquid crystal spatial light modulators loaded with random phase screens to generate partially coherent Gaussian-Schell model beams is related to the theoretical model, and the properties of the generated beams are easy to adjust. Therefore, they have attracted extensive research interest.

In ordinary light sources, the process of atomic light emission is a spontaneous radiation process, and the radiation of atoms is spontaneous and independent; therefore, the photons emitted by each atom have different frequencies, emission direction, and initial phase. Thus ordinary light sources are not coherent [78]. Lasers are stimulated by radiation processes, and the emitted laser light has strong coherence.

1.8.1.1 Coherence of Fundamental Mode Gaussian Beams

In this case, a laser beam output was generated using an optical resonator. One of the main functions of a resonator is to reflect light back and forth many times in an optical cavity to increase the working length of the active medium and improve the working density of the resonator. Any paraxial light in the cavity is transmitted back and forth many times without overflowing out of the cavity. The resonator enables a laser to stably emit light, and is called a stable cavity. The light emitted by a laser with a stable cavity is transmitted in space in the form of a Gaussian beam. The structure of the laser is shown in Fig. 1.15.

The fundamental-mode Gaussian beam that propagates along the z-axis, regardless of the structure of the stable cavity, can be expressed in the following general form:

$$\psi_{00}(x, y, z) = \frac{c}{w(z)} e^{-\frac{r^2}{w^2(z)}} e^{-i[k(z+\frac{r^2}{2R})arctg\frac{z}{f}]} \tag{1.63}$$

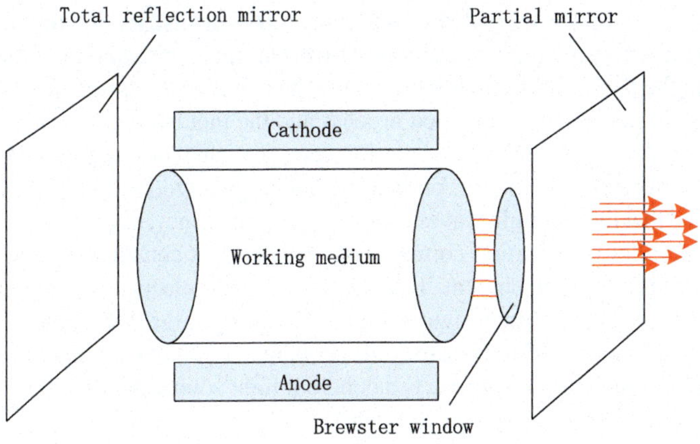

Fig. 1.15 Schematic diagram of the internal structure of the laser

where C is a constant factor, and the other variables are defined as

$$r^2 = x^2 + y^2, k = \frac{2\pi}{\lambda}, w(z) = w_0\sqrt{1 + \left(\frac{z}{f}\right)^2},$$

$$R = R(z) = z\left[1 + \left(\frac{f}{z}\right)^2\right] = f\left(\frac{z}{f} + \frac{f}{z}\right) = z + \frac{f^2}{z},$$

$$f = \frac{\pi w_0^2}{\lambda}, w_0 = \sqrt{\frac{\lambda f}{\pi}}$$

w_0 is the waist radius of the fundamental mode Gaussian beam; f is the confocal parameter of Gaussian beam; $R(z)$ is the curvature radius of the equal phase plane of the Gaussian beam intersecting the transmission z-axis at the point; $w(z)$ is the spot radius on the equal phase plane of the Gaussian beam intersecting the transmission z-axis at the point [79].

The Gaussian beam has a uniform amplitude distribution on the wavefront, exhibits a Gaussian amplitude distribution, and the light energy is concentrated at the center of the beam. The field amplitude and intensity distribution of the fundamental transverse mode of the Gaussian beam are given by

$$U_{00} = \exp\left(\frac{-2}{1+\zeta}\frac{x^2+y^2}{w_s^2}\right) \tag{1.64}$$

$$I_{00} = U_{00}^2 = \exp\left(\frac{-4}{1+\zeta^2}\frac{x^2+y^2}{w_s^2}\right) \tag{1.65}$$

The constant term [80] is ignored in the equation.

The Gaussian beams exhibited a Gaussian amplitude distribution on the isophase plane. The laser beam output is generated by the stimulated radiation of the working medium in the laser cavity under the action of an external electromagnetic field. The photons generated by the stimulated process and the incident photons that cause the stimulated radiation have the same photon state, i.e., they have the same frequency, phase wave vector, and other properties, including polarization [80].

Provided that the two light waves have the same frequency, a constant phase difference and parallel polarization components, the beam is considered to be coherent light, otherwise, it is incoherent. For light waves, the electric and magnetic field oscillate. The origin of human sight and film sensitivity is the light vector. If the light vectors of two beams of light satisfy the coherence condition, they are coherent, and the corresponding light source is a coherent light source [81]. Coherence can be divided into temporal and spatial coherence.

(1) Temporal coherence

The light sources used are not strictly monochromatic, and the light waves they radiate always have a certain frequency width centered at a specific frequency. According to the light-emitting mechanism of the light source, the light wave emitted by any light source consists of a series of finite-length wave trains that are separated from each other by discontinuous phase changes. These phase changes are the result of a random process, by which excited atoms transition between energy levels in the light source, producing short, random wave trains of radiation. A given light source has a specific mean wave train length, which is the coherence length. The time required for light to pass traverse the coherence length is called the coherence time, and the relationship between the two is given by $L_c = c\tau_c$, where c is the speed of light in a vacuum, and τ_c is approximately inversely proportional to the spectral width of the light wave emitted by the light source; i.e., the longer the coherence time, the better the monochromaticity of the light source [82].

(2) Spatial coherence

The spatial coherence of light refers to the coherence of the wave field at the same moment as the spatial point on the cross-section perpendicular to the transmission direction of the light. Waves may have variable phase differences in different regions of space, and only light waves within a certain spatial range have relatively fixed phase differences. Therefore, only light waves within a certain space are coherent. This property is known as the spatial coherence of a wave. Spatial coherence is also known as lateral coherence. In the laser, the light waves were initially not spatially coherent. When light waves are reflected back and forth many times in an optical cavity, they diffuse owing to diffraction. The diffractive aperture causes the light emitted from each point on the beam cross-section to mix with each other, and the space of the beam is established. Coherence causes the originally incoherent light to be spatially coherent [83]. The coherence of the laser is the same across the entire beam cross-section; therefore, the fundamental-mode laser has perfect spatial coherence [84].

1.8.1.2 Coherence of Gaussian-Schell Model Beams

In 1978, Collet and Wolf proposed a GSM beam model. This beam has good directivity and a limited transverse coherence length, and the transverse section has a Gaussian light intensity and coherence distribution. Partially coherent light sources can also produce the same far-field intensity distribution as lasers, with good directionality, but only partial spatial coherence [85]. Partially coherent beams have a wide range of applications in free-space optical communications, optical imaging, nonlinear optics, optical trapping, and laser scanning. Gaussian-Schell model beams are quasi-monochromatic planar light sources with an intensity distribution and spatial coherence that are both Gaussian functions.

Suppose that the cross-spectral density function of partially coherent light at the source can be expressed as

$$J(r_1, r_2, 0) = \sqrt{I(r_1, 0)I(r_2, 0)}\mu(r_1 - r_2, 0) \tag{1.66}$$

In Eq. (1.76), the complex spatial coherence $\mu(r_1 - r_2, 0)$ is only related to the distance between r_1 and r_2 two points. If both the light intensity and complex space coherence are characterized by a Gaussian function, we have:

$$I(r, 0) = I_0 \exp\left(-\frac{2r^2}{w_0}\right) \tag{1.67}$$

$$u(r_1 - r_2, 0) = \exp\left(-\frac{(r_1 - r_2)^2}{2\sigma_0}\right) \tag{1.68}$$

The beam described by Eq. (1.76) is called an isotropic Gaussian-Schell model beam, where I_0 is a constant, w_0 and σ_0 are the beam waist width and correlation length of the Gauss-Schell model beam at the exit surface, respectively [85]. The isotropic Gaussian-Schell model beam has a Gaussian light intensity distribution and Gaussian spatial coherence distribution. The model play an important role in modern optics as a physical model for describing partially coherent beams. Their properties and transmission characteristics have been extensively investigated.

1.8.2 Phase Screen Theory

The intensity distribution and coherence of the Gaussian-Schell model beam are Gaussian. Therefore, to convert the fundamental-mode Gaussian beam emitted by the laser into a Gaussian-Schell model beam with similar properties to the theoretical model, we first ensure that the fundamental-mode Gaussian beam propagate through the phase screen. The intensity distribution was Gaussian. After adjusting the phase distribution of the fundamental-mode Gaussian beam, The coherence of the beam needs to meet the characteristics of Gaussian type. After the aforementioned two conditions are satisfied, the beam emitted from the spatial light modulator is the required Gaussian Schell-model beam.

The phase function of the random phase screen is denoted as $\xi(p, q)$, and the transfer function of the phase screen can be expressed as:

$$t(p, q) = \exp(j\xi(p, q)) \tag{1.69}$$

After the fundamental-mode Gaussian beam described by Eq. 1.79 propagates through the phase screen, the obtained light field is expressed as follows:

$$U'(p, q) = U(p, q)t_A(p, q) \tag{1.70}$$

where $U(p, q)$ denotes the field of the original fundamental mode Gaussian beam.

According to Schell's theorem, the light intensity of the beam after propagation through the phase screen can be expressed as Eq. (1.81)

$$I(x, y) = \frac{1}{(\lambda z)^2} \iint u(\Delta p, \Delta q)R(\Delta p, \Delta q) \exp\left[j\frac{2\pi}{\lambda z}(\Delta px + \Delta qy)\right]d\Delta pd\Delta q \tag{1.71}$$

where $u(\Delta p, \Delta q)$ is the self-coherence function of the fundamental-mode Gaussian beam, and $R(\Delta p, \Delta q)$ is the self-coherence function of the random phase screen [86].

The light emitted by the laser is quasi-perfectly coherent. The self-coherence function of the fundamental-mode Gaussian beam in Eq. (1.76) can be considered to be a constant. Therefore, when the autocorrelation function of the random phase screen in Eq. (1.76) is a Gaussian function, the intensity distribution of the beam after propagation through the phase screen is also Gaussian, which is consistent with the light intensity distribution characteristics of the Gaussian-Schell model beam of the theoretical model.

The phase function of the random phase screen, $\xi(p, q)$, is obtained using the spatial convolution of a random Gaussian function $f(p, q)$ and a Gaussian filter function $r(p, q)$, as shown in the following equation:

$$\xi(p, q) = r(p, q) \otimes f(p, q) \tag{1.72}$$

where $\otimes$ represents a spatial convolution, p and q represent the coordinates of a point in space, $r(p, q)$ is a Gaussian two-dimensional random function with zero mean that can be generated using a computer, and the Gaussian filter function is $f(p, q)$, which is defined in the following equation:

$$f(p, q) = A_1 \exp\left(-\frac{p^2 + q^2}{\delta_f}\right) \tag{1.73}$$

By discretizing $f(p, q)$,$r(p, q)$ we obtain $f(m, n)$, and $r(m, n)$, and if a spatial convolution is performed on the discretized $f(m, n)$ and $r(m, n)$, the real part of the convolution is displayed in the form of a grayscale image, from which the random phase screen can be obtained.

Given that $r(m, n)$ is a random function, the correlation function between any two points on the phase screen can be obtained as follows:

$$r(m_1, n_1) \otimes r(m_2, n_2) = \delta(m_1 - m_2, n_1 - n_2) \tag{1.74}$$

Based on the convolution property of the Gaussian function, the convolution of the two Gaussian functions is a Gaussian function, which can be obtained as follows:

$$f(m_1, n_1) \otimes f(m_2, n_2) = A_2 \exp\left[-\frac{A_3\rho^2}{2\delta_f^2}\right] \tag{1.75}$$

where A_2 is a constant and ρ is the distance between points in space. From the above argument, we obtain the following

$$\xi(m_1, n_1) \otimes \xi(m_2, n_2) = A_3 \exp\left[-\frac{\rho^2}{2\delta_\xi^2}\right] \tag{1.76}$$

From (1.76), it can be shown that the autocorrelation function of the phase function introduced by the phase screen is Gaussian. The autocorrelation function of the phase screen is the same as the autocorrelation function of the beam after it propagates through the phase screen. Therefore, the random phase function meets the requirement of generating a Gaussian-Schell partial coherent beam, and the coherence length is the sum of the variance of the Gaussian random function δ_r and the Gaussian function δ_f. The relationship between the two is given by $\delta_\xi^2 = \left(8\pi.\delta_f^4/\delta_r^2\right)$.

1.8.3 Generation of Random Phase Screen and Result Analysis

According to the theory of phase screens, a random phase screen is generated by the spatial convolution of a random Gaussian function and a Gaussian function. First, the Fourier transform of the discretized random Gaussian and Gaussian functions was obtained.

$$r(m, n) \to^F R(M, N) \tag{1.77}$$

$$f(m, n) \to^F F(M, N) \tag{1.78}$$

According to the convolution theorem, a convolution in the space domain is equivalent to the product in the frequency domain. Thus, we obtain the following:

$$\xi(M, N) = R(M, N)F(M, N) \tag{1.79}$$

An inverse Fourier transform is performed using the results from Eq. (1.89) to obtain the value of $\xi(m, n)$. By displaying the obtained real part of this term in grayscale, the random phase screen required to generate a Gaussian-Schell model beam can be obtained. When δ_f and δ_r When taking different values, and the sampling interval is selected as 12.3 μm. Figure 1.16 shows the result for a random phase screen with a size of 1024 × 1024.

As shown in Fig. 1.16, when δ_r is constant and δ_f decreases, the number of speckles in the grayscale image increases. The voltage loaded by the liquid crystal spatial light modulator corresponds to an 8-bit grayscale value. The loading of different grayscale values cause the liquid crystal to produce different phase delays. The corresponding relationship between phase delay and gray value is the phase modulation characteristics of liquid crystal spatial light modulator. Since the fundamental-mode Gaussian beam emitted by the laser is approximately coherent, the constant phase difference condition is satisfied. When the fundamental mode Gaussian beam passes through the gray-scale image, the phase adjustment of the beam is the same where the color of the gray-scale image is the same. Therefore, the phase difference of the beam

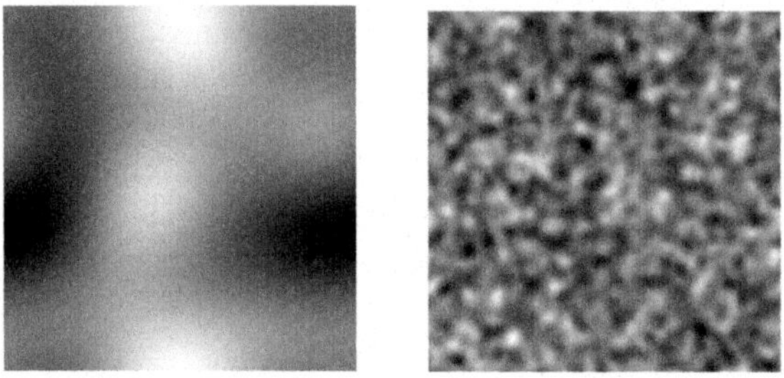

(a) $\delta_f = 0.18\ cm$, $\delta_r = 4\ cm$ (b) $\delta_f = 0.018\ cm$, $\delta_r = 4\ cm$

(c) $\delta_f = 0.0018\ cm$, $\delta_r = 4\ cm$ (d) $\delta_f = 0.00018\ cm$, $\delta_r = 4\ cm$

Fig. 1.16 Radom phase screen for different values of the Gaussian random function and constant Gaussian function after propagation through the phase screen at points with the same intensity value is still constant, and this part of the beam is still correlated after the phase screen. As coherence decreases, the number of grayscale speckles in the phase screen increases. After the Gaussian beam propagates through the spatial light modulator loaded with the phase screen, the phase of the beam changes, thereby influencing the coherence of the beam. This is the working principle of a liquid–crystal spatial light modulator loaded with a random phase screen to generate partially coherent light.

1.8.4 Experiment and Result Analysis

Spatial light modulator contains many independent units, which are arranged in a one-dimensional or two-dimensional array in space. Each unit can independently receive the control of optical signal or electrical signal, and change the properties of optical itself such as transmittance, reflectivity, refractive index according to this signal. Add light or electric signal to the spatial light modulator to control the spatial light modulator. The amplitude is modulated by absorption, the phase is modulated by refractive index, and the polarization state is modulated by the rotation of the polarization plane. The voltage loaded by the spatial light modulator corresponds to the gray value. Loading different gray values will change the arrangement of the liquid crystal, resulting in the transformation of the refractive index of light, so as to adjust the beam. By converting the change of the beam into an electrical signal and loading it on the spatial light modulator, the beam can be controlled and the phase of the beam can be adjusted accordingly. When using spatial light modulator to generate Gaussian Schell beam, the change of LCD array of spatial light modulator is controlled by loading the gray image of phase screen generated by computer.

The Gaussian-Schell model beam is generated by loading a random phase screen onto a spatial light modulator, and the properties of the generated partially coherent light can be adjusted by changing only the relevant parameters in the phase screen. When the spatial light modulator is loaded with a random phase screen to generate partially coherent light, the core components are the spatial light modulator, the phase screen, and a laser. A schematic of the experimental setup is shown in Fig. 1.17.

The laser used in the experiment was a helium–neon gas laser with a center wavelength of 632.8 nm. The lens group was composed of lenses 1 and 2. The function of the lens group is to expand and collimate the fundamental-mode Gaussian beam emitted by the laser. The collimated fundamental-mode Gaussian beam is incident on a liquid crystal spatial light modulator (SLM) loaded with a random phase screen. Lens 3 focuses the beam reflected from the spatial light modulator. A white screen was used to observe the spot formed by the focused beam. The two-dimensional light intensity distribution at $\delta_f = 0.18\,\text{cm}$, $\delta_r = 4\,\text{cm}$, is shown in Fig. 1.18.

As shown in Fig. 1.18, the intensity distribution of the beam emitted from the liquid crystal spatial light modulator loaded with the random phase screen is Gaussian, which is as expected considering the Gaussian distribution of the intensity of the Gaussian-Schell model beam.

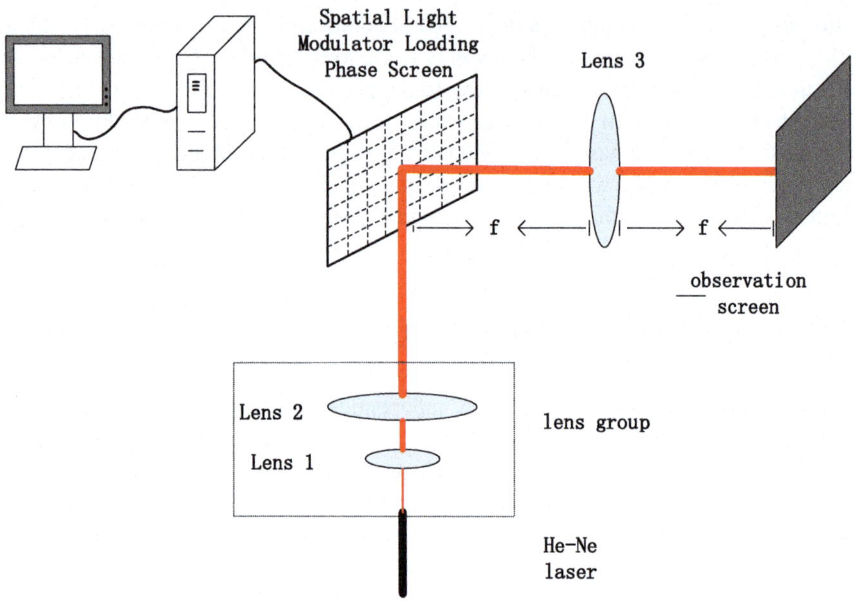

Fig. 1.17 Schematic diagram of the experimental setup

Fig. 1.18 Experimentally obtained two-dimensional light intensity distribution

The three-dimensional light intensity distribution measured by the beam analyzer in the experiment is the three-dimensional light intensity distribution after the light reflected by the spatial light modulator travels a certain distance in space. The specific parameters used in the experiment are: transmission distance of the beam is $z = 15\,\text{m}$, the wavelength of the laser is $\lambda = 632.8\,\text{nm}$, $\delta_f = 0.18\,\text{cm}$, $\delta_r = 4\,\text{cm}$, $\delta_f = 0.00018\,\text{cm}$, and $\delta_r = 4\,\text{cm}$. The measured three-dimensional light intensity distribution of the experiment is shown in (a), (b) and (c) of Fig. 1.19. Based on the change of the curve in Fig. 1.19, it can be concluded that as the coherence decreases, the maximum value of the intermediate light intensity of the light spot gradually decreases.

Fig. 1.19 The experimentally measured three-dimensional light intensity distribution

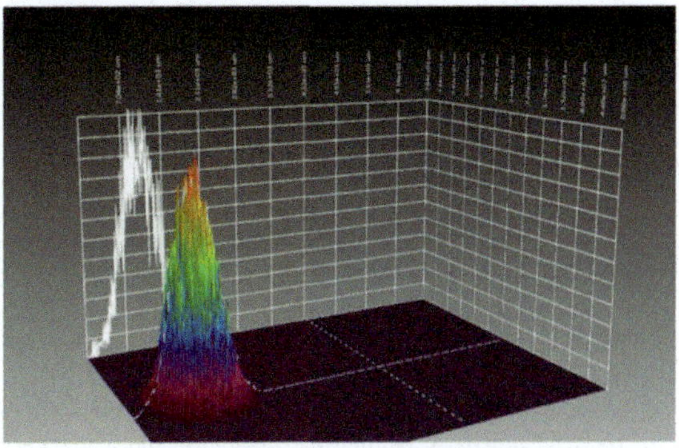

(a) $\delta_f = 0.18\ cm,\ \delta_r = 4\ cm$

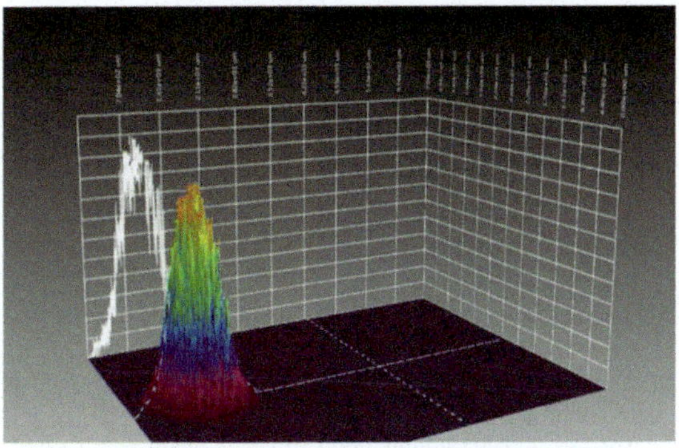

(b) $\delta_f = 0.0018\ cm,\ \delta_r = 4\ cm$

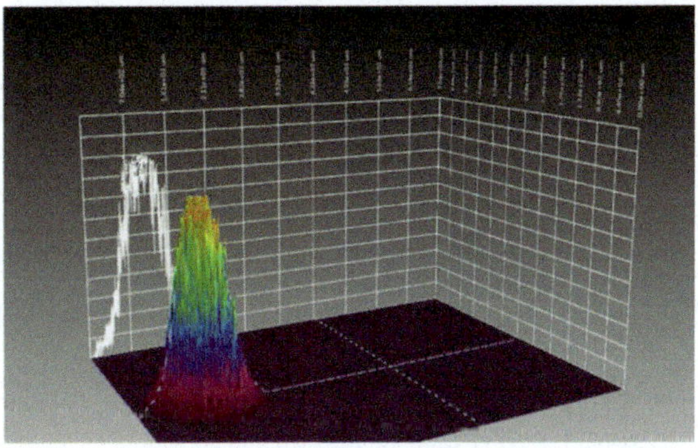

(c) $\delta_f = 0.00018\ cm, \delta_r = 4\ cm$

References

1. Wang H (2001) Laser communication technology in the atmosphere. Infrared Laser Eng 30(2):123–127
2. Ke X (2009) Coding theory in wireless laser communication system. Science Press, Beijing, pp 8–10
3. Zijuan J (2005) Research on atmospheric laser communication system. Changchun University of Science and Technology, Changchun, p 7
4. Lu W (1997) Overview of optical communication. Opt Commun Technol 21(3):171–180
5. Wu X (2006) Atmospheric optical turbulence, models and measurement technology. J Anhui Normal Univ 29(2):104–107
6. Ishimaur A (1978) Wave propagation and scattering in random media. Academic Press, New York (written by A. Ishimaru, Propagation and scattering of waves in random Media, translated by Huang Runheng et al., proofread by Zhou Xiuji, Beijing: Science Press, 1986)
7. Wu J, Yang C, Liu J (2005) The theory of light transport in the atmosphere. Beijing University of Posts and Telecommunications Press, Beijing
8. Zhang Y, Chi Z (1997) Transmission and imaging of light waves in the atmosphere. National Defense Industry Press, Beijing
9. Rao R (2005) Propagation of light in turbulent atmosphere. Anhui Science and Technology Press, Hefei
10. Chiba T (1971) Spot dancing of the laser beam propagated through the turbulent atmosphere. App I Opt 10(11):2456–2461
11. Andrews LC, Phillips RL (2005) Laser beam propagation through random media. SPIE Press, 83–99
12. Chunrside JH, Lataitis RJ (1990) Wander of an optical beam in the turbulent atmosphere. Appl Opt 29(7):926–930
13. Xiao X, Voelz DG (2012) Beam wander analysis for focused partially coherent beams propagating in turbulence. Opt Eng 51(2):601–607
14. Tofsted H (1992) Outer-scale effects in beam-wander and angle-of-arrival variances. Appl Opt 31(27):5865–5870
15. Zhensen W, Xiang N, Wang M (2013) The broadening and drift of partially coherent Gauss-Shell beams in atmospheric turbulence. Infrared Laser Eng 42(3):658–662
16. Recolons J, Andrews LC, Phillips RL (2007) Analysis of beam wander effects for a horizontal-path propagating Gaussian-beam wave: focused beam case. Opt Eng 46(8):86–92
17. Recolons J, Andrews LC, Phillips RL (2007) Analysis of beam wander effects for a horizontal-path propagating Gaussian-beam wave: focused beam case. Opt Eng 46(8):88–92
18. Gao C, Siyuan Y (2006) Beam drift in strong turbulent regions. Intense Laser Particle Beam 18(10):1597–1601
19. Tan Q, Huang X (1987) Calculation of laser beam drift in turbulent atmosphere. J Wuyi Univ (Soc Sci Edn) 03(1):69–74
20. Fante RL (1980) Electromagnetic beam propagation in turbulent media: an update. Proc IEEE 68(11):1424–1443
21. Mironov, Nosov V (1977) On the theory of spatially limited light beam displacements in a randomly inhomogeneous medium. J Opt Soc Am 67(8):1073–1080
22. Wang L (2012) Study on the drift and spreading characteristics of light beams in atmospheric turbulence. Xidian University, Xi'an
23. Hill RJ (1982) Theory of saturation of optical scintillation by strong turbulence: plane-wave variance and covariance and spherical-wave covariance. JOSA 72(2):212–222
24. Hill RJ, Clifford SF (1981) Theory of saturation of optical scintillation by strong turbulence for arbitrary refractive-index spectra. J Soc Am A 71(6):675–686
25. Wu Z (1986) Intensity correlation and scintillation of saturation state of wave propagation in random media. J Radio Wave Propag 1(1):72–83

26. Flatte SM, Gerber JS (2000) Irradiance-variance behavior by numerical simulation for plane-wave and spherical-wave optical propagation through strong turbulence. J Opt Soc Am A 17(6):1092–1097
27. Li X, Ji X, Zhu J (2013) Higher-order intensity moments of light beams in atmospheric turbulence. Acta Physica Sinica 62(4):044217-1-7
28. Brown Jr WP (1971) Second moment of a wave propagating in a random medium. J Opt Soc Am 61(8):1051–1059
29. Filinov VS (1995) Monte Carlo simulation in the theory of wave propagation in random media: II. Scintillation index, second and fourth moments. Waves Random Media 5(3):277–281
30. Wu J (1999) Analytic solutions of the two-frequency and two-position mutual coherence function for wave propagating through random media. Chin J Electron 8(3):313–316
31. Lee SY, Liu CH, Yeh KC (1980) Research topics in electromagnetic wave theory. Wiley, New York
32. Ito S, Furutsu K (1982) Theoretical analysis of the high-order irradiance moments of light waves observed in turbulent air. J Opt Soc Am 72(6):760–764
33. Jian W, Liu R (1989) The solution of forward multiple scattering moment equation and its application in ionospheric scintillation. Chin Sci 19(12):1322–1332
34. Li G, Jian W, Huang J (2003) The solution of two-frequency and two-point mutual coherence function suitable for random ionospheric media. J Radio Wave Sci 18(2):157–160
35. Li X, Ji X, Zhu J (2013) Higher-order intensity moments of light beams in atmospheric turbulence. Acta Physica Sinica 62(4):217–223
36. Wang SCH, Plonus MA (1979) Optical beam propagation for a partially coherent source in the turbulent atmosphere. Opt Soc Am 69(9):1297–1304
37. Zhang Y, Song Z (1986) Fluctuation of laser beam propagation direction in turbulent atmosphere. Acta Opt 6(12):1111–1117
38. Yi X (2005) Research on the propagation characteristics of Gaussian beams in turbulent atmosphere. Xi'an University of Electronic Science and Technology, Xi'an
39. Whitman AM, Beran MJ (1970) Beam spread of laser light propagating in a random medium. J Opt Soc Am A 60(12):1595–1602
40. Poirier JL (1972) Beam spreading in a turbulent medium. J Opt Soc Am A 62(7):893–898
41. Fante RL (1975) Electromagnetic beam propagation in turbulent media. Proc IEEE 63(12):1669–1688
42. Fan C (1999) Atmospheric turbulence effect on oblique propagation of Gaussian beam waves. Chin J Quant Electron 16(6):519–525
43. Andrews LC (1995) Single-pass and double-pass propagation through complex paraxial optical systems. J Opt Soc Am A 12(1):137–150
44. Andrews LC, Miller WB (1995) The mutual coherence function and the backscatter amplification effect for a reflected Gaussian-beam wave in atmospheric turbulence. Wave Random Media 5(2):167–182
45. Andrews LC, Philips RL, Al-Halash MA (1999) Theory of optical scintillation. J Opt Soc Am A 16(6):1417–1429
46. Andrews LC, Phillips RL, Hopen CY et al (2001) Theory of optical scintillation: Gaussian-beam wave model. Wave Random Media 11(2):271–291
47. Andrews LC, Phillips RL, Miller WB Mutual coherence function for a double-passage retroreflected atmospheric turbulence. App Opt 36(3):698–708
48. Rao R (2005) Propagation of light in turbulent atmosphere. Anhui Science and Technology Press, Hefei, pp 4–9
49. Li X (2007) Principle and technology of satellite-to-ground laser communication link. National Defense Industry Press, Beijing, pp 118–121
50. Liu S, Liang F, Liu SI (2006) Atmospheric turbulence. Peking University Press, Beijing, 1–9
51. Wang L (2012) Study on the drift and expansion characteristics of light beams in atmospheric turbulence. Xidian University, Xi'an, 17–19
52. Cheng F (2012) Transmission characteristics of partially coherent flat-topped beams in a turbulent atmosphere. Soochow University, Suzhou, pp 9–11

53. Wu J, Le S (1988) Theory of light propagation in random media. Chengdu Institute of Telecommunications Engineering Press, Chengdu, 128–131
54. Yi X (2005) Research on the propagation characteristics of Gaussian beams in turbulent atmosphere. Xidian University, Xi'an, 14–17
55. Fei H (1995) Turbulence, intermittency and atmospheric boundary layer. Science Press, Beijing, pp 5–8
56. Huang S (2012) Research on the oblique transmission characteristics of partially coherent array beams. Beijing University of Posts and Telecommunications, Beijing, pp 27–28
57. Jian W, Chunping Y, Jianbin L (2005) Optical transport theory in atmosphere. Beijing University of Posts and Telecommunications Press, Beijing, pp 142–147
58. Dou L (2012) Research on the propagation characteristics of partially coherent optical ring beams along horizontal and oblique paths in atmospheric turbulence. Sichuan Normal University, Chengdu, pp 22–24
59. Zhang X, Quanquan D, Bin Z (2012) Beam expansion of partially coherent light propagating obliquely in a turbulent atmosphere. Acta Optics Sinica 32(12):1201001-1-7
60. Wei H, Zhensen W (2008) The broadening and drift characteristics of laser beam oblique propagation in atmospheric turbulence. J Radio Wave Propag 23(4):611–615
61. Yi X, Guo L, Wu Z (2005) Research on the scintillation problem of Gaussian beams in turbulent atmospheric sloping transmission. Acta Optics Acta Sinica 25(4):433–438
62. Zhimin Luo W, Zhensen GL (2002) Research on oblique propagation of light wave scintillation considering internal scale effect. J Xidian Univ 29(4):455–459
63. Haddon J, Vilar E (1986) Scattering induced microwave scintillations from clear air and rain on earth space paths and the influence of antenna aperture. IEEE Trans Antennas Propag 34(5):646–657
64. Shaik KS (1988) Atmospheric propagation effects relevant to optical communications. 34(5):646–657
65. Pearson JE (1976) Atmospheric turbulence compensation using coherent optical adaptive techniques. Appl Opt 15(3):622–631
66. Kaimal JC, Wyngaard JCJ, Izumi Y et al (1972) Spectral characteristics of surface-layer turbulence. Q J R Meteorol Soc 98(417):563–589
67. Kaimal JC, Eversole RA, Lenschow DH et al (1982) Spectral characteristics of the convective boundary layer over uneven terrain. J Atmosph Sci 39(5):1098–1114
68. Flatté SM (1983) Wave propagation through random media: contributions from ocean acoustics. Proc IEEE 71(11):1267–1294
69. Tatarskii VI (1961) Wave propagation in a turbulent medium. McGraw-Hill, New York
70. Baribeau R, Rioux M (1991) Influence of speckle on laser range finders. Appl Opt 30(20):2873–2878
71. Östlund P, Friberg AT (2001) Radiometry and far-zone polarization for partially coherent electromagnetic planar sources. Opt Commun 197(03):1–8
72. Fei W (2008) Theoretical and experimental research on partially coherent light. Zhejiang University, Zhejiang
73. Wolf E, Collett E (1978) Partially coherent sources which produce the same far-field intensity distribution as a laser. Opt Commun 25(3):293–296
74. De Santis P, Gori F, Guattari G et al (1979) An example of a Collett-Wolf source. Opt Commun 29(3):256–260
75. Deschamps J, Courjon D, Bulabois J (1983) Gaussian Schell-model sources: an example and some perspectives. JOSA 73(3):256–261
76. Tervonen E, Friberg AT, Turunen J (1992) Gaussian Schell-model beams generated with synthetic acousto-optic holograms. JOSA A 9(5):796–803
77. Shirai T, Korotkova O, Wolf E (2005) A method of generating electromagnetic Gaussian Schell-model beams. J Opt A Pure Appl Opt 7(5):232–242
78. Su X (2003) Different expressions of coherence of light. J Yuncheng Univ 21(5):12–14
79. Thiel KH, Wehr A (2004) Performance capabilities of laser scanners–an overview and measurement principle analysis. Int Arch Photogramm Remote Sens Spatial Inf Sci 36(8):14–18

80. Hennessy RT, Leibowitz HW (1972) Laser optometer incorporating the Badal principle. Behavior Res Methods Instrum 4(5):237–239
81. Haken H (1964) Theory of coherence of laser light. Phys Rev Lett 13(11):329
82. Noe R, Heidrich H, Hoffmann D (1988) Endless polarization control systems for coherent optics. J Lightwave Technol 6(7):1199–1208
83. Yang Y (2008) About the coherence of lasers. Discuss Phys Teach 26(307):36–42
84. Anfu F (1983) The spatial coherence of light. Sichuan Laser J 4(1):32–34
85. Jacobs SD, Cerqua KA, Marshall KL et al (1988) Liquid-crystal laser optics: design, fabrication, and performance. JOSA B 5(9):1962–1979
86. Takeda M, Wang W, Naik DN et al (2014) Spatial statistical optics and spatial correlation holography: a review. Opt Rev 21(6):849–861

Chapter 2
Light Intensity Distribution and Beam Expansion

Atmospheric turbulence can cause laser intensity scintillation, beam expansion, and spot drift, resulting in an increase in the bit error rate of the communication systems. Partially coherent light can suppress atmospheric turbulence compared to fully coherent light. This section discusses the intensity distribution and expansion of a laser beam after it travels through atmospheric turbulence.

2.1 Introduction

The transmission of laser light in a turbulent atmosphere has been studied extensively, and its theoretical foundation is becoming increasingly well-established [1]. Initially, researchers mainly focused on turbulent effects such as light intensity distribution and beam expansion of light waves in the horizontal transmission mode. In 1967, Feizulin investigated beam expansion when a laser beam was transmitted along an atmospheric channel and focused on the analysis of the effect of scattering and diffraction caused by inhomogeneous media on beam expansion [2]. In 1970, Whitma et al. found that when a light beam was transmitted in a random medium, its intensity distribution always obeyed a Gaussian distribution, and the size of its beam diameter was related to the transmission distance [3]. In the late 1970s, Leader investigated the transmission characteristics of partially coherent light in atmospheric turbulence in a study on the scattering characteristics of laser beams from rough surfaces [4, 5]. In 1978, Foley used a partially coherent Gaussian-Schell (GSM) beam as a model to study the effects of the initial beam radius and beam coherence on beam directionality [6]. In 1990, Wu Jian and Boradman discovered that "partially coherent light is more capable of suppressing turbulence than fully coherent light" in an investigation of the intensity distribution and coherence of partially coherent Gauss-Schell beams in the atmosphere [7]. In 1999, Santarsiero derived an expression for the Schell source (beam quality factor) factor and analyzed the far-field intensity distribution and beam

X. Ke, *Partially Coherent Optical Transmission Theory in Optical Wireless Communication*, Optical Wireless Communication Theory and Technology,
https://doi.org/10.1007/978-981-97-7158-5_2

distribution of a GSM beam and the first type of zero-order Bessel-Schell model (JSM) beam during transmission are analyzed [8]. In 2002, Greg and Emil studied the mean square beam width of partially coherent light in atmospheric turbulence and explained the physical mechanism by which partially coherent light is better than fully coherent light in suppressing turbulence [9]. In 2003, Tomohiro et al. deduced the beam expansion and angular expansion expressions for an equivalent GSM light source for transmission through atmospheric turbulence and found that a partially coherent Gauss-Schell beam can generate the same radiation as a fully coherent laser beam. In the same year, Ricklin used a partially coherent GSM beam as a laser source to study the effects of turbulence intensity and spatial coherence on wireless optical communication links, and under the condition of weak turbulence, the expressions of intensity fluctuation covariance and logarithmic intensity fluctuation of Gauss beam are derived [11]. In 2006, Zhang analyzed the variation law of beam expansion for different transmission conditions by numerically calculating the short-term turbulent expansion radius of Yura and Fante beams [12]. In 2008, Motlagh analyzed the influence of atmospheric turbulence on the signal-to-noise ratio and bit error rate of free-space optical communication systems under different turbulent conditions [13]. In 2009, Yang Ailin et al. used the Hermitian-hyperbolic Cosine-Gaussian beam as an example to study the beam width expansion and angular expansion of a beam during horizontal transmission in atmospheric turbulence. In addition, the transmission distance, atmospheric refractive index structure constant, beam, and the effects of order and beam waist width on the beam width and angular spread were analyzed, and physical explanations were provided for the obtained results [14]. In 2010, Borah et al. analyzed the relationship between the coherence length of a beam, the radius of curvature of the wavefront, the beam width, and other parameters, and the performance of a communication system. It was determined that when the initial beam width was large, the transmission distance and beam wavelength were short, and the use of partially coherent light is conducive to improving communication performance [15].

In satellite laser communication, aerial surveying, mapping, and other applications, laser beams are primarily transmitted along oblique paths. In 1999, Fan Chengyu analyzed the changes of logarithmic amplitude fluctuation variance, expansion radius, drift angle variance and other parameters of the Gaussian collimated beam during oblique propagation at different target heights according to the distribution of the near earth atmospheric structure constant with height measured in Kunming [16]. In 2006, Zhang deduced an expression for the intensity distribution of a finite beam as it propagated obliquely in atmospheric turbulence. The results revealed that the average intensity was not only related to the emission aperture half-plane, transmission distance, and long-term lateral coherence length, but also to the emission aperture radius [17]. In 2007, Zhang et al. analyzed the relationship between the equivalent radius of a Gaussian beam and the transmission distance, initial beam radius, and beam wavelength, considering the influence of the turbulent scale [18]. In In 2008, Qian XianMei et al. Used the multi-layer phase screen theory to analyze the variation of statistical parameters such as the effective radius of the beam, the root mean square of the spot centroid drift, and the power density with the zenith angle, laser wavelength, and initial radius when the laser is transmitted in the atmosphere at an oblique path [19]. In 2008, Chu et al. investigated the average light intensity and beam expansion of multi-Gaussian beams as they traveled obliquely through atmospheric turbulence and analyzed the changes in light intensity and expansion with parameters such as altitude, beam coherence length, and beam waist width [20]. In 2009, Qian Xianmei conducted a comparative study of uniformly distributed phase screens, equal Rytov index spaced phase screens, and equal Fried parameter spaced phase screens, and discovered that the equal Fried parameter spaced phase screen was the best solution when the beam travels along a non-uniform turbulent path [21]. In 2010, Chu et al. deduced an expression for the light intensity distribution of a partially coherent Gauss–Schell beam in a turbulent atmosphere and analyzed two cases of relay transmission and direct transmission. The light intensity distribution was superior to that of direct transmission, and the variation in the peak light intensity of the beam with the height of the target was also investigated [22]. In 2012, Zhang et al. used a partially coherent flat-top beam as an example to analyze the effects of the beam order, spatial coherence, zenith angle, and other parameters on the beam width as it propagated along an oblique path [23]. In 2013, Duan Meiling et al. studied the average intensity and mean square beam width of a GSM beam for transmission in the upward, downward, and horizontal paths in atmospheric turbulence, and discovered that when the beam was transmitted in the downlink in atmospheric turbulence, the influence of turbulence on the beam is the least [24]. In 2014, Yongtao et al. studied the intensity distribution of non-paraxial polychromatic partially coherent Laguerre Gaussian beams propagating in free space and found that the intensity distribution characteristics of the beam were determined by the initial parameters. When transmitted in space, the spectral shift was related to the initial parameters of the beam and the transmission distance [25].

2.2 Light Intensity Distribution and Beam Expansion

2.2.1 Geometric Model of Beam Propagation in Atmospheric Turbulence

A transmission model for a light beam propagating through the atmosphere is shown in Fig. 2.1. The main difference between oblique and horizontal transmission is the difference in the structure constants of the atmospheric refractive index. The atmospheric structure constant of the inclined transmission path changes with a change in altitude and is also affected by factors such as wind speed, temperature, humidity, and air pressure. In horizontal transmission, the atmospheric structure constant can generally assume a typical value. Therefore, in the case of inclined transmission, the situation is more complicated.

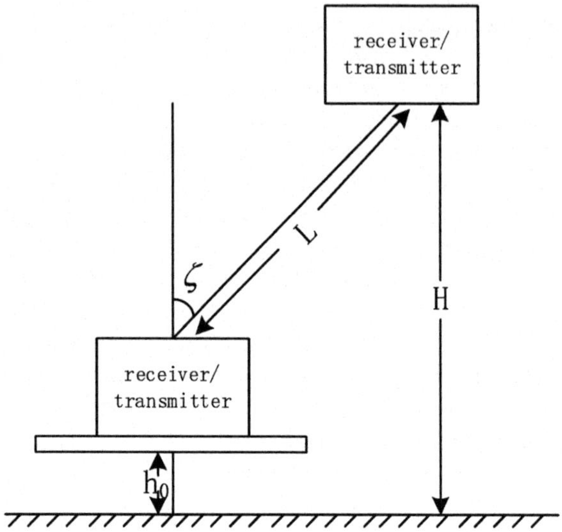

Fig. 2.1 Transmission model of light beams in the atmosphere

2.2.2 *Mathematical Model of Beam Propagation in Atmospheric Turbulence*

The cross-spectral density of a partially coherent GSM beam at z = 0 is

$$W^{(0)}(\boldsymbol{\rho}_{s1},\boldsymbol{\rho}_{s2},0)=A\exp\left(-\frac{|\boldsymbol{\rho}_{s1}|^2+|\boldsymbol{\rho}_{s2}|^2}{4\sigma_s^2}\right)\times\exp\left(-\frac{|\boldsymbol{\rho}_{s1}-\boldsymbol{\rho}_{s2}|^2}{2\sigma_g^2}\right) \tag{2.1}$$

where $\boldsymbol{\rho}_{s1}$, $\boldsymbol{\rho}_{s2}$ are the coordinate vectors of two points in the source plane, and the parameters A, σ_s and σ_g represent the light intensity, beam waist width, and coherence length of the light source, respectively.

Based on the generalized Huygens-Fresnel principle, the cross-spectral density at z after the GSM beam passes through atmospheric turbulence is given by [26]

$$W(\boldsymbol{\rho}_1,\boldsymbol{\rho}_2,z)=\left(\frac{k}{2\pi z}\right)^2\iint d^2\boldsymbol{\rho}_{s1}\iint d^2\boldsymbol{\rho}_{s2}W^{(0)}(\boldsymbol{\rho}_{s1},\boldsymbol{\rho}_{s2},0)\times\exp\left\{-\frac{ik}{2z}\left[(\boldsymbol{\rho}_1-\boldsymbol{\rho}_{s1})^2-(\boldsymbol{\rho}_2-\boldsymbol{\rho}_{s2})^2\right]\right\}\times\left\langle\exp\left[\phi*(\boldsymbol{\rho}_1,\boldsymbol{\rho}_{s1})+\phi(\boldsymbol{\rho}_2,\boldsymbol{\rho}_{s2})\right]\right\rangle \tag{2.2}$$

In the equation, $\boldsymbol{\rho}_1$, $\boldsymbol{\rho}_2$ are the coordinate vectors of the two points in the z-plane, and $\langle\cdot\rangle$ is the ensemble statistical average, which is defined as:

$$\langle \exp[\phi * (\boldsymbol{\rho}_1, \boldsymbol{\rho}_{s1}) + \phi(\boldsymbol{\rho}_2, \boldsymbol{\rho}_{s2})] \rangle = \exp\left[-\frac{1}{2} D_\phi(\boldsymbol{\rho}_{s1} - \boldsymbol{\rho}_{s2}, \boldsymbol{\rho}_1 - \boldsymbol{\rho}_2)\right] \quad (2.3)$$

where $D_\phi(\boldsymbol{\rho}_{s1} - \boldsymbol{\rho}_{s1}, \boldsymbol{\rho}_1 - \boldsymbol{\rho}_2)$ represents the wave structure function, which can be expressed as [27, 28] for oblique transmission.

$$D_\phi(\boldsymbol{p}, \boldsymbol{Q}, z) = 8\pi^2 k^2 \sec\zeta \int_{h_0}^{H} \int_{0}^{\infty} \kappa \phi_n(\kappa, h)\{1 - J_0[|(1-\xi)\boldsymbol{p} + \xi\boldsymbol{Q}|\kappa]\} \mathrm{d}\kappa \mathrm{d}h \quad (2.4)$$

There, ζ represents the zenith angle of the beam emission, and is the angle between the transmission path and the vertical direction. $\boldsymbol{J}_0$ is the zero-order Bessel function of the first type, which can be expanded as follows:

$$J_0(x) = 1 - \left(\frac{x}{2}\right)^2 + \frac{1}{(2!)^2}\left(\frac{x}{2}\right)^4 - \cdots \quad |x| < \infty \quad (2.5)$$

$\boldsymbol{p} = \boldsymbol{\rho}_1 - \boldsymbol{\rho}_2, \boldsymbol{Q} = \boldsymbol{\rho}_{s1} - \boldsymbol{\rho}_{s2}$, P is the difference between the coordinate vectors of two points in the z plane Q is the difference between the coordinate vectors of two points in the source plane $\phi_n(\kappa, h) = \phi_n(\kappa)C_n^2(h)$, and $\phi(\kappa)$ represent the refractive index power spectrum, and $C^2(h)$ is the atmospheric structure constant model, which are independent of each other. When the beam is transmitted upward, $\xi = 1 - \frac{h-h_0}{H-h_0}$, H represents the vertical height of the receiver above the ground, and h_0 is the vertical height of the transmitter above the ground. When the beam is transmitted downward, $\xi = \frac{h-h_0}{H-h_0}$.

Under strongly turbulent conditions, the Bessel function in Eq. (2.5) can be approximated using the first two terms [29] as follows:

$$J_0[|(1-\xi)\boldsymbol{p} + \xi\boldsymbol{Q}|\kappa] = 1 - \frac{1}{4} \times [(1-\xi)\boldsymbol{p} + \xi\boldsymbol{Q}]^2 \kappa^2 \quad (2.6)$$

Substituting Eq. (2.6) into Eq. (2.4), after integration and simplification, we obtain

$$D_\phi(\boldsymbol{p}, \boldsymbol{Q}, Z) = B_0(B_1\boldsymbol{p}^2 + B_2\boldsymbol{p}\cdot\boldsymbol{Q} + B_3\boldsymbol{Q}^2) \quad (2.7)$$

where,

$$B_0 = 2\pi^2 k^2 \sec\zeta \int_0^\infty 0.033\kappa^3(-\kappa^2/\kappa_m^2)(\kappa^2 + \kappa_0^2)^{-11/6} \mathrm{d}\kappa \quad (2.7a)$$

$$B_1 = \int_{h_0}^{H} C_n^2(h)(1-\xi)^2 dh \quad (2.7b)$$

$$B_2 = 2\int_{h_0}^{H} C_n^2(h)(1-\xi)\xi dh \tag{2.7c}$$

$$B_3 = \int_{h_0}^{H} C_n^2(h)\xi^2 dh \tag{2.7d}$$

Substituting (2.1) and (2.7) into (2.2), we obtain

$$\begin{aligned} W(\boldsymbol{\rho}_1, \boldsymbol{\rho}_2, z) = & \left(\frac{k}{2\pi z}\right)^2 \iint d^2\boldsymbol{\rho}_{s1} \iint d^2\boldsymbol{\rho}_{s2} A \\ & \exp\left(-\frac{|\boldsymbol{\rho}_{s1}|^2 + |\boldsymbol{\rho}_{s2}|^2}{4\sigma_s^2}\right) \times \exp\left(-\frac{|\boldsymbol{\rho}_{s1} - \boldsymbol{\rho}_{s2}|^2}{2\sigma_g^2}\right) \\ & \times \exp\left\{-\frac{ik}{2z}\left[(\boldsymbol{\rho}_1 - \boldsymbol{\rho}_{s1})^2 - (\boldsymbol{\rho}_2 - \boldsymbol{\rho}_{s2})^2\right]\right\} \\ & \times \exp\left[-\frac{1}{2}B_0(B_1\boldsymbol{p}^2 + B_2\boldsymbol{p}\cdot\boldsymbol{Q} + B_3\boldsymbol{Q}^2)\right] \end{aligned} \tag{2.8}$$

In Eq. 2.8, let $\boldsymbol{\rho}_1 = \boldsymbol{\rho}_2 = \boldsymbol{\rho}$. After the integral is performed, the intensity distribution of the GSM beam during the oblique transmission can be obtained

$$I(\boldsymbol{\rho}, z) = W(\boldsymbol{\rho}, \boldsymbol{\rho}, z) = \frac{A}{\Delta^2(z)} \exp\left[-\frac{|\boldsymbol{\rho}|^2}{2\sigma_s^2 \Delta^2(z)}\right] \tag{2.9}$$

where,

$$\Delta^2(z) = 1 + \left[\frac{z}{k\sigma_s\delta}\right]^2 + \frac{2Mz^2}{k^2\sigma_s^2} \tag{2.9a}$$

$$\frac{1}{\delta^2} = \frac{1}{4\sigma_s^2} + \frac{1}{\sigma_g^2} \tag{2.9b}$$

$$M = \frac{1}{2}B_0B_3 \tag{2.9c}$$

When the atmospheric refractive index structure constant $C_n^2(h)$ is constant (horizontal transmission) $M = \frac{1}{3}\pi^2k^2z\int_0^\infty \kappa^3\phi_n(\kappa)\mathrm{d}\kappa$ is constant (horizontal transmission), and for propagation through free space, $M = 0$.

The beam intensity is related to the atmospheric refractive index structure constant, beam coherence length, and beam waist width. Equation (2.9) is similar to the expression for the light intensity distribution when the beam is transmitted horizontally, as derived from the literature [10].

For the turbulent power spectrum model [30], the modified Von Karmon spectrum is used considering both the inner and outer scales, and the expression is given by:

$$\phi_n(\kappa, h) = 0.033C_n^2(h)\exp(-\kappa^2/\kappa_m^2) \times (\kappa^2 + \kappa_0^2)^{-11/6} \tag{2.10}$$

In the equation, $\kappa_m = 5.92/l_0$, $\kappa_0 \approx 2\pi/L_0$, where l_0 and L_0 represent the inner and outer scales of turbulence, respectively. $C_n^2(h)$ represents the atmospheric refractive index structure model, wherein the model proposed by ITU-R in 2001 that describes the variation with height is used, which is expressed as follows:

$$\begin{aligned} C_n^2(h) &= 8.148 \times 10^{-56} v_{RMS}^2 h^{10} e^{-h/1000} \\ &\quad + 2.7 \times 10^{-16} e^{-h/1500} + C_0 e^{-h/100} \end{aligned} \tag{2.11}$$

where $v_{RMS}^2 = \sqrt{v_g^2 + 30.69v_g + 348.91}$ is the vertical-path wind speed, v_g represents the near-surface wind speed, and C_0 is the near-surface atmospheric refractive index structure constant. Moderate turbulence C_0 lies between $10^{-17}\text{m}^{-2/3}$ and $10^{-13}\text{m}^{-2/3}$. In this study, the light intensity and expansion of the beam transmission are discussed under the assumption that the near-surface atmospheric refractive index structure constant is $C_0 = 1.7 \times 10^{-14}\text{m}^{-2/3}$. For horizontal transmission, h in Eq. (2.11) is the transmitter height.

To analyze the influence of atmospheric turbulence on a beam's intensity distribution, the normalized light intensity of the beam is defined as [31]

$$I^N(\boldsymbol{\rho}, z) = \frac{I_{ts}(\boldsymbol{\rho}, z)}{I_{fs}(\boldsymbol{\rho}, z)|_{\rho=0,\sigma_g\to\infty,M=0}} \tag{2.12}$$

where $I_{ts}(\boldsymbol{\rho}, z)$ is the intensity distribution of the beam when it travels obliquely in atmospheric turbulence, and $I_{fs}(\boldsymbol{\rho}, z)|_{\rho=0,\sigma_g\to\infty,M=0}$ is the on-axis light intensity value of the completely coherent beam when it travels obliquely in free space.

The normalized light intensity of the beam can be obtained as follows [32]:

$$I^N(\boldsymbol{\rho}, z) = \frac{1 + (z/2k\sigma_s^2)^2}{\Delta^2(z)} \exp\left[-\frac{\rho^2}{2\sigma_s^2\Delta^2(z)}\right] \tag{2.13}$$

Furthermore, the expression for the transmission of partially coherent light in free space can be derived [32]

$$I_N(\rho, z) = \frac{1 + \left(\frac{z}{2k\sigma_s^3}\right)^2}{1 + \left(\frac{z}{k\sigma_s^2\delta}\right)^2} \exp\left(-\frac{\rho^2}{2\sigma_s^2\left[1 + (\frac{z}{k\sigma_s^2\delta})^2\right]}\right) \tag{2.14}$$

The expression for transmission of partially coherent light in atmospheric turbulence is given by [32]:

$$I_N(\rho, z) = \frac{1 + \left(\frac{z}{2k\sigma_s^3}\right)^2}{1 + \left(\frac{z}{k\sigma_s^2\delta}\right)^2 + \frac{2M_3z^2}{k^2\sigma_s^2}} \exp\left(-\frac{\rho^2}{2\sigma_s^2\left[1 + \left(\frac{z}{k\sigma_s^2\delta}\right)^2 + \frac{2M_3z^2}{k^2\sigma_s^2}\right]}\right) \tag{2.15}$$

The equation for transmission of perfectly coherent light in free space is given by [32]:

$$I_N(\rho, z) = \exp\left(-\frac{\rho^2}{2\sigma_s^2\left[1 + \left(\frac{z}{2k\sigma_s^3}\right)^2\right]}\right) \tag{2.16}$$

The transmission of perfectly coherent light in atmospheric turbulence can be described by the equation [32]:

$$I_N(\rho, L) = \frac{1 + \left(\frac{z}{2k\sigma_s^3}\right)^2}{1 + \left(\frac{z}{2k\sigma_s^3}\right)^2 + \frac{2M_3z^2}{k^2\sigma_s^2}} \exp\left(-\frac{\rho^2}{2\sigma_s^2\left[1 + \left(\frac{z}{2k\sigma_s^3}\right)^2 + \frac{2M_3z^2}{k^2\sigma_s^2}\right]}\right) \tag{2.17}$$

According to the beam expansion RMS beam width definition [33],

$$\omega(z) = \sqrt{\frac{\int I(\boldsymbol{\rho}, z)|\boldsymbol{\rho}|^2 d^2\boldsymbol{\rho}}{\int I(\boldsymbol{\rho}, z) d^2\boldsymbol{\rho}}} \tag{2.18}$$

Substitute (2.9) into (2.18) to get [34]

$$\omega(z) = \left[2\sigma_s^2\Delta^2(z)\right]^{\frac{1}{2}} = \left[2\sigma_s^2 + \frac{2}{k^2\delta^2}z^2 + \frac{4M}{k^2}z^2\right]^{\frac{1}{2}} \tag{2.19}$$

To intuitively analyze the influence of turbulence on beam expansion, the relative beam width is defined as the ratio of the beam width in turbulence to the free-space beam width, and is expressed as [23]

$$\omega_r(z) = \left[1 + \frac{\frac{4M}{k^2}z^2}{2\sigma_s^2\frac{2}{k^2\delta^2}z^2}\right]^{\frac{1}{2}} \tag{2.20}$$

Moreover, the angular spread of the beam can also be obtained based on the definition of angular spread [35]

$$\theta = \lim_{z\to\infty}\frac{\omega(z)}{z} = \left[\frac{2}{k^2\delta^2} + \frac{4M}{k^2}\right]^{\frac{1}{2}} \tag{2.21}$$

Similar to Eq. (2.19), the first term of Eq. (2.21) represents the angular spread that occurs in free space, which is related to the beam parameter term, and the second term represents the angular spread caused by turbulence, where M is related to the turbulence term.

The relative angular expansion expression of the beam can then be obtained [36]

$$\theta_r = \frac{\theta_{turb}}{\theta_{free}} = \left[1 + 2M\delta^2\right]^{1/2} \tag{2.22}$$

In the equation, θ_{turb} is the angular spread of the light beam when it propagates through turbulent flow, and θ_{free} represents the angular spread of the light beam when it travels in through free space.

2.3 Numerical Analysis of Beam Transmission in Atmospheric Turbulence

2.3.1 *Numerical Analysis of Light Intensity Distribution*

The transmission mode of a light beam in the atmosphere can be divided into two types: horizontal and oblique transmissions, according to the zenith angle. Figure 2.2 analyzes the variation of the on-axis light intensity with the transmission distance when the beam travels along three different paths: the horizontal link, uplink, and downlink. From the figure, it is evident that when the beam is transmitted through the atmosphere, the light intensity value along the axis decreases with an increase in the transmission distance. At the same transmission distance, the light intensity value of the downlink is the largest, followed by the uplink. The light intensity of the horizontal transmission path is the lowest. The results show that when the beam is transmitted obliquely in atmospheric turbulence, the downlink transmission is less affected by the turbulence than the uplink transmission, and the horizontal transmission is the most affected.

When a light beam propagates in atmospheric turbulence, the light intensity on the receiving surface is weakened owing to random fluctuations in the atmospheric refractive index. Likewise, when a beam of light travels in free space, the intensity of the light is reduced owing to the effects of vacuum diffraction. Moreover, the coherence of the beam and the light intensity at the receiving surface are different. The normalized intensity distributions of partially coherent light and fully coherent light in atmospheric turbulence and free space are analyzed in Fig. 2.3 according to Eqs. (2.14)–(2.17).

As shown in this figure, whether the GSM beams are fully coherent or partially coherent, the intensity distribution is the same for transmission through free space for two scenarios: oblique transmission and horizontal transmission, i.e., when the

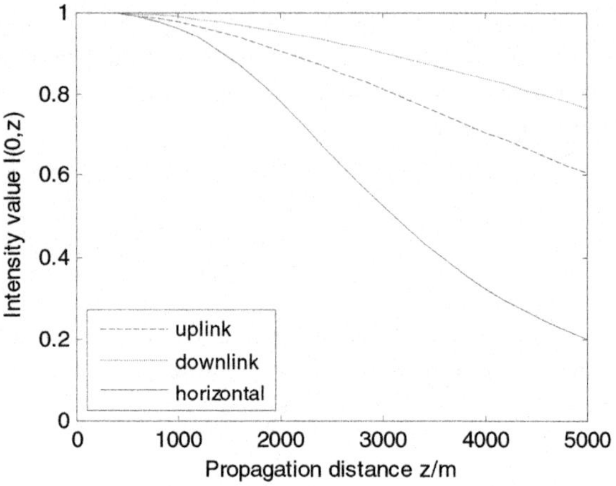

Fig. 2.2 Variation of on-axis light intensity with transmission distance for different transmission paths

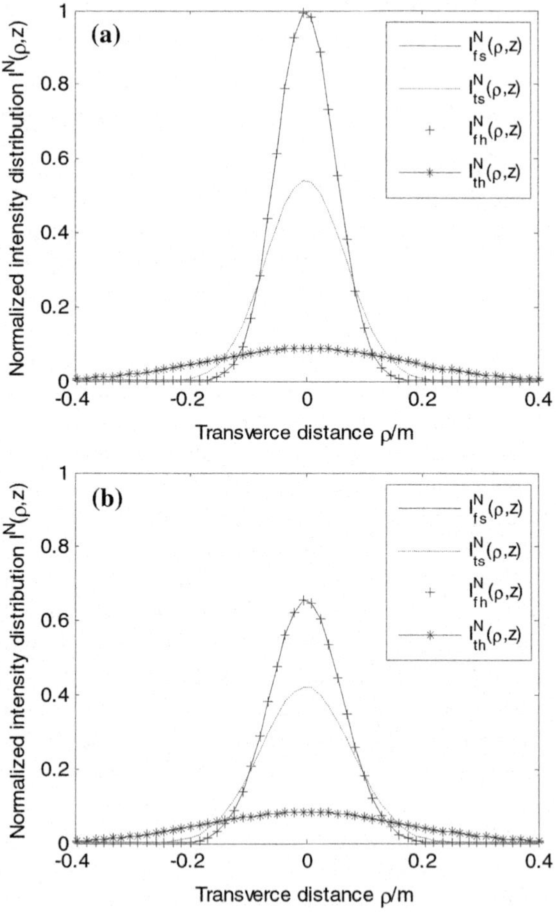

Fig. 2.3 Shows the normalized intensity distribution of fully coherent and partially coherent GSM beams during propagation in atmospheric turbulence. **a** Fully coherent GSM beam, **b** partially coherent GSM beam

beam is transmitted in free space. The intensity distribution is unrelated to the transmission path, which can be explained based on Eq. (2.9). When the beam is transmitted through atmospheric turbulence, the expression for the turbulence term varies according to the atmospheric refractive index structure constant in both oblique and horizontal transmission modes. For transmission in free space, the intensity does not differ according to the transmission path. It is evident that during transmission in atmospheric turbulence, the light intensity distribution during oblique transmission is more concentrated compared to the horizontal mode. For transmission through free space or atmospheric turbulence, the fully coherent GSM beam is more concentrated than the partially coherent GSM beam. Comparing the light intensity distributions of partially coherent GSM light and fully coherent GSM beams in a turbulent atmosphere to their respective light intensity distributions in free space, the intensity of partially coherent GSM light exhibits minimal difference. Therefore, it can be concluded that partially coherent GSM light has better anti-turbulence characteristics than fully coherent GSM light. This is consistent with the result for the partially coherent and the fully coherent light intensity distribution obtained when a beam propagates horizontally in the atmosphere [10].

Figure 2.4 shows the distribution of the normalized light intensity of partially coherent GSM light at different zenith angles. Consider the case of ramp transmission $h_0 = 2\,\text{m}$, $C_0 = 1.7 \times 10^{-14}\,\text{m}^{-2/3}$, $v_g = 2.8\,\text{m/s}$; in the case of horizontal transmission, $C_n^2(2) = 1.7 \times 10^{-14}\,\text{m}^{-2/3}$; $z = 7071\,\text{m}$; $l_0 = 0.01\,\text{m}$ $L_0 = 10\,\text{m}$; $\sigma_g = 0.02\,\text{m}$; $\sigma_s = 0.05\,\text{m}$; $\lambda = 650\,\text{nm}$. The zenith angle of the day is 0°, i.e., the light intensity distribution is most concentrated during vertical transmission, and the beam expansion is the smallest. The light intensity distribution during horizontal transmission is the most scattered, and the expansion is the most pronounced; the more dispersed the light intensity distribution, the more pronounced the beam spread. For other conditions, the light intensity distribution of the beam along the inclined path is always better compared to that of the horizontal transmission. The main physical reason for this result is that when the height is $C_n^2(h)$ lower than 10 km, the structure constant of the refractive index of the atmosphere decreases with an increase in height. Moreover, when it is greater than 10 km, it increases slightly, then decreases owing to the influence of the upper-air wind speed. The closer the beam is transmitted to the ground, the greater the influence of turbulence, and the beam quality is the worst.

According to Eq. (2.9c), the size of the turbulence term is related to the atmospheric structure constant near the ground and the inner and outer scales of the turbulence. Therefore, in Fig. 2.5 and Fig. 2.6, partially coherent GSM beams are transmitted through atmospheric turbulence and the atmospheric structure constants near the ground, and the internal and external scales of turbulence are analyzed.

For the same turbulence conditions, the light intensity distribution during oblique transmission of a beam is more concentrated compared to the case of horizontal transmission, and the peak light intensity on the axis is higher. The greater the turbulence intensity, the more scattered the light intensity distribution, and the more pronounced the beam expansion. This is because when a beam travels through the atmosphere, the random fluctuation of the atmospheric refractive index causes phase distortion, which

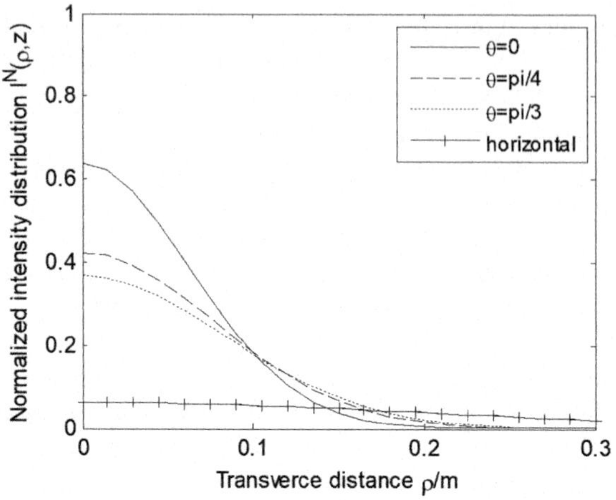

Fig. 2.4 Normalized light intensity distribution of partially coherent GSM light transmitted through atmospheric turbulence at different zenith angles

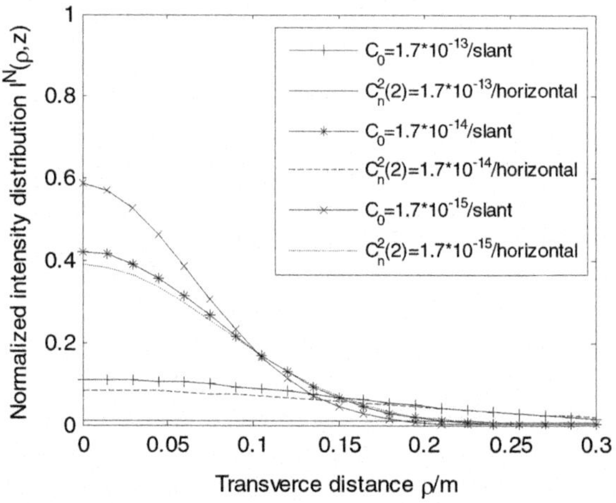

Fig. 2.5 Normalized light intensity distribution of partially coherent GSM beams for different turbulence intensities

causes the beam to expand beyond a certain distance, resulting in a more scattered intensity distribution and a lower peak intensity. The larger the random fluctuation, the lower the light intensity, and the more pronounced is the beam expansion.

Figure 2.6 shows the influence of the inner and outer scales of turbulence on the beam intensity distribution. As shown in Fig. 2.6a, the light intensity distribution of the inclined transmission mode is more concentrated than that of the horizontal mode, and the beam expansion is smaller. Moreover, the smaller the internal scale, the more scattered the intensity distribution, and the more pronounced the beam expansion. If the inner scale is assumed to be $l_0 = 0.01$ m, the beam-normalized intensity distribution is as shown in Fig. 2.6b. At different outer scales, the normalized intensity distribution curves are very close, and the intensity distribution of the beam is slightly affected by the change in the outer scale of the turbulent flow. When analyzing the influence of atmospheric turbulence on the beam intensity distribution,

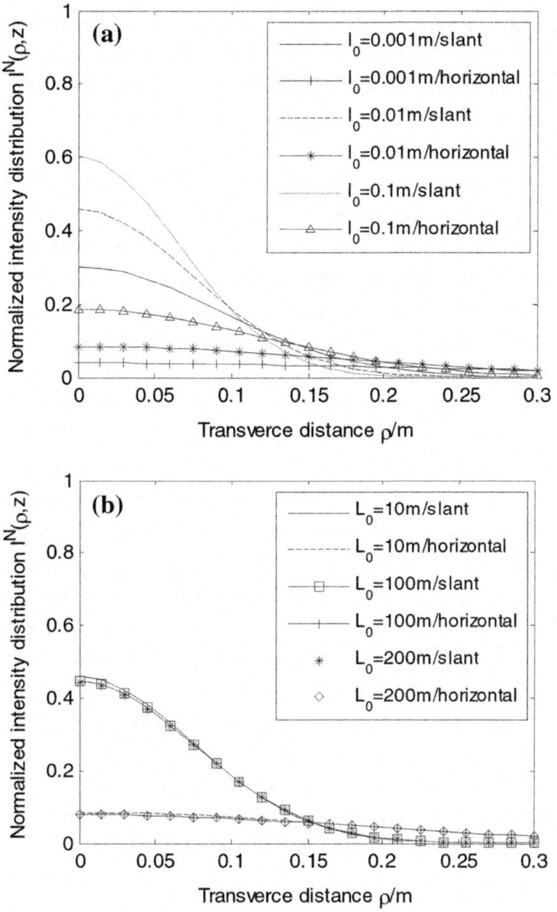

Fig. 2.6 **a** Normalized light intensity at different internal scales, **b** normalized light intensity at different outer scales

only the inner scale may be considered, and the outer scale can be ignored. The main physical reason for this result is that the smaller the inner scale of the turbulence, the smaller the turbulent vortices contained in the beam cross-section, and the more significant the diffraction of the beam that irradiates these vortices, which causes the intensity of the beam to vary in space and time. The random distribution appears to be more pronounced, causing the light to become more dispersed. The larger the outer scale of turbulence, the more significant the drift of the beam on the receiving surface and the less the effect on beam expansion.

It can be seen from Fig. 2.7 that the smaller the coherence length during oblique transmission, the lower the peak intensity, the more scattered the intensity distribution, and the more pronounced the beam expansion. The smaller the coherence length of the beam, the closer the transmission curve in turbulence and that in free space, which indicates that the smaller the coherence length of the beam, the smaller the influence of turbulence on the light intensity.. The intensity of light is determined by both free-space diffraction and turbulence. The lower the coherence, the less the light intensity is affected by turbulence. Moreover, the light intensity is weakened instead of enhanced, which is caused by the increased influence of Vacuum diffraction on the beam.

Figure 2.8 shows the variation of the light intensity with beam waist width for a partially coherent GSM beam at the center of the receiving plane for different turbulence intensities. The results show that the greater the turbulence intensity during oblique transmission, the greater the change in the intensity at the center of the receiving plane with the width of the beam waist. As the width of the beam waist increases, the intensity initially decreases rapidly, then gradually increases, and the light intensity is the weakest at the waist. Equation 2.15 indicates that the light intensity is related to the width of the beam waist. Thus, by taking its partial derivative, it can be determined that there is an extreme point of 0.019, which can explain why the light intensity is the weakest at this point. This finding can aid in the selection of an appropriate beam waist width for different practical applications.

The wavelength of the beam also affects its intensity distribution. Figure 2.9 is an analysis of the normalized distribution of light intensity when partially coherent GSM beams are transmitted at different wavelengths along an oblique path. It can be observed from the figure that the normalized light intensity of the beam is reduced with an increase in the wavelength during oblique transmission, indicating that the longer the wavelength, the weaker the normalized light intensity, and the shorter the wavelength, the higher the light intensity. This occurs because the atmosphere has different absorption lines for light beams of different wavelengths, resulting in different attenuation of light waves of different wavelengths. In addition, the intensity distribution is also different. Considering Eq. 2.9, the normalized light intensity of the beam is proportional to the wave number, and according to the relationship between the wave number and the wavelength, it can be deduced that the normalized light intensity $k = 2\pi/\lambda$ is inversely proportional to the wavelength.

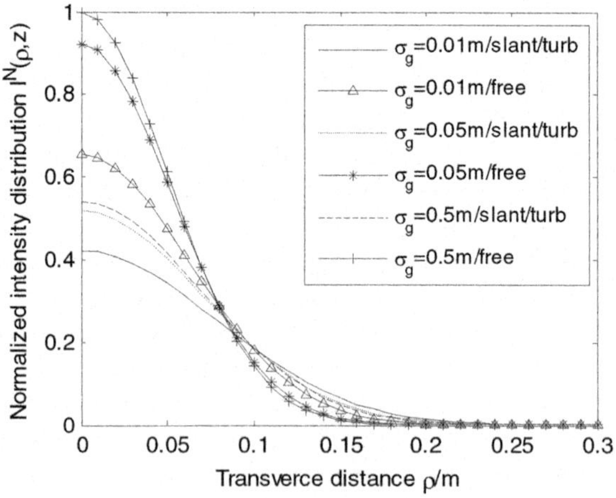

Fig. 2.7 Normalized light intensity distribution of partially coherent GSM beams with different coherence lengths

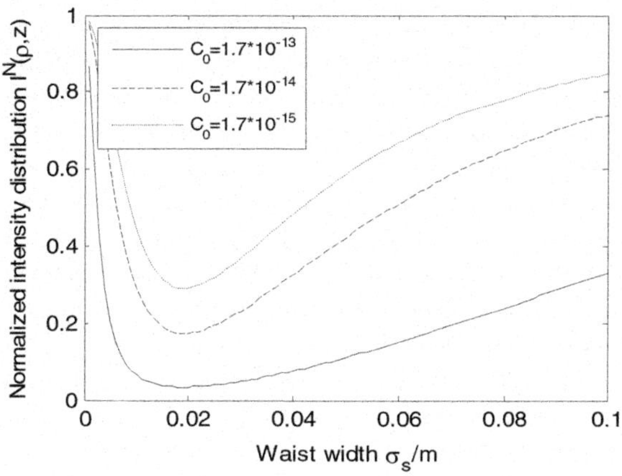

Fig. 2.8 Variation of normalized light intensity with beam waist width during partially coherent GSM beam transmission for different turbulence intensities

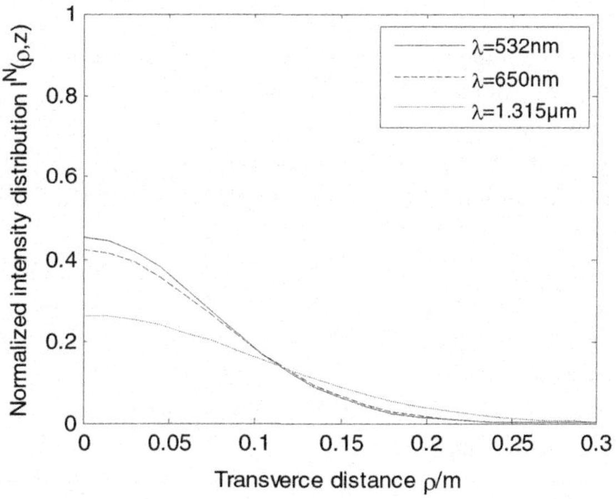

Fig. 2.9 Normalized light intensity distribution of partially coherent GSM beams at different wavelengths during oblique transmission

2.3.2 *Numerical Analysis of Beam Expansion*

Using Eqs. 2.18, 2.19, 2.20 and 2.21, the expansion and angular expansion of the GSM beam were systematically analyzed for propagation along the slant and horizontal transmission paths in atmospheric turbulence. Figure 2.10 shows the variation of the beam spread with transmission distance at different zenith angles. When considering oblique transmission, the transmitter height $h_0 = 2\,\text{m}$, near-surface atmospheric refractive index structure constant $C_0 = 1.7 \times 10^{-14}\,\text{m}^{-2/3}$,near-surface wind speed $v_g = 2.8\,\text{m/s}$; atmospheric refractive index structure constant for horizontal transmission $C_n^2(2) = 1.7 \times 10^{-14}\,\text{m}^{-2/3}$; inner scale $l_0 = 0.01\,\text{m}$; outer scale $L_0 = 10\,\text{m}$; coherence length $\sigma_g = 0.02\,\text{m}$; beam waist width $\sigma_s = 0.05\,\text{m}$, and wavelength $\lambda = 650\,\text{nm}$. It can be seen from Fig. 2.10a that when the zenith angle is constant, the beam expansion increases with the increase in the transmission distance, and when the transmission distance is constant, the larger the zenith angle, the more pronounced the beam expansion. At the top angle of the sky $\zeta < \pi/3$, the change curve of the beam expansion with transmission distance is very close to that of vertical transmission; within the range of the top angle of the sky $\zeta < \pi/3$, the beam expansion is relatively small and significantly smaller than that of horizontal transmission. From Fig. 2.10b, it can be seen that for a certain transmission distance, the larger the zenith angle, and the larger the relative beam width, and the greater the influence of atmospheric turbulence on the beam expansion. For vertically upward transmission, the relative beam width is the smallest, and the turbulence suppression ability of the beam is best. When the beam is transmitted horizontally, the relative beam width is the largest, and the turbulence suppression ability of the beam is at a minimum. This is caused by the decrease in the atmospheric refractive index structure constant with increasing height. The closer to the ground, the larger the atmospheric structure constant, the greater the influence of atmospheric turbulence on a beam during transmission, and the more pronounced the beam expansion. However, owing to the influence of the atmospheric refractive index, the wavefront phase distortion generated during the transmission of the beam becomes more significant with an increase in the transmission distance, which leads to an increase in the influence of atmospheric turbulence and a more pronounced beam expansion. These conclusions are consistent with the results in the literature [32] for the beamwidth and a relative beamwidth of partially coherent flat-topped (PCFT) beams as a function of the zenith angle.

Figure 2.11 shows the beam angular spread and relative angular spread as a function of the zenith angle. The parameters used in the calculation for oblique transmission include the total transmission distance $L = 7500\,\text{m}$, the refractive index structure constant of the near-surface atmosphere $C_0 = 1.7 \times 10^{-14}\,\text{m}^{-2/3}$, the coherence length $\sigma_g = 0.02\,\text{m}$, the beam waist width $\sigma_s = 0.05\,\text{m}$, the wavelength $\lambda = 650\,\text{nm}$, the near-surface wind speed $v_g = 2.8\,\text{m/s}$, the turbulent inner scale $l_0 = 0.01\,\text{m}$, the turbulent outer scale $L_0 = 10\,\text{m}$, and the transmitter height $h_0 = 2\,\text{m}$. It can be seen from the figure that when the beam is transmitted in a turbulent atmosphere, its angular spread and relative angular spread increase with the increase of the zenith angle when the zenith angle is less than $\pi/3$. When this angle is greater than $\pi/3$, i.e., close to the horizontal transmission, the angular spread and the relative angular spread increase rapidly with the increase of the zenith angle. This can be explained based on the variation of the atmospheric refractive index structure constant with height.

Figure 2.12 shows the variation of the beam angular spread and relative angular spread with transmission distance for oblique transmission with a zenith angle $\zeta = \pi/4$, near-surface atmospheric refractive index structure constant $C_0 = 1.7 \times 10^{-14}\,\text{m}^{-2/3}$, horizontal transmission, atmospheric refractive index structure constant $C_n^2(2) = 1.7 \times 10^{-14}\,\text{m}^{-2/3}$, coherence length $\sigma_g = 0.02\,\text{m}$, beam waist width $\sigma_g = 0.05\,\text{m}$, wavelength $\lambda = 650\,\text{nm}$,near-surface wind speed $v_g = 2.8\,\text{m/s}$, turbulent inner scale $l_0 = 0.01\,\text{m}$, turbulent outer scale $L^0 = 10\,\text{m}$, and emission machine height $h_0 = 2\,\text{m}$. As shown in Fig. 2.12a, b, when a beam is transmitted via the horizontal and oblique transmission modes, its angular spread and relative angular

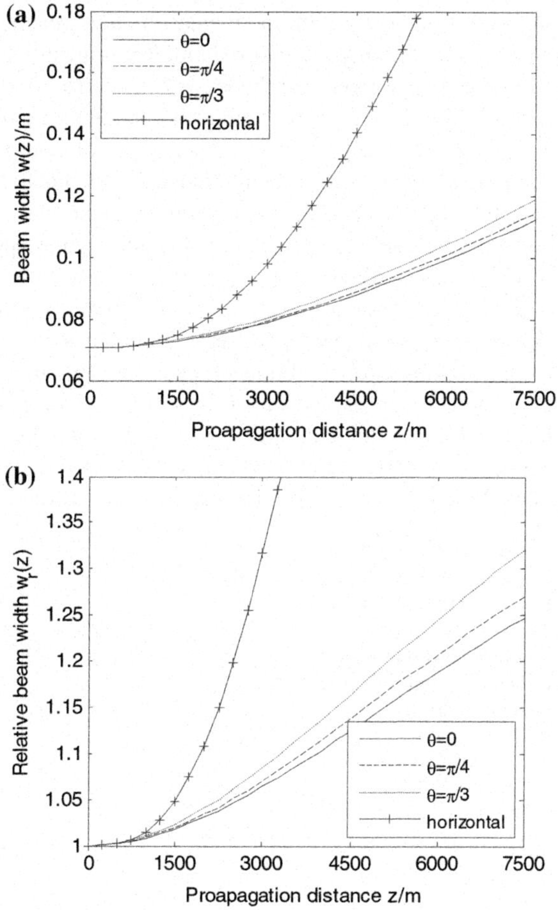

Fig. 2.10 Variation of beam spread with transmission distance for different zenith angles. **a** Beam width, **b** beam relative beam width

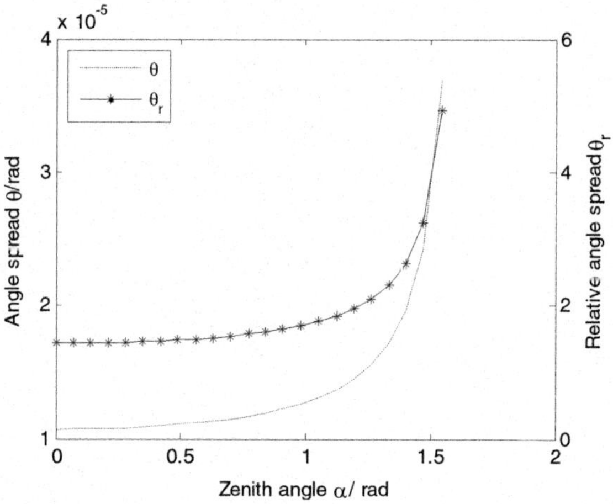

Fig. 2.11 Angular spread and relative angular spread of a beam as a function of the zenith angle

spread both increase with the increase in the transmission distance; the greater the transmission distance, the more significant is the influence of atmospheric turbulence on the angular spread, and the greater the corresponding angular spread. When the transmission distance is constant, the angular spread of the beam is less affected by turbulence during oblique transmission compared to horizontal transmission by approximately an order of magnitude. This is due to the deterioration of coherence and phase distortion as the beam propagates through atmospheric turbulence. As the transmission distance increases, the beam angular spread is affected by turbulence and becomes significant.

Figure 2.13 shows the variation of the partially coherent GSM beam spread with transmission distance under different turbulence intensities. The parameters selected for calculation during oblique transmission include the zenith angle $\zeta = \pi/4$, transmitter height $h_0 = 2\,\text{m}$, near-surface wind speed $v_g = 2.8\,\text{m/s}$, inner scale $l_0 = 0.01\,\text{m}$, outer scale $L_0 = 10\,\text{m}$, coherence length $\sigma_g = 0.02\,\text{m}$, beam waist width $\sigma_s = 0.05\,\text{m}$, and wavelength $\lambda = 650\,\text{nm}$. It can be seen from Fig. 2.13a that for the same turbulence intensity, the longer the transmission distance, the more

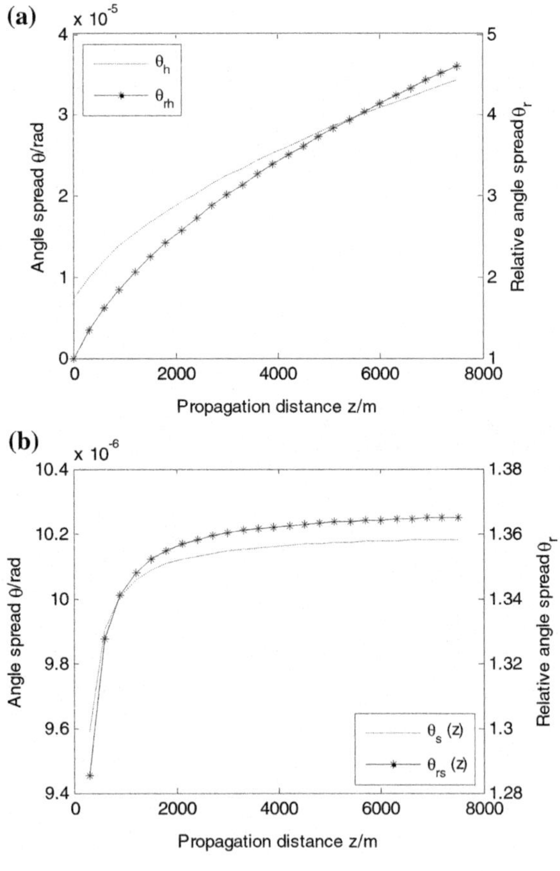

Fig. 2.12 Beam angular spread and relative angular spread as a function of transmission distance. **a** Horizontal transmission, **b** incline transmission

pronounced the beam expansion. When the transmission distance is constant, the higher the turbulence intensity, the more significant the beam expansion, and the transmission is superior. Figure 2.13b shows that for different turbulence intensities, the relative beam width increases with an increase in the transmission distance. The greater the turbulence intensity, the faster the growth rate, and the beam expansion is affected by the turbulence. Owing to the influence of the atmospheric refractive index, wavefront phase distortion occurs during the beam transmission process, and the larger the random fluctuation, the more significant the phase distortion. Thus, the effect of turbulence on beam expansion will increase. Moreover, the relative beam width for oblique transmission is always smaller compared to that of horizontal transmission, i.e., the anti-turbulence ability of oblique transmission is greater. This is because the atmospheric structure constants for oblique transmission are smaller than those for horizontal transmission close to the ground. Therefore, the turbulent effects of the oblique transmission of the beam are reduced.

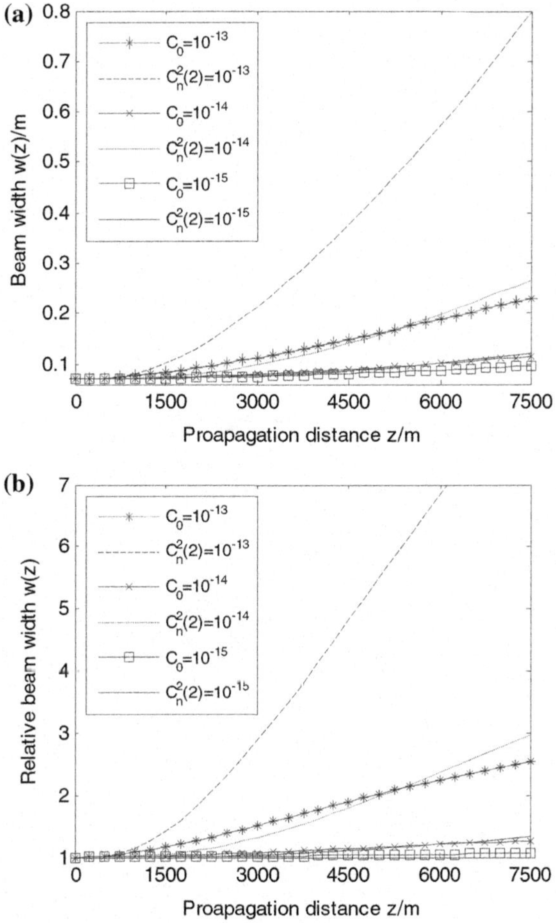

Fig. 2.13 Variation of beam spread with transmission distance for different turbulence intensities **a** beam width, **b** beam relative beam width

Figure 2.14 shows the beam angular spread and relative angular spread as a function of turbulence intensity for a total transmission distance $L = 7500\,\text{m}$, zenith angle for oblique transmission $\zeta = \pi/4$, coherence length $\sigma_g = 0.02\,\text{m}$, beam waist width $\sigma_s = 0.05\,\text{m}$, wavelength $\lambda = 650\,\text{nm}$, near-surface wind speed $v_g = 2.8\,\text{m/s}$, turbulent inner scale $l_0 = 0.01\,\text{m}$, turbulent outer scale $L_0 = 10\,\text{m}$, and transmitter height $h_0 = 2\,\text{m}$. It can be seen from Fig. 2.14 that the beam angular spread and relative angular spread increase with the increase of the turbulence intensity, and the stronger the turbulence, the greater the rate of increase. For a given turbulence intensity, the influence of turbulence on the angular spread of a beam for oblique transmission is much smaller compared to horizontal transmission. This is because the greater the fluctuation of the refractive index of the atmosphere, the greater the influence of turbulence on the transmission of the beam, and the greater the angular spread.

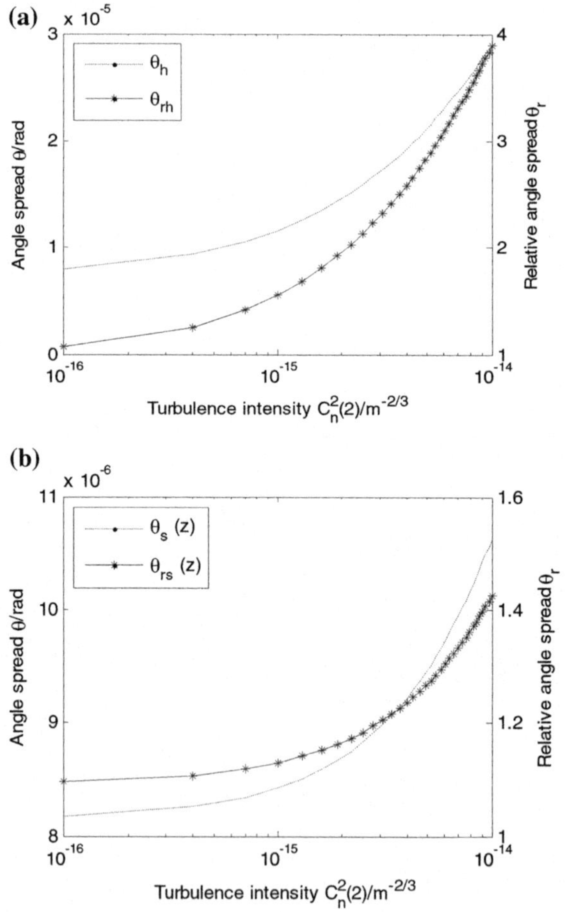

Fig. 2.14 Beam angular spread and relative angular spread as a function of turbulence intensity. **a** Horizontal transmission, **b** Incline transmission

Figure 2.15 shows the change in the beam expansion transmission distance under different inner (a) and outer scale (b) conditions for oblique transmission and a zenith angle $\zeta = \pi/4$, transmitter height $h_0 = 2\,\text{m}$, near-surface atmospheric refractive index structure constant $C_0 = 1.7 \times 10^{-14}\,\text{m}^{-2/3}$, near-surface wind speed $v_g = 2.8\,\text{m/s}$, atmospheric refractive index structure constant for horizontal transmission $C_n^2(2) = 1.7 \times 10^{-14}\,\text{m}^{-2/3}$, coherence length $\sigma_g = 0.02\,\text{m}$, beam waist width $\sigma_s = 0.05\,\text{m}$, and wavelength $\lambda = 650\,\text{nm}$. Figure 2.15a, b show the variation in the relative beam width with the transmission distance when the outer $L_0 = 10\,\text{m}$ and the inner scales $l_0 = 0.01\,\text{m}$ are assumed. Comparing Fig. 2.15a, b, it is evident that the relative beam width for oblique beam transmission is always smaller compared to that associated with horizontal transmission. As such, the anti-turbulence ability of the beam for oblique transmission is always superior to that of the horizontal mode beam. Moreover, the smaller the internal scale, the larger the relative beam width, the greater the influence of turbulence on the beam expansion, and the influence of the outer scale on the relative beam width is small and can be ignored. This is because the smaller the turbulent internal scale, the smaller the turbulent vortices contained in the beam cross-section, and the more significant the diffraction of the beam irradiated on these turbulent vortices. This results in an increased angular spread of the beam. The larger the outer scale of the turbulence, the more significant the drift of the beam on the receiving surface, and the less its influence on expansion and angular expansion.

Figure 2.16 shows the variation of the beam expansion with transmission distance for different coherence lengths for oblique transmission and a zenith angle $\zeta = \pi/4$, transmitter height $h_0 = 2\,\text{m}$, near-surface atmospheric refractive index structure constant $C_0 = 1.7 \times 10^{-14}\,\text{m}^{-2/3}$, near-surface wind speed $v_g = 2.8\,\text{m/s}$, atmospheric refractive index structure constant for horizontal transmission $C_n^2(2) = 1.7 \times 10^{-14}\,\text{m}^{-2/3}$, coherence length $\sigma_g = 0.02\,\text{m}$, beam waist width $\sigma_s = 0.05\,\text{m}$, and wavelength $\lambda = 650\,\text{nm}$. As shown in Fig. 2.16a, when the beam travels a certain distance, the larger the coherence length, the smaller the beam expansion, i.e., the worse the beam's coherence and the more pronounced the expansion. As shown in Fig. 2.16b, for a certain transmission distance, the larger the coherence length, the larger the relative beam width, which means that the beam's coherence is better and the beam expansion is more affected by turbulence. When the coherence is constant, beam expansion during oblique transmission is significantly smaller compared to that during horizontal transmission, and the corresponding turbulence suppression is also significantly greater. As the transmission distance increases, the beam becomes less coherent, and is less affected by turbulence.

Figure 2.17 shows the beam angular spread and relative angular spread as a function of coherence length for the total transmission distance $L = 7500\,\text{m}$, zenith angle $\zeta = \pi/4$, near-surface atmospheric refractive index structure constant for oblique transmission $C_0 = 1.7 \times 10^{-14}\,\text{m}^{-2/3}$, atmospheric refractive index structure constant for horizontal transmission $C_n^2 = 1.7 \times 10^{-14}\,\text{m}^{-2/3}$, beam waist width $\sigma_s = 0.05\,\text{m}$, wavelength $\lambda = 650\,\text{nm}$, near-surface wind speed $v_g = 2.8\,\text{m/s}$, turbulent inner scale $l_0 = 0.01\,\text{m}$, turbulent outer Scale $L_0 = 10\,\text{m}$, and transmitter height $h_0 = 2\,\text{m}$. It can be seen from the figure that when the coherence length is less than approximately

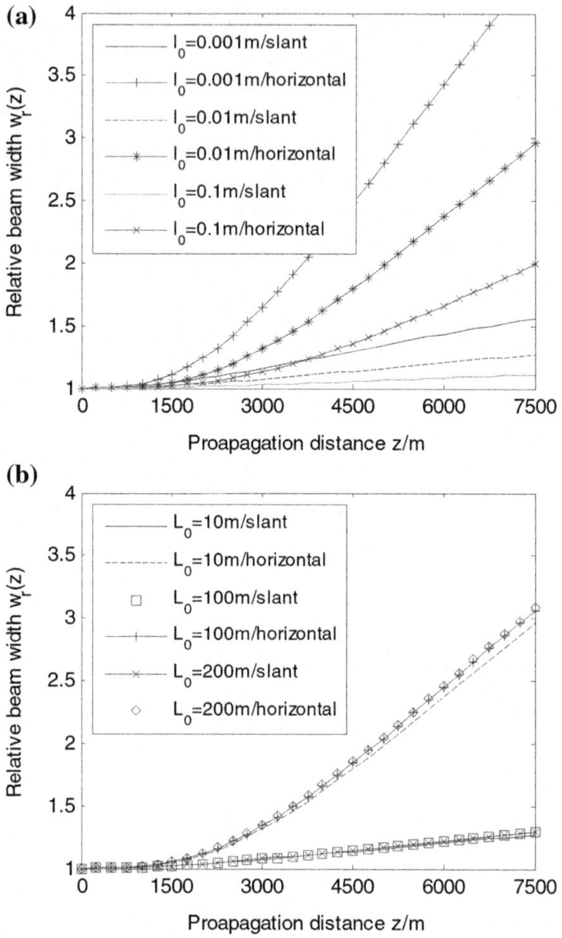

Fig. 2.15 Variation of relative beam width with transmission distance for different inner (**a**) and outer scale (**b**) conditions. **a** Relative beamwidth versus transmission distance at different internal scales, **b** relative beamwidth versus transmission distance at different outer scales

0.02 m, the beam angular spread decreases rapidly with the increase of this parameter. The smaller the angular spread, the more significant the angular spread, and the relative angular spread of the beam increases with the increase of the coherence length, i.e., the greater the coherence length of the beam, the greater the influence of turbulence. Comparing Fig. 2.17a, b, it is evident that for a certain coherence length, the effect of influence on the angular spread during oblique transmission of the beam is much smaller compared to the case of horizontal transmission.

Figure 2.18 shows the variation in the relative beam width with the beam waist width for different turbulence intensities when the beam is transmitted obliquely with a zenith angle $\zeta = \pi/4$, transmitter height $h_0 = 2\,\text{m}$, near-surface wind speed $v_g = 2.8\,\text{m/s}$, transmission distance $L = 7500\,\text{m}$, inner scale $l_0 = 0.01\,\text{m}$, outer scale $L_0 = 10\,\text{m}$, coherence length $\sigma_g = 0.02\,\text{m}$, and wavelength $\lambda = 650\,\text{nm}$. It can be seen from the figure that for the same turbulence intensity, the relative

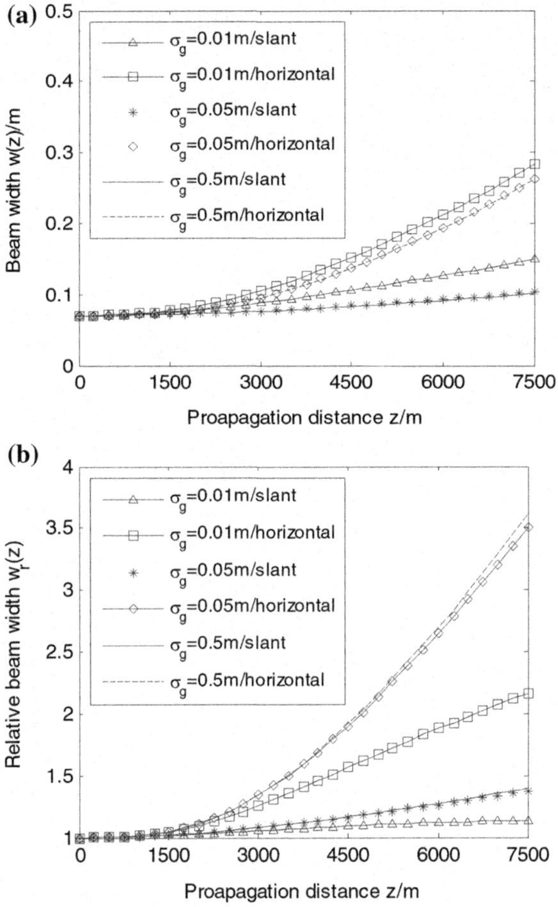

Fig. 2.16 Variation of relative beam width with transmission distance for different coherence lengths. **a** Beam expansion, **b** relative beam expansion

beam width first increases, then decreases with an increase in the beam waist width. Moreover, the greater the turbulence, the greater the maximum relative beam width, and the corresponding increase and decrease speeds are also greater. These results are consistent with the variation in the relative beam width of PCFT beams of different orders with the beam waist width of the fundamental mode Gaussian beam that has been reported in the literature [32].

Figure 2.19 shows the beam angular spread and relative angular spread as a function of the beam waist width for a total transmission distance $L = 7500\,\text{m}$, zenith angle $\zeta = \pi/4$, near-surface atmospheric refractive index structure constant for oblique transmission $C_0 = 1.7 \times 10^{-14}\,\text{m}^{-2/3}$, atmospheric refractive index structure constant for horizontal transmission $C_n^2 = 1.7 \times 10^{-14}\,\text{m}^{-2/3}$, coherence length $\sigma_g = 0.02\,\text{m}$, wavelength $\lambda = 650\,\text{nm}$, near-surface wind speed $v_g = 2.8\,\text{m/s}$, turbulent inner scale $l_0 = 0.01\,\text{m}$, turbulent flow Outer dimension $L_0 = 10\,\text{m}$, and transmitter height $h_0 = 2\,\text{m}$. It can be seen from the figure that when the beam waist

Fig. 2.17 Beam angular spread and relative angular spread as a function of coherence length.
a Horizontal transmission,
b incline transmission

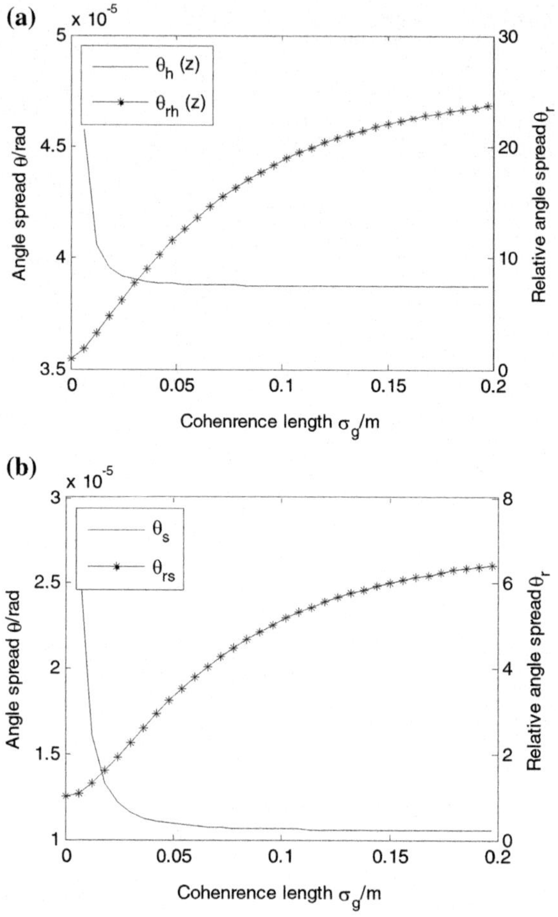

Fig. 2.18 Variation of relative beam width with beam waist width for different turbulence intensities

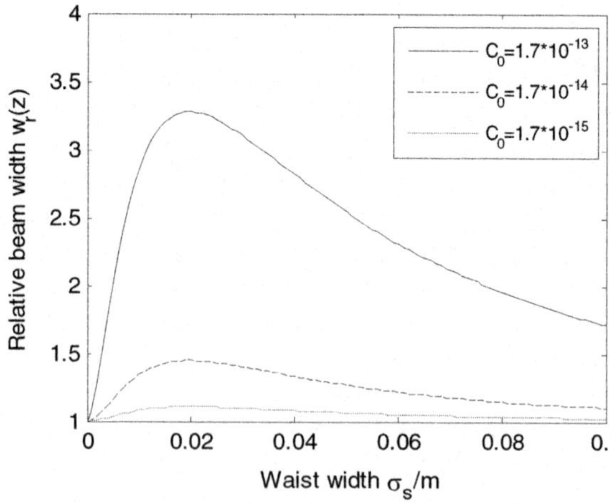

width is approximately 0.02 m, the beam angular spread decreases rapidly, whereas the relative angular spread increases rapidly with the increase of the beam waist width. When the beam waist width is greater than 0.02 m, the angular spread and the relative angle spread rapidly. The expansion is minimal with the increase of the beam waist width. In Fig. 2.19a, b, At the same beam waist width, compared with the horizontal propagation, the angular spread of the beam in oblique propagation is less affected by turbulence. The obtained results are consistent with the variation of the angular spread and the relative angular spread of the partially coherent Hermitian-Hyperbolic Cosine-Gaussian beam with the width of the beam waist in [17], which can be mutually confirmed.

Figure 2.20 shows the variation in the relative beam width with the transmission distance when the beam is transmitted along an oblique path for different wavelengths for a zenith angle $\zeta = \pi/4$, transmitter height $h_0 = 2\,\text{m}$, near-surface atmospheric refractive index structure constant $C_0 = 1.7 \times 10^{-14}\,\text{m}^{-2/3}$, near-surface wind speed $v_g = 2.8\,\text{m/s}$, inner scale $l_0 = 0.01\,\text{m}$, outer scale $L_0 = 10\,\text{m}$, coherence length $\sigma_g = 0.02\,\text{m}$, and beam waist width $\sigma_s = 0.05\,\text{m}$. It can be seen from the figure

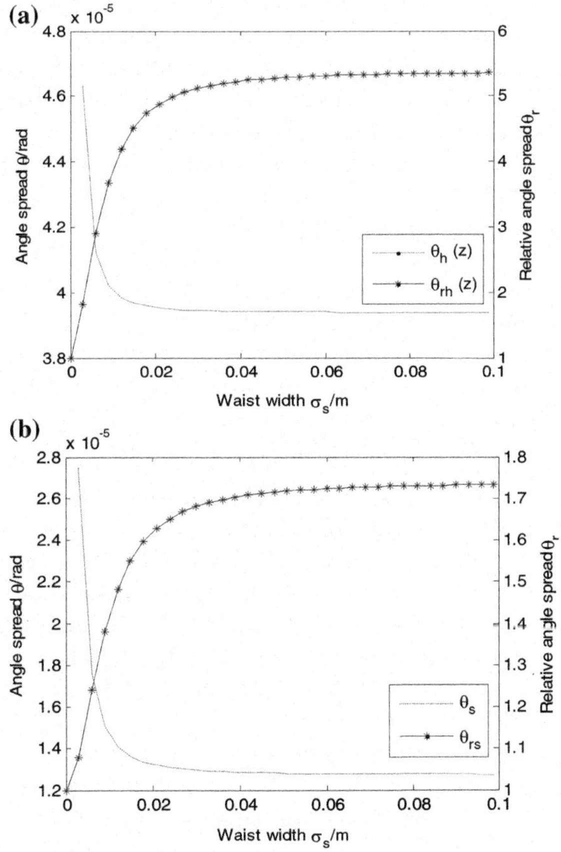

Fig. 2.19 Beam angular spread and relative angular spread as a function of beam waist width. **a** Horizontal transmission, **b** incline transmission

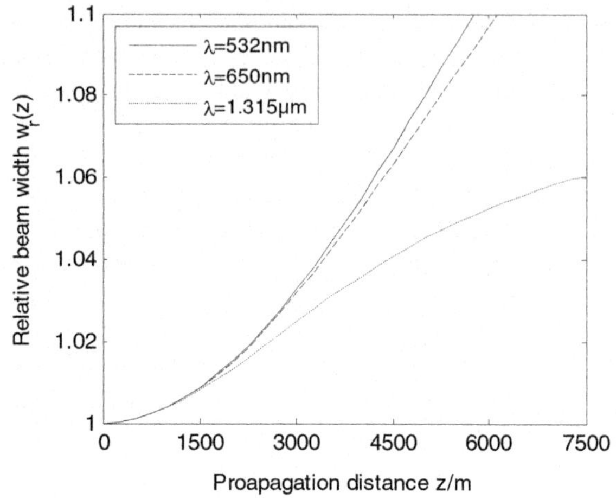

Fig. 2.20 Variation of relative beam width with transmission distance at different wavelengths

that for a certain transmission distance, the longer the wavelength, the smaller the relative beam width, and the less the effect of turbulence on the beam expansion, which means that the longer the wavelength, the greater the turbulence suppression characteristic. This is because of the different attenuation of light beams of different wavelengths when they are transmitted in the atmosphere. The corresponding beam expansion is different, and the relative beam width is also affected. Moreover, based on Eq. 2.20, the relative beam width is inversely proportional to the wavelength. Therefore, the shorter the wavelength and the greater the relative beam width, the greater the influence of turbulence.

In this chapter, the normalized intensity distributions of fully coherent and partially coherent GSM light propagating in atmospheric turbulence and free space were compared. In addition, the light intensity distribution and beam expansion of GSM beams traveling along different paths were systematically analyzed to determine the influence of parameters such as the vertex angle, transmission distance, turbulence intensity, and beam coherence length on the beam intensity, spread, and angular spread. Based on the findings, the following can be concluded:

(1) When a beam is transmitted through atmospheric turbulence, the turbulence suppression characteristics of partially coherent light are greater than those of fully coherent light, and the normalized light intensity distribution of a partially coherent GSM beam for horizontal transmission in free space is the same as that for the case of inclined transmission. The normalized light intensity distribution is the same and is most affected by turbulence when the beam travels along a horizontal path, followed by upward transmission, and downward transmission.
(2) The influence of turbulence on the intensity distribution, beam spread and angular spread of the oblique propagation of the beam in atmospheric turbulence is less than that of the horizontal propagation; The smaller the zenith angle, the more concentrated the normalized intensity distribution of the beam,

the stronger the peak intensity, the smaller the beam spread and angle spread, and the smaller the influence of turbulence.

(3) The expansion coefficient of the beam increases with an increase in the transmission distance, and the growth rate of the expansion coefficient in the case of horizontal transmission is much greater than that of the inclined transmission. Moreover, the stronger the turbulence intensity, the greater the effect of turbulence on the beam width and angular expansion. As the beam width and angular spread become more pronounced, the greater the beam coherence length, the smaller the beam width and angular spread, and the greater the turbulence effect on the beam.

(4) The influence of parameters such as the zenith angle, transmission distance, and turbulence intensity on the beam width and angular spread is essentially the same.

References

1. Qian X, Zhu W, Rao R (2008) Research progress of partially coherent light transmission in turbulent atmosphere. J Atmosph Environ Opt 3(2):81–90
2. Feizulin ZI, Kravtsov YA (1967) Broadening of a laser beam in a turbulent medium. Radio Phys Quant Electron 10(12):33–35
3. Whitma AM, Beran MJ (1970) Beam spread of laser light propagating in a random medium. J Opt Soc Am 60(02):1595–1602
4. Leader JC (1976) An analysis of the spatial coherence of laser light scattered from a surface with two scales of roughness. J Opt Soc Am 66(6):536–546
5. Leader JC (1978) Atmospheric propagation of partially coherent radiation. J Opt Soc Am 68(05):175–185
6. Foley JT, Zubairy MS (1978) The directionality of Gaussian Schell-model beams. Optics Commun 26(3):297–300
7. Wu J (1990) Propagation of Gaussian-Schell beam through turbulent media. J Mod Opt 37(4):671–684
8. Santarsiero M, Gori F, Borghi R et al (1999) Spreading properties of beams radiated by partially coherent Schell-model sources. J Opt Soc Am 16(02):106–112
9. Greg G, Emil W (2002) Spreading of partially coherent beams in random media. J Opt Soc Am A Opt Image Sci Vis 19(8):1592–1598
10. Tomohiro S, Aristide D, Emil W (2003) Mode analysis of spreading of partially coherent beams propagating through atmospheric turbulence. J Opt Soc Am A Opt Image Sci Vis 20(06):1094–1102
11. Ricklin JC, Davidson FM (2003) Atmospheric optical communication with a Gaussian Schell beam. J Opt Soc Am A 20(5):856–866
12. Zhang Y (2006) Comparison of two short-term expansion theories of turbulent atmospheric beams. Infrared Laser Eng 35(02):394–398
13. Motlagh AC, Ahmadi V, Ghassemlooy Z et al (2008) The effect of atmospheric turbulence on the performance of the free space optical communications. IEEE Trans Antennas Propag 12(05):540–543
14. Yang A, Li J, Lv B et al (2009) A comparative study of beam width and angular spread in atmospheric turbulence. Acta Physica Sinica 58(4):2451–2460
15. Borah DK, Voelz DG (2010) Spatially partially coherent beam parameter optimization for free space optical communications. Opt Express 20(18):20746–20758

16. Fan C (1999) Atmospheric turbulence effect of Gaussian beam wave slope transmission. Chin J Quantum Electron 18(12):519–525
17. Zhang Y, Wang G (2006) Slant path average intensity of finite optical beam propagating in turbulent atmosphere. Chin Opt Lett 4(10):559–562
18. Zhang Y, Wang G (2007) The average light intensity and short-term beam spread of a laser beam propagating in a slanted atmosphere. Infrared Laser Eng 36(2):167–171
19. Qian X, Zhu W, Rao R (2008) Numerical simulation analysis of turbulence effect of ground-to-air laser atmospheric sloping transmission. Infrared Laser Eng 37(5):787–792
20. Chu X, Liu Z, Wu Y (2008) Propagation of a general multi-Gaussian beam in turbulent atmosphere in a slant path. J Opt Soc Am A 25(1):74–79
21. Qian X, Zhu W, Rao R (2009) Phase screen distribution for numerical simulation of light propagation over non-uniform turbulent paths. Acta Physica Sinica 58(09):6633–6639
22. Chu XX, Liu ZJ, Wu Y (2010) Comparison between relay propagation and direct propagation of Gaussian-Schell-model beam in turbulent atmosphere along a slant path. Chin Phys B 19(9):94–102
23. Zhang X, Dan Q, Zhang B (2012) Beam expansion of partially coherent light propagating obliquely in a turbulent atmosphere. Acta Optics Sinica 32(12):1201–2127
24. Duan ML, Li JH, Wei JL (2013) Influence of different propagation paths on the propagation of laser in atmospheric turbulence. Opt Elec Lett 9(6):0477–0480
25. Zhang Y, Liu L, Wang F et al (2014) Average intensity and spectral shifts of a partially coherent standard or elegant Laguerre-Gaussian beam beyond paraxial approximation. Opt Quant Electron 46(12):365–379
26. Dan YQ, Zhang B (2008) Beam propagation factor of partially coherent flat-topped beams in a turbulent atmosphere. Opt Express 16(20):15563–15575
27. Andrews LC, Phillips RL (2005) Laser beam propagation through random media, 45(09). SPIE Optical Engineering Press, Bellingham, 195
28. Shirai T, Dogariu A, Wolf E (2003) Directionality of Gaussian Schell-model beams propagating in atmospheric turbulence. Opt Lett 28(8):610–612
29. Wei L, Liu LR, Sun JF, Yang QG, Zhu YJ (2007) Change in degree of coherence of partially coherent electromagnetic beams propagating through atmosphere turbulence. Opt Commun 27(1):1–8
30. Hua W, Xiangchao W, Aijun Z et al (2007) Influence of atmospheric turbulence on the temporal coherence of obliquely propagating quasi-monochromatic Gauss-Shell beams. Acta Optics Sinica 27(9):1–9
31. Xiang J, Wu Z, Wang M (2013) Broadening and drift of partially coherent Gauss-Shell beams in atmospheric turbulence. Infrared Laser Eng 42(3):658–662
32. Ke X, Wang W (2015) Light intensity and expansion of partially coherent light in oblique and horizontal atmospheric turbulence. J Appl Sci 33(2):142–154
33. Meiling D, Jinhong L, Jilin W (2013) Extension of partially coherent Hermitian Gaussian beams in oblique atmospheric turbulence. Intense Laser Particle Beam 25(9):2252–2256
34. Wang W (2015) Light intensity distribution and beam expansion of partially coherent light in atmospheric turbulence, 12(03). Xi'an University of Technology, Xi'an, 33–35
35. Yang AL, Zhang ET, Ji XL et al (2008) Angular spread of partially coherent Hermite-Cosh-Gaussian beams propagating through atmospheric turbulence. Opt Express 16(12):8366–8380
36. Ke X, Wang W (2015) Beam expansion and angular expansion of partially coherent light in atmospheric turbulence. Chin J Infrared Laser Eng 44(7):1–14

Chapter 3
Beam Drift

The effect of atmospheric turbulence on light waves is essentially the disruption of their coherence. Beam drift creates spot noise on the detector surface, thereby reducing the quality of communication and increasing the BER, as well as causing the spot to drift out of the detection range, which requires relocation searching. This section analyzes the drift and extension of a beam propagating through atmospheric turbulence.

3.1 Introduction

The theory of partially coherent optical atmospheric transport was developed during the early 1970s and the 1980s. Leader [1, 2] first observed variations in spatial coherence in a study of laser scattering on rough surfaces. In the early 1990s, Wu et al. [3, 4] investigated a partially coherent Gauss-Schell model for transport in turbulent media. In 2002, the transport properties of partially coherent light in atmospheric turbulence were explored by Ponomarenk [5] and Gbur et al. [6], who examined the radius and diffusion angle of a beam, respectively. In 2003, Salem et al. [8] also theoretically demonstrated that partially coherent light is more advantageous for transmission in a turbulent atmosphere. In the same year, Dogari et al. [9] experimentally demonstrated that "partially coherent light is less affected by turbulence than fully coherent light." Shirai et al. [10] provided a physical explanation for the resilience of partially coherent light to the influence of turbulence. In the same year, Borghi et al. [11] investigate the coherence properties of partially coherent beams by measuring the light intensity of partially coherent quasi-monochromatic light.

In 1960, Chernov [12] and Beckmann [13] simultaneously analyzed the drift of beams in turbulent atmospheres using the geometrical-optical approximation method (GOM), but without considering the finite size of the beams. In 1975, Richard [14] used Ehrenfest's law to obtain an expression for the arrival angle of a beam in a weakly

X. Ke, *Partially Coherent Optical Transmission Theory in Optical Wireless Communication*, Optical Wireless Communication Theory and Technology,
https://doi.org/10.1007/978-981-97-7158-5_3

undulating inhomogeneous turbulent medium and applied the von Karman spectrum to analyze the arrival angle undulation drift of Gaussian-focused and collimated beams. In 1990, Chumside and Lataitis [15] used the GOM to develop an expression including the drift variance of collimated and focused beams In 2005, Andrews [16, 17] proposed a drift model for weak turbulence conditions based on the Rytov approximation and the GOM, introduced a Gaussian filter function, and attempted to extend it to strong turbulence modes. In 2007, Berman [18] used the photon distribution function to analyze spot drift characteristics. Qian et al. [19] numerically simulated the expansion and drift of laser beams of different wavelengths using the multilayer phase screen method, and showed that the larger the zenith angle, the greater the intensity of the turbulence effect and the greater the influence of turbulence [19]. It was shown that the shorter the wavelength, the greater the influence of turbulence, the drift of the spot decreases with the increase in the initial radius of the beam, and is independent of the wavelength. Voelz et al. [20] investigated the spot drift of a focused partially coherent beam in atmospheric turbulence based on geometrical optics approximation and re-examined the beam size, mean light intensity distribution, and scintillation index of the beam.

Zhang and Song [21] discussed the effects of the degree of spatial coherence of the light source, the turbulence spectrum, and the wavelength on the spot drift, and analyzed the drift variation for different turbulence intensities and transmission distances. In the same year, in the literature [22], they also used the Markov approximation to derive general expressions for the beam drift angle and arrival angle undulation for plane mirror reflection, and investigated the strong and weak turbulence undulation regions. In 1987, Tan and Huang [23] calculated the drift variance of laser beams in turbulent atmospheres. A specific expression for calculating the drift variance was derived, and the results were discussed. Rao [24] conducted an experimental study on the spot drift of focused laser beams using an array detection system and found that the vertical drift is slightly larger than the horizontal drift, and the beam drift in different cases is compared and analyzed., In 2005, Yi [25] derived an expression for the drift of the horizontal transmission of a Gaussian beam in a turbulent atmosphere in the weakly undulating region based on the Markov approximation, assuming that the complex amplitude of the light wave satisfied the random field of the normal distribution, and is axisymmetric about the observation point. Gao et al. [26] expressed the drift of the horizontal transmission of a Gaussian beam in a weakly undulating region based on the Markov approximation, [26] and developed a beam drift variance that was suitable for application to the strong undulation region based on the weak undulation theory. Their results were in good agreement with the experimental data. They also conducted a numerical analysis of the effects of the spatial coherence length, wavelength, and turbulence intensity of the light source on the beam drift and discussed the factors that affected the beam drift. Wang [27] et al. designed a spot drift detection system to realize close-range detection of a beam from a He–Ne laser. In 2008, Hongyan [28, 29] used the Mellin transform to derive the beam drift variance of a collimated laser beam transmitted through a ground-to-air turbulent atmosphere in the weak undulation range without considering the turbulent inner and outer scales, but considering the turbulent outer scale. Qian and

Rao [30] investigate beam expansion, spot drift, and light intensity scintillation of partially coherent light using numerical simulation, further verifying the conclusion that "turbulence has less effect on partially coherent light". In 2011, Wu et al. [31] investigated the beam drift variance of partially coherent electromagnetic Gaussian light. Li and Wu [32, 33] analyzed the drift of partially coherent light based on the light intensity distribution and compared the results with those of their predecessors. In 2014, Gao et al. [34] discussed the drift and polarization properties of beams in oblique-range atmospheric turbulence, using electromagnetic Gauss-Schell beams as an example.

3.2 Beam Drift of Partially Coherent Beams in Atmospheric Turbulence

The beam drift is mainly due to the refractive effect of large-scale turbulent vortices. When the beam diameter is much smaller than the turbulence size, the main effect of turbulence is the random deflection of the beam away from its original transmission direction. This occurs when the transmission direction of the beam or the projected position on the receiving surface drifts randomly, i.e., when beam drift occurs, as shown in Fig. 3.1. The change in the drift of a partially coherent GSM beam in atmospheric turbulence is discussed in the following section [35].

Beam drift is usually described by a change in the position of the center of mass of the spot. The coordinates of the center-of-mass of the spot are defined as [35].

$$\rho = \frac{\iint \rho I(\rho) d\rho}{\iint I(\rho) d\rho} \tag{3.1}$$

i.e.

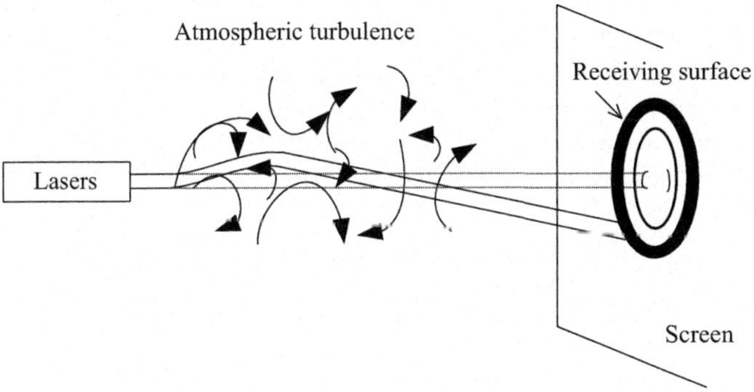

Fig. 3.1 Beam drift of partially coherent beams in atmospheric turbulence [35]

$$x_c = \frac{\iint xI(x, y)dxdy}{\iint I(x, y)dxdy}, y_c = \frac{\iint yI(x, y)dxdy}{\iint I(x, y)dxdy} \quad (3.2)$$

The drift variance of the center of mass is then given by:

$$\sigma_\rho^2 = \langle \rho_c^2 \rangle = \frac{\iint \iint (\rho_1 \cdot \rho_2) I(\rho_1) I(\rho_2) d^2\rho_1 d^2\rho_2}{\left[\iint I(\rho) d\rho\right]^2} \quad (3.3)$$

3.3 Beam Drift and Beam Extension for Horizontal Transmission

3.3.1 *Beam Drift*

Andrews used the Rytov approximation to obtain the mean light-intensity distribution along an axis [16].

$$\langle I(0, L) \rangle = \frac{w_0^2}{W_{LT}^2} = \frac{w_0^2}{W^2(1+T)} \quad (3.4)$$

where $\langle \cdot \rangle$ represents tethered averaging, W_{LT} is long-term beam extension, and W_{LT}^2 can be expressed as [16],

$$W_{LT}^2 = \underbrace{W^2}_{diffrac} + \underbrace{W^2 T_{ss}}_{\substack{sm-scale \\ spread}} + \underbrace{W^2 T_{Ls}}_{\substack{l\arg e-scale \\ beam\ wander}} \quad (3.5)$$

Decomposing T in Eq. 3.4 as $T = T_{ss} + T_{LS}$, T_{ss} and T_{LS} yield the beam expansion due to small-scale turbulence and the beam drift due to large-scale turbulence, respectively. Therefore, the last term of Eq. 3.5 can be interpreted as the beam drift, ρ_c, and the sum of the first and second terms are the short-term beam expansion.

Andrews and Philips used the Rytov approximation to analyze the last term in Eq. 3.5 to obtain a classical drift variance model [16],

$$\langle \rho_c^2 \rangle = W^2 T_{LS} = 4\pi^2 k^2 w^2(L) \int_0^L \int_0^\infty \kappa \Phi_n(\kappa) H_{LS}(\kappa, z) \times \left[1 - e^{-\Lambda_p L\kappa^2 (1 - z/L)^2 / k}\right] d\kappa dz \quad (3.6)$$

where Λ_P is the beam output parameter, $\Lambda_P = 2L/kw^2(L)$, and $w(L)$ is the radius of the beam at the receiving end in free space. Since the main factor that causes beam drift is large-scale turbulence, a large-scale filter function was introduced, which contributed to beam expansion when the size of the random inhomogeneous turbulent vortex was equal to or larger than the beam size. This function can be expressed as:

$$H_{\mathrm{LS}}(\kappa, z) = \exp[-\kappa^2 w^2(z)] \tag{3.7}$$

To emphasize the natural properties of beam drift, applying the geometric optics approximation to the last term in Eq. 3.6 yields

$$1 - e^{-\Lambda_p L\kappa^2(1-z/L)^2/k} \cong \frac{\Lambda_\mathrm{p} L\kappa^2(1-z/L)^2}{k}, \quad L\kappa^2/k << 1 \tag{3.8}$$

Using Eqs. 3.7 and 3.8 to simplify Eq. 3.6 yields

$$\langle \rho_c^2 \rangle = 8\pi^2 L^2 \int_0^L \int_0^\infty \kappa^3 \Phi_\mathrm{n}(\kappa) \exp\left[-k^2 w^2(z)\right] \times (1 - \frac{z}{L})^2 \mathrm{d}\kappa \mathrm{d}z \tag{3.9}$$

Since the beam drift is mainly due to large-scale turbulence, the exponential spectrum was chosen to consider the large-scale [37]

$$\Phi_n(\kappa) = 0.033 C_n^2 \kappa^{-11/3}\left[1 - \exp(-\kappa^2/\kappa_0^2)\right] \tag{3.10}$$

In Eq. 3.10, κ_0 is the spatial wave number corresponding to the outer scale L_0, $\kappa_0 = 1/L_0$, and C_n^2 is the refractive index structure constant, which characterizes the strength of the turbulence and is constant for horizontal transmission.

After substituting Eqs. 3.7 and 3.10 into Eq. 3.6 and integrating the result, we obtain:

$$\langle \rho_c^2 \rangle = 7.25 C_n^2 L^2 \int_0^L \left(1 - \frac{z}{L}\right)^2 w^{-1/3}(z) \left[1 - \left(\frac{\kappa_0^2 w^2(z)}{1 + \kappa_0^2 w^2(z)}\right)\right]^{1/6} dz \tag{3.11}$$

In Eq. 3.11, $w(z)$ is the radius of expansion of the partially coherent GSM beam at the reception point in atmospheric turbulence. To derive an expression for drift in the case of partially coherent GSM light and numerical analysis of this phenomenon, it is necessary to analyze $w(z)$ and demonstrate that it can be applied to various turbulence models including moderate, strong, and weak turbulence.

3.3.2 Beam Extension Radius

The complex amplitude distribution of the Gaussian beam in free space at z = 0, for the evening axis condition, is [36]

$$U(\mathbf{r},0) = \exp\left[-\left(\frac{1}{w_0^2} + \frac{jk}{2F_0}\right)\mathbf{r}^2\right] \tag{3.12}$$

In Eq. 3.12, $r = |\mathbf{r}| = (x^2 + y^2)^{1/2}$ denotes a two-dimensional radial vector that is perpendicular to the transmission direction, w_0 and F_0 are the initial beam radius and phase-front radius of curvature, respectively, and $k = 2\pi / \lambda$ is the number of light waves in free space.

According to the generalized Huygens-Fresnel principle [37], the cross-spectral density function at the receiving end of a partially coherent GSM beam after transmission in atmospheric turbulence is given by [20]

$$\begin{aligned} W(\rho_1,\rho_2;L) &= \langle U(\rho_1,L)U^*(\rho_2,L)\rangle \\ &= \frac{1}{(\lambda L)^2}\iint d^2\mathbf{r}_1 d^2\mathbf{r}_2 W(\mathbf{r}_1,\mathbf{r}_2,0) \times \langle \exp[\psi(\mathbf{r}_1,\rho_1) + \psi^*(\mathbf{r}_2,\rho_2)]\rangle \\ &\quad \times \exp\left\{\frac{ik}{2L}\left[(\rho_1 - \mathbf{r}_1)^2 - (\rho_2 - \mathbf{r}_2)^2\right]\right\} \end{aligned} \tag{3.13}$$

where $W(\mathbf{r}_1,\mathbf{r}_2,0)$ is the Cross-Spectral Density Function (CSDF) at the emitting end. Considering the phase spread at the emission aperture, the emission field can be expressed as follows:

$$\widetilde{U}(\mathbf{r},0) = U(\mathbf{r},0)\exp[i\varphi_d(\mathbf{r})] \tag{3.14}$$

where $\exp[i\varphi_d(\mathbf{r})]$ describes the random disturbance caused by phase diffusion. Assuming that the spatial random phase is an independent Gaussian function, the CSDF of the GSM beam at the transmitting end can be expressed as:

$$\begin{aligned} W_0(\mathbf{r}_1,\mathbf{r}_2,0) &= \langle \widetilde{U}(\mathbf{r}_1,0)\widetilde{U}^*(\mathbf{r}_2,0)\rangle \\ &= U(\mathbf{r}_1,0)U^*(\mathbf{r}_2,0) \\ &\quad \times \langle \exp[i\psi_{d1}(\mathbf{r}_1)]\exp[i\psi_{d2}(\mathbf{r}_2)]\rangle \\ &= U(\mathbf{r}_1,0)U^*(\mathbf{r}_2,0)\exp\left[-(\mathbf{r}_1 - \mathbf{r}_2)^2 / 2\sigma_g^2\right] \end{aligned} \tag{3.15}$$

The parameter σ_g^2 is the Gaussian variance of the ensemble mean that describes the random phase that characterizes the spatial coherence of the emitting light source.

$$\begin{aligned} \mathbf{r_c} &= (\mathbf{r}_1 + \mathbf{r}_2)/2,\ \mathbf{r}_d = \mathbf{r}_1 - \mathbf{r}_2 \\ \rho_c &= (\rho_1 + \rho_2)/2,\ \rho_d = \rho_1 - \rho_2 \end{aligned} \tag{3.16}$$

Substituting Eqs. 3.14 and 3.16 into Eq. 3.15 yields

$$W_0(\mathbf{r}_c, \mathbf{r}_d, 0) = \exp\left\{-\frac{1}{w_0^2}\left[\frac{1}{2}\left(\mathbf{r}_d^2 + 4\mathbf{r}_c^2\right)\right] - \frac{ik}{2F_0}(2\mathbf{r}_d \cdot \mathbf{r}_c) - \frac{\mathbf{r}_d^2}{2\sigma_g^2}\right\} \tag{3.17}$$

According to Eqs. 3.16, 3.17 can also be written as

$$\exp\left\{\frac{jk}{2L}\left[(\rho_1 - \mathbf{r}_1)^2 - (\rho_2 - \mathbf{r}_2)^2\right]\right\} = \exp\left\{\frac{jk}{L}\left[(\mathbf{r}_c - \rho_c)\cdot(\mathbf{r}_d - \rho_d)^2\right]\right\} \tag{3.18}$$

The complex phase cross-correlation function of a spherical wave can be expressed as [37]

$$\left\langle \exp\left(\psi(\mathbf{r}_1, \rho_1) + \psi^*(\mathbf{r}_2, \rho_2)\right)\right\rangle = \exp\left[-D_\psi / 2\right] \tag{3.19}$$

where D_ψ is the phase structure function, which can be expressed as

$$D_\psi(\mathbf{r}_c, \rho_c) \cong 8\pi^2 k^2 L \int_0^1 \int_0^\infty \kappa \Phi_n(\kappa) \times [1 - J_0(|(1-\xi)\mathbf{r}_d + \xi\rho_d|\kappa)] d\kappa d\xi \tag{3.20}$$

where $\Phi_n(\kappa)$ is the refractive-index fluctuation power spectrum model. By substituting the Kolmogorov spectrum into Eq. 3.20, we have:

$$\begin{aligned} D_\psi(\mathbf{r}_d, \rho_d) &\cong 2.92 k^2 L \int_0^1 dt C_n^2 |\mathbf{r}_d t + (1-t)\rho_d|^{5/3} \\ &\cong 2\left\{\left[\mathbf{r}_d^2 + \mathbf{r}_d \cdot \rho_d + \rho_d^2\right] / \rho_0^2\right\} \end{aligned} \tag{3.21}$$

From Eqs. 3.19 and 3.21, we have,

$$\left\langle \exp\left(\psi(\mathbf{r}_1, \rho_1) + \psi^*(\mathbf{r}_2, \rho_2)\right)\right\rangle = \exp\left[\frac{-1}{\rho_0^2}\left(r_d^2 + \mathbf{r}_d \cdot \rho_d + \rho_d^2\right)\right] \tag{3.22}$$

where $\rho_0 = (0.55 C_n^2 k^2 z)^{-3/5}$ is the coherent length of a spherical wave in atmospheric turbulence, and C_n^2 is the atmospheric refractive index structure constant that describes the intensity of turbulence, and is a constant for horizontal transmission. When the transmission distance $z = L$, the cross-spectral density function at the receiver can be written as

$$\begin{aligned} W(\rho_1, \rho_2; L) = \frac{1}{(\lambda L)^2} \iint d^2\mathbf{r_d} \iint d^2\mathbf{r}_c W_0(\mathbf{r}_1, \mathbf{r}_2; 0) \exp\left\langle \exp\left[\psi(\mathbf{r}_1, \rho_1) + \psi^*(\mathbf{r}_2, \rho_2)\right]\right\rangle \\ \times \exp\left\{\frac{ik}{2L}\left[(\rho_1 - \mathbf{r}_1)^2 - (\rho_2 - \mathbf{r}_2)^2\right]\right\} \end{aligned} \tag{3.23}$$

Substituting Eq. 3.18 into 3.22, we have:

$$W(\rho_c, \rho_d; L) = \frac{1}{(\lambda L)^2} \iint d^2\mathbf{r_d} \iint d^2\mathbf{r}_c \exp\left(\frac{-2\mathbf{r_c}^2}{w_0^2}\right) \times \exp\left[\frac{-ik\mathbf{r}_c \cdot \mathbf{r}_d}{F_0} + \frac{ik\mathbf{r}_c \cdot (\mathbf{r}_d - \rho_d)}{L}\right] \times \exp\left[-\frac{\mathbf{r}_d^2}{2w_0^2} - \frac{\mathbf{r}_d^2}{2\sigma_g^2} - \frac{r_d^2 + \mathbf{r}_d \cdot \rho_d + \rho_d^2}{\rho_0^2} - \frac{ik\rho_c \cdot (\mathbf{r}_d - \rho_d)}{L}\right] \tag{3.24}$$

By integrating Eq. 3.24, we have:

$$W(\rho_c, \rho_d, L) = \langle I(\rho_c, \rho_d, L)\rangle = \frac{w_0^2}{w_\zeta^2(L)} \exp\left\{-\rho_d^2\left(\frac{1}{\rho_0^2} + \frac{1}{2w_0^2\Lambda_0^2}\right) + \frac{2j\rho_c \cdot \rho_d}{w_0^2\Lambda_0}\right\} \exp\left[\frac{-2\rho_c^2}{w_\zeta^2(L)}\right] \times \exp\left[\frac{-(j\phi)^2\rho_d^2}{2w_\zeta^2(L)}\right] \exp\left[\frac{-2j\phi\rho_c \cdot \rho_d}{w_\zeta^2(L)}\right] \tag{3.25}$$

where $\phi \equiv \frac{\Theta_0}{\Lambda_0} - \Lambda_0 \frac{w_0^2}{\rho_0^2}$, $w_\zeta(z)$, and $F_\zeta(z)$ are the effective radii of the beam in atmospheric turbulence, i.e., the expanding radius and curvature radius before the phase, respectively. $w_\zeta(z)$ and $F_\zeta(z)$ can be expressed as:

$$w_\zeta(L) = w_0(\Theta_0^2 + \zeta\Lambda_0^2)^{1/2}, \ \zeta = \zeta_s + \frac{2w_0^2}{\rho_0^2} \tag{3.26}$$

$$F_\zeta(L) = \frac{L(\Theta_0^2 + \zeta\Lambda_0^2)}{\phi\Lambda_0 - \zeta\Lambda_0^2 - \Theta_0^2}, \ \phi = \frac{\Theta_0}{\Lambda_0} - \Lambda_0\frac{w_0^2}{\rho_0^2} \tag{3.27}$$

In Eqs. 3.26 and 3.27, $\Theta_0 = (F_0 - L)/F_0$, $\Lambda_0 = 2L/kw_0^2$ is an effective beam parameter. ζ is the global coherence parameter, and $\zeta_s = 1 + w_0^2/\sigma_g^2$ is the light source coherence parameter, where [38] $2\sigma_g^2 = l_c^2$. When the transmission distance $z = L$, (3.26) is an expression of $w(z)$.

3.3.3 Analysis of Beam Spreading Characteristics

When calculating the facula expansion and the other parameters at the receiving end, $z = L$ was used. In [16], Andrew and Philips derived the beam expansion radius under weak turbulence based on the traditional Rytov approximation as:

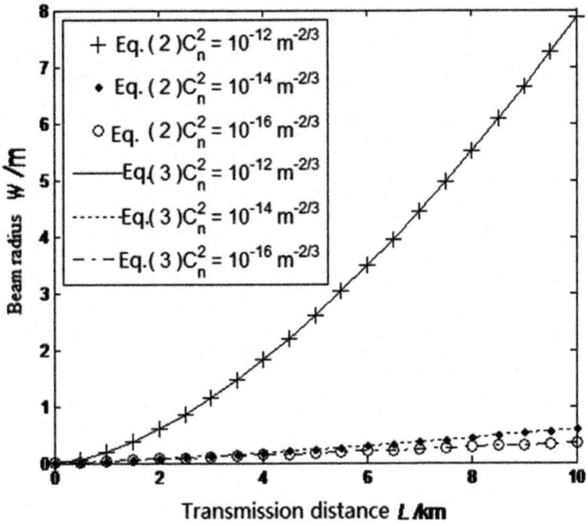

Fig. 3.2 Variation of the spreading radii of two beams with transmission distance

$$w(z) = w_0 \left[\left(1 - \frac{z}{F}\right)^2 + \xi_s \left(\frac{2z}{kw_0^2}\right)^2 \right]^{1/2} \tag{3.28}$$

Figure 3.2 shows the results for wavelength λ=0.63 μm, initial radius $w_0 = 0.01$ m, light source coherence parameter $\varsigma_s = 3$, and transmission distance $L = 10$ km for Eqs. 3.26 and 3.28. From Fig. 3.2, in the case of weak turbulence, i.e., $C_n^2 = 10^{-16}$, the curves of the two equations are the same, whereas, in the case of medium strong turbulence, only the curve for Eq. 3.26 is obtained. This establishes that this equation can be applied to the cases of medium strong and weak turbulence.

Using Eq. (3.26), as shown in Fig. 3.3, we take the wavelength as λ=0.63 μm, the light source coherence parameter $\varsigma_s = 3$, and the transmission distance $L = 2$ km. The smaller the initial radius of the beam, the greater the influence of turbulence on the extended radius. The spreading radius is almost constant for weak turbulence. As the intensity of the turbulence increases, the beam spreads faster. By changing the transmission distance, The parameters such as wavelength and coherence parameters of light source remain unchanged, and the initial radius is $w_0 = 0.01$ m. Figure 3.4 shows the spread of the GSM beam under turbulence ($C_n^2 = 10^{-14}\ m^{-2/3}$) and free space ($C_n^2 = 0$) for 2 km. For an increase in ς_s (fully coherent when $\varsigma_s = 1$), i.e., when the coherence of the light source is poor, the transmission distance is larger, and the two curves are closer. This indicates that the beam spread of partially coherent light is less affected by atmospheric turbulence than that of fully coherent light, which is consistent with the conclusion obtained using partially coherent Hermite Gaussian beams [39]. The result of this expansion is consistent with previously published findings [40].

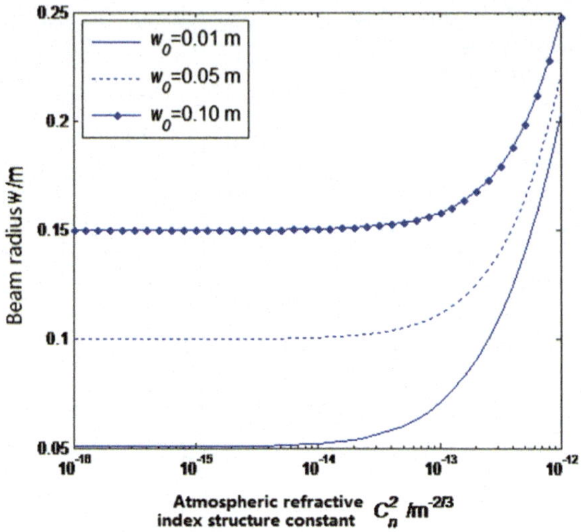

Fig. 3.3 Variation of beam radius with refractive index structure constant

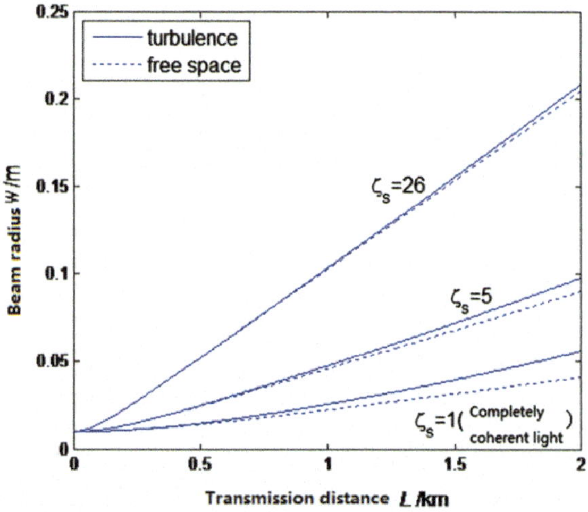

Fig. 3.4 Variation of beam radius with transmission distance

3.3.4 Analysis of Beam Drift Characteristics

Since Eq. 3.11 contains a hypergeometric function, to facilitate analysis, it is numerically integrated to study the influence of wavelength and the coherence parameters of the light source on beam drift. The parameters are set as $\kappa_0 = 0.3\,\mathrm{m}^{-1}$ and $\varsigma_s = 26$, as shown in Fig. 3.5, which compares the beam drift characteristics under different wavelengths. As shown in Fig. 3.5a, when the transmission distance $L = 1$ km, the wavelength has little effect on the beam drift. An increase in the initial beam radius w_0 causes a reduction of the drift difference caused by different wavelengths. In Fig. 3.5b, $w_0 = 0.01$ m. After comparison with Fig. 3.5a, it is determined that when the transmission distance increases, the drift variance in the medium- and strong-fluctuation areas increases rapidly. The wavelength is related to the drift; the shorter the wavelength, the more pronounced the drift.

In Fig. 3.6, we set the simulation parameters to $\kappa_0 = 0.3\,\mathrm{m}^{-1}$, $\varsigma_s = 3$, and $w_0 = 0.05\,\mathrm{m}$. As shown in Fig. 3.6a, within a transmission distance of 10 km, the spatial coherence length of the light source under weak turbulence has little effect on the beam drift, which is consistent with the conclusion in reference [42]. Figure 3.6b, c show that, in the range of $l_c \leq 0.005$, the change in the spatial coherence length of a beam propagating in medium turbulence affects the drift. The difference in the magnitude of the drift was within the range of the 0.01–0.1 m. As the degree of

Fig. 3.5 Variation of drift variance with refractive index structure constant

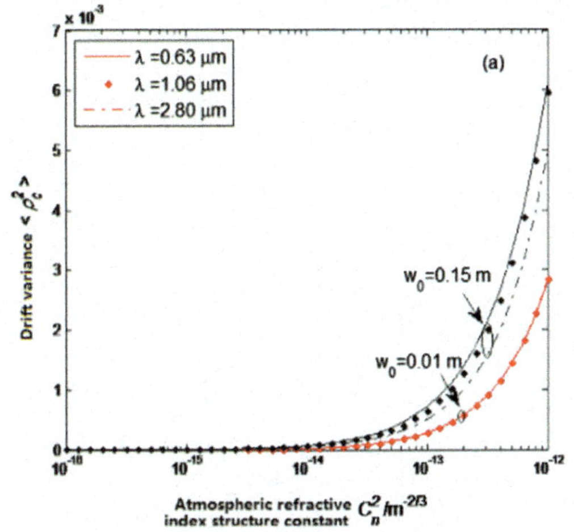

(a) Comparison of initial beam radius at different wavelengths

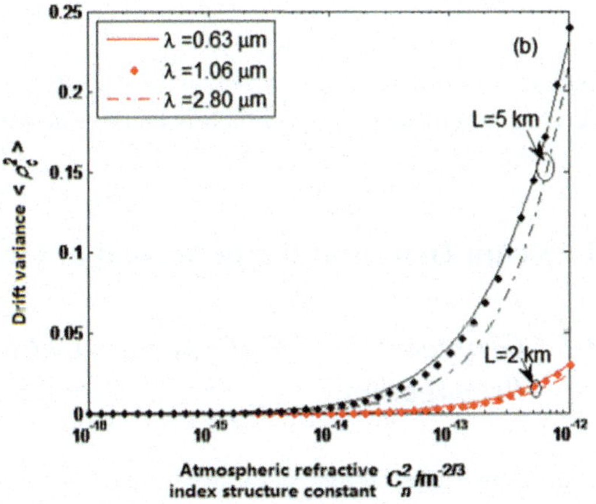

(b) Beam drift at different transmission distances

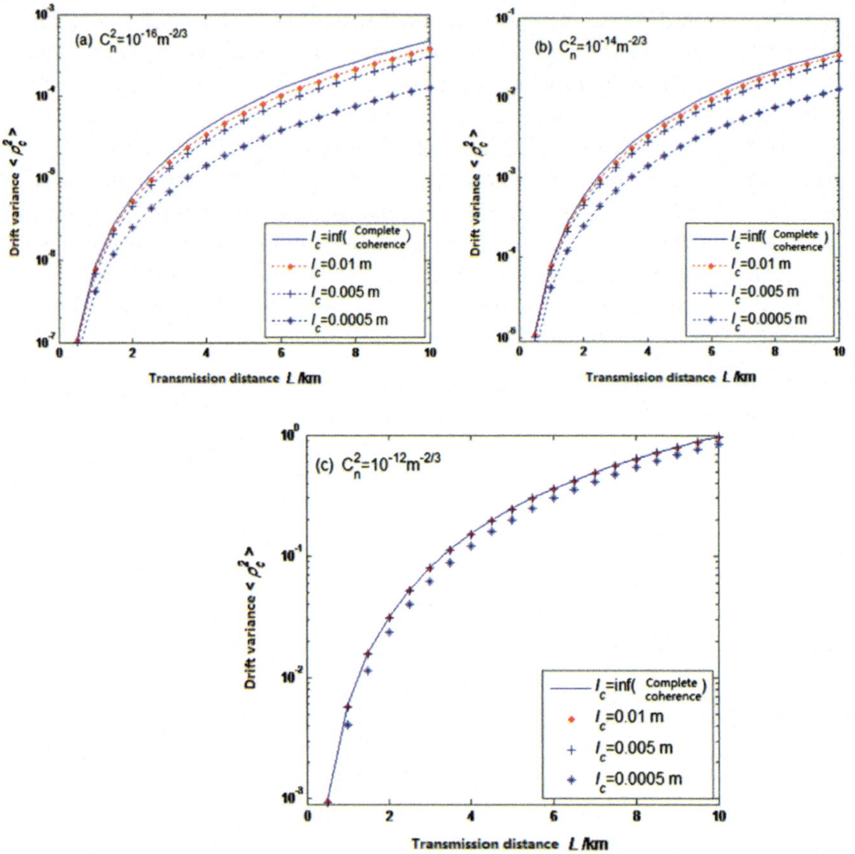

Fig. 3.6 Relationship between drift variance and beam spatial coherence length

turbulence increases, the difference in the drift fluctuation caused by different spatial coherence lengths decreases, which is consistent with the results in [41].

3.4 Beam Drift and Beam Spread of Oblique Propagation

3.4.1 Geometric Model of Partially Coherent Optical Slant Propagation

The atmospheric refractive index structure constant and the turbulent outer scale can be assumed to be constant when the beam propagates horizontally in the atmosphere, but the change in this parameter must be considered when the beam propagates

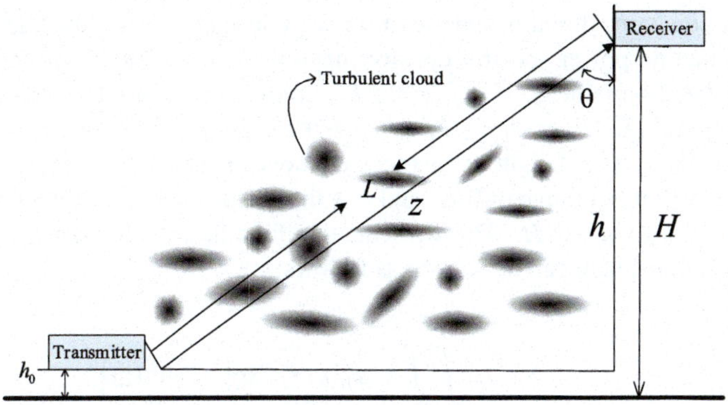

Fig. 3.7 Schematic diagram of slant transmission [35]

obliquely. Next, we analyzed the beam broadening and drift characteristics from the angle of oblique propagation.

Assuming that the height of the transmitter from the ground is 0, a schematic diagram of the oblique transmission path of a light beam in atmospheric turbulence is shown in Fig. 3.7 [35].

In Fig. 3.7, According to the geometric relationship shown in Fig. 3.7, it can be seen that

$$\cos\theta = H/L = h/z(h \le H, z \le L) \tag{3.29}$$

where θ is the zenith angle, H is the vertical height of the receiver above the ground, h_0 is the vertical height of the transmitting end from the ground, L is the distance between the transmitter and receiver, z is the transmission distance which has values in the range of 0 to L, and h is the vertical height of the transmission which has a range of 0 to H .

3.4.2 *Theoretical Derivation of Beam Drift*

In oblique transmission, we set $h_0 = 0$, and use the classical drift variance model for horizontal transmission [16]:

$$\begin{aligned}\left\langle \rho_c^2 \right\rangle = W^2 T_{LS} = 4\pi^2 k^2 w^2(L) \int_0^L \int_0^\infty \kappa \phi_n(\kappa) H_{LS}(\kappa, z) \\ \times \left[1 - e^{-\Lambda_p L \kappa^2 \xi^2 / k}\right] d\kappa \, dz\end{aligned} \tag{3.30}$$

In this equation, the atmospheric structure constant, $C_{\mathrm{n}}^2(h)$, and the refractive index fluctuation power spectral density function $\phi_{\mathrm{n}}(\kappa, h)$ vary with height. First, the normalized variable $\xi = h/H{=}z/L$ is introduced. Based on the geometric relationship in Fig. 3.7, $\cos\theta = H/L = h/z(h \le H, z \le L)$ where θ is the zenith angle, H is the vertical height of the receiver above the ground, z is the transmission distance, which has a range of 0–L, and h is the vertical height of the transmission, which has a range of 0 to H . The drift equation for a laser with a slant transmission path in the atmosphere can be written as follows:

$$\langle \rho_c^2 \rangle = 8\pi^2 L^2 \int_0^L \int_0^\infty \kappa^3 \phi_{\mathrm{n}}(\kappa, z) \exp\left[-k^2 w^2(z)\right] \times \left(1 - \frac{z}{L}\right)^2 \mathrm{d}\kappa \mathrm{d}z \tag{3.31}$$

Equation (3.31) describes the classical drift variance model for laser beam propagation in the atmosphere, which is consistent with horizontal propagation. Provided that the atmospheric refractive index fluctuation power spectrum and the spread radius of the beam are known, the drift characteristics of the beam can be investigated.

Similarly, the large-scale index spectrum [35] can be selected, i.e.

$$\Phi_n(\kappa) = 0.033 C_n^2(z) \kappa^{-11/3}\left[1 - \exp(-\kappa^2/\kappa_0^2)\right] \tag{3.32}$$

In Eq. 3.32, $C_{\mathrm{n}}^2(z)$ is the refractive index structure constant that describes a turbulent atmosphere. In this case, the variation of the atmospheric structure constant with height H-V data published by ITU-R of the International Telecommunication Union–Ministry of Radio Communication was adopted the model H-V of the atmospheric structure constant varying with height published by the International Telecommunication Union radio communication department (ITU-R) is adopted,. Using the distance normalization variable $\xi = h/H{=}z/L$, this can be written as follows:

$$\begin{aligned} C_{\mathrm{n}}^2(z) &= C_{\mathrm{n}}^2(\xi L){=}C_{\mathrm{n}}^2(\xi H/\cos\theta) \\ &= 8.148 \times 10^{-56} v_{\mathrm{RMS}}^2 (\xi H/\cos\theta)^{10} \\ &\quad \times \exp(-\xi H/\cos\theta/1000) + 2.7 \times 10^{-16} \\ &\quad \times \exp(-\xi H/\cos\theta/1500) \\ &\quad + C_{n0}^2 \exp(-\xi H/\cos\theta/100) \end{aligned} \tag{3.33}$$

where $v_{\mathrm{RMS}}^2 = \sqrt{v_{\mathrm{g}}^2 + 30.69 v_{\mathrm{g}} + 348.91}$ is the wind speed along the vertical path, v_g is the wind speed near the ground (generally $v_g{=}21\ \mathrm{m/s}$, and C_{n0}^2 is the atmospheric structure constant near the ground (generally, $C_{\mathrm{n0}}^2 = 1.7 \times 10^{-14}\ m^{-2/3}$).

Substituting Eq. 3.32 into Eq. 3.31, we have:

$$\langle\rho_c^2\rangle = 7.25L^3 \int_0^1 C_n^2(\xi L)(1-\xi)^2 w^{-1/3}(\xi L)\left[1-\left(\frac{\kappa_0^2 w^2(z)}{1+\kappa_0^2 w^2(z)}\right)^{1/6}\right]d\xi \quad (3.34)$$

Without losing generality, the exponential spectrum is still selected to study slant transmission.

3.4.3 Beam Spreading Radius

In oblique transmission, owing to the change in the refractive index fluctuation power spectrum, the phase structure function D_ψ in Eq. 3.19 becomes

$$D_\psi(\rho_c, \rho_d) \cong 8\pi^2 k^2 L \int_0^1 \int_0^\infty \kappa \Phi_n(\kappa, z) \times [1 - J_0(|(1-\xi)\mathbf{r}_d + \xi\rho_d|\kappa)]d\kappa d\xi \quad (3.35)$$

By substituting the Kolmogorov spectrum into Eq. 3.35, we have:

$$D_\psi(\mathbf{r}_d, \rho_d) \cong 2.92k^2 L \int_0^1 dt C_n^2(\xi L)|\mathbf{r}_d t + (1-t)\rho_d|^{5/3} \cong 2\{[\mathbf{r}_d^2 + \mathbf{r}_d \cdot \rho_d + \rho_d^2]/\rho_0^2\} \quad (3.36)$$

Similarly, from Eqs. 3.34 and 3.36, we have:

$$\langle\exp(\psi(\mathbf{r}_1, \rho_1) + \psi^*(\mathbf{r}_2, \rho_2))\rangle = \exp\left[\frac{-1}{\rho_0^2}(r_d^2 + \mathbf{r}_d \cdot \rho_d + \rho_d^2)\right] \quad (3.37)$$

For oblique propagation, the coherent length of a spherical wave in atmospheric turbulence is given by:

$$\rho_0' - [1.46k^2 \int_0^L C_n^2(z)(1 - z/L)^{5/3} dz]^{-3/5} \quad (3.38)$$

where $C_n^2(z)$ is the atmospheric refractive index structure constant that describes the turbulence intensity. Substituting Eqs. 3.35 and 3.38 into Eq. 3.24 and integrating the result, we obtain:

$$W(\rho_c, \rho_d, L) = \langle I(\rho_c, \rho_d, L)\rangle$$

$$
= \frac{w_0^2}{w_\zeta^2(L)} \exp\left\{-\rho_d^2\left(\frac{1}{\rho_0'^2} + \frac{1}{2w_0^2\Lambda_0^2}\right) + \frac{2j\rho_c \cdot \rho_d}{w_0^2\Lambda_0}\right\} \exp\left[\frac{-2\rho_c^2}{w_\zeta^2(L)}\right]
$$

$$
\times \exp\left[\frac{-(j\phi)^2\rho_d^2}{2w_\zeta^2(L)}\right] \exp\left[\frac{-2j\phi\rho_c \cdot \rho_d}{w_\zeta^2(L)}\right] \tag{3.39}
$$

where $\phi \equiv \frac{\Theta_0}{\Lambda_0} - \Lambda_0 \frac{w_0^2}{\rho_0'^2}$, $w_\zeta(z)$, and $F_\zeta(z)$ are the beam spreading radius and pre-phase curvature radius in oblique atmospheric turbulence, respectively. This can be expressed as follows:

$$
w_\zeta(L) = w_0(\Theta_0^2 + \zeta\Lambda_0^2)^{1/2}, \quad \zeta = \zeta_s + \frac{2w_0^2}{\rho_0'^2} \tag{3.40}
$$

$$
F_\zeta(L) = \frac{L(\Theta_0^2 + \zeta\Lambda_0^2)}{\phi\Lambda_0 - \zeta\Lambda_0^2 - \Theta_0^2}, \quad \phi = \frac{\Theta_0}{\Lambda_0} - \Lambda_0 \frac{w_0^2}{\rho_0'^2} \tag{3.41}
$$

In the preceding two equations, consistent with horizontal transmission, $\Theta_0 = (F_0 - L)/F_0$, $\Lambda_0 = 2L/kw_0^2$ is an effective beam parameter. ζ is the global coherence parameter, and $\zeta_s = 1 + w_0^2/\sigma_g^2$ is the light source coherence parameter, where [37] $2\sigma_g^2 = l_c^2$.

When the transmission distance $z = L$, Eq. 3.40 can be simplified as:

$$
w(z) = w_0\left[\left(1 - \frac{z}{F_0}\right)^2 + \left(\varsigma_s + \frac{2w_0^2}{\rho_0'^2}\right)\left(\frac{2z}{kw_0^2}\right)^2\right]^{1/2}
$$

$$
= w_0\left[\left(1 - \frac{z}{F_0}\right)^2 + \varsigma_s\left(\frac{2z}{kw_0^2}\right)^2 + \frac{2w_0^2}{\rho_0'^2}\left(\frac{2z}{kw_0^2}\right)^2\right]^{1/2} \tag{3.42}
$$

Equation 3.42 describes the beam-spreading radius of a partially coherent GSM beam in atmospheric turbulence during oblique path propagation. The first two terms in the equation quantify the amount of the free-space diffraction, and the last term describes the amount of turbulent spread. F_0 is the wavefront phase curvature, z is the transmission distance variable, ς_s is the coherent parameter of the light source, which always satisfies $\varsigma_s \geq 1$ (the beam is completely coherent when $\varsigma_s = 1$, partially coherent when $\varsigma_s > 1$, and completely incoherent when $\varsigma_s \to \infty$), and l_c is the spatial coherence length. The smaller l_c, the worse the coherence. When $l_c \to \infty$, the light is completely coherent, whereas $l_c \to 0$ describes complete incoherence.

3.4.4 *Analysis of Ramp Transmission Expansion Characteristics*

As shown in Fig. 3.8, using Eq. 3.42, the parameters are set as follows: $\lambda = 1.06$ μm, H = 500 m, $K_n = 1$, and the curve of the spread radius of the collimated beam with the propagation distance under different initial radii is shown. In Fig. 3.8, when the propagation distance is within 10 km and $w_0 = 0.5$ m, i.e., the light wave is approximately a plane wave, the effective radius of the beam is almost unchanged and the expansion effect is very small, which is consistent with the conclusion that the expansion effect of a plane wave is very small in theory. A comparison of Fig. 3.8a, b shows that when the receiver height is constant, the partially coherent light expands more rapidly compared to the fully coherent light. When the initial radius of the beam is small, such as in $w_0 = 0.01$ m, the spread change is the most pronounced. This result is consistent with the results of Ref. [43].

Taking $w_0 = 0.01$ m in Fig. 3.9, the broadening characteristics of fully coherent light and partially coherent light under oblique path transmission at different receiver heights are analyzed. As shown in Fig. 3.9, under the same conditions, the beam spread of partially coherent light is more pronounced, and the effective radius is larger than that of fully coherent light. When the transmission distance is fixed, a higher receiver height results in a smaller spreading effect, indicating that oblique propagation is more conducive to the propagation of light beams in a turbulent atmosphere. This is consistent with the results obtained by Duan Meiling using Hermite Gaussian beams [44]. Moreover, in Fig. 3.9, it is evident that the spread change curve of partially coherent light at different receiver heights is closer to that of fully coherent light, i.e., atmospheric turbulence has less impact on the beam spread of partially coherent light. The physical mechanism leading to this result is described below.

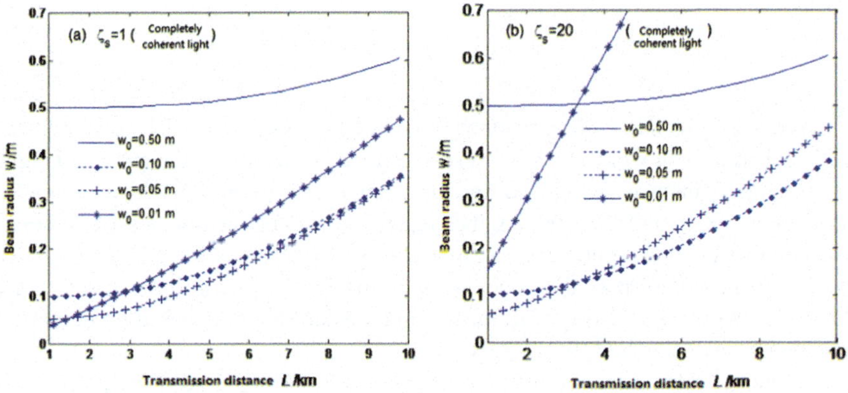

(a) Fully coherent light (b) Partially coherent light

Fig. 3.8 Variation of effective beam radius with transmission distance at different initial radii

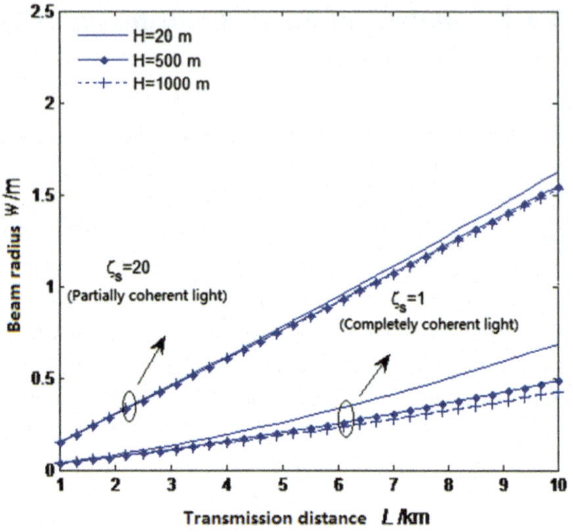

Fig. 3.9 Variation of effective beam radius with transmission distance at different receiver heights

According to Eq. 3.42, the expansion associated with free space diffraction is given by:

$$\Delta w_{\text{atm}} \sim w_0 \left[\frac{2w_0^2}{\rho_0^2} \left(\frac{2z}{kw_0^2} \right)^2 \right]^{1/2} \tag{3.43}$$

where $\rho_0 = [1.46k^2 \int_0^L C_n^2(\xi L)(1 - z/L)^{5/3} dz]^{-3/5}$

The expansion of the turbulence effect is given by:

$$\Delta w_{\text{diff}} \sim w_0 \left[\left(1 - \frac{z}{F}\right)^2 + \varsigma_s \left(\frac{2z}{kw_0^2} \right)^2 \right]^{1/2} \tag{3.44}$$

From Eq. 3.44, it is evident that the turbulence expansion of partially coherent and fully coherent light yields the same result, and for an increase in the receiver height, the turbulence intensity is reduced, and the change in the expansion becomes smaller. When h = 500 m and h = 1000 m, the change in the expansion decreased. When ζ_s increases, that is, the coherence decreases, and Δw_{diff} increases, the expansion caused by the diffraction of partially coherent light and fully coherent light in free space increases, which is consistent with the of Qian Xian Mei et al. based on numerical simulation in document [19]; it is also with [42].

Figure 3.10 describes the drift variance curve for oblique beam propagation in atmospheric turbulence under different conditions. In Fig. 3.10a, the parameters are set as: λ=0.63 μm, H = 500 m, w_0 = 0.01 m. Based on Fig. 3.10, the drift change of fully coherent light during oblique transmission at different wavelengths is very small, which is consistent with the results of Qian Xian Mei using numerical simulation [28]. The longer the wavelength of partially coherent light, the more pronounced the drift, which is consistent with the conclusion of Li Yaqing using the spot centroid distribution map [31]. Other parameters are the same as above, and the wavelength λ=0.63 μm is used. As shown in Fig. 3.10b, when the beam slant propagation (h = 500 m) occurs at the same distance, the greater the receiver height, the smaller the beam drift. Compared to horizontal propagation (h = 0 m), slant propagation is less affected by turbulence. When the receiver height h is constant and the transmission distance is within 5 km, the coherence is worse as the drift variance becomes smaller. Moreover, compared to Fig. 3.8a, the short-range transmission drift is smaller, which is consistent with the research results in [26]. In Figs. 3.10c, d, the other conditions in Fig. 3.10a are unchanged, κ_0 and w_0 are changed, and the drift variance curves of the light beam at different turbulence external scales and initial beam radius are shown. In Fig. 3.10c, the smaller the value of κ_0, the larger the outer scale of turbulence, the more pronounced the drift, and the drift effect of partially coherent light is smaller than that of fully coherent light. In Fig. 3.10d, when the propagation distance is greater than 2 km, the smaller the initial radius of the beam, the greater the drift change. When the initial waist radii are 0.1 and 0.5 m, the difference between the drift variance change curves of the partially coherent and fully coherent light is very small.

In this chapter, the beam drift characteristics of partially coherent GSM beams were investigated for horizontal and oblique propagation paths in atmospheric turbulence. First, the drift variance equation was derived for different atmospheric refractive index fluctuation power spectra based on the classical drift variance model under weak turbulence, using the Rytov approximation and geometrical optics approximation. Secondly, according to the generalized Huygens-Fresnel principle and the cross-spectral density function of partially coherent GSM beams, the unified beam propagation equation of partially coherent beams propagating in medium, strong and weak turbulence is derived, Finally, using the extended radius, the drift variance model for weak turbulence is extended to strong turbulence, and the broadening and drift characteristics of some GSM beams in medium, strong and weak turbulence are investigated. The main conclusions are as follows:

(1) When propagating horizontally, compared to fully coherent light, the beam spread of partially coherent light is less affected by atmospheric turbulence, whereas the beam spread caused by diffraction is exacerbated in free space. Therefore, in practice, the advantages and disadvantages of the two should be evaluated to determine the optimal beam according to actual need.
(2) In the optical and near-infrared bands, the horizontal transmission distance is short and the wavelength had little effect on the drift. Therefore, to reduce the impact of atmospheric turbulence, we should attempt to select an appropriately large initial radius, long wavelength, and high distance from the ground.
(3) Under the general conditions of a laser engineering emission system, the change in the spatial coherence length of the light source for horizontal transmission has

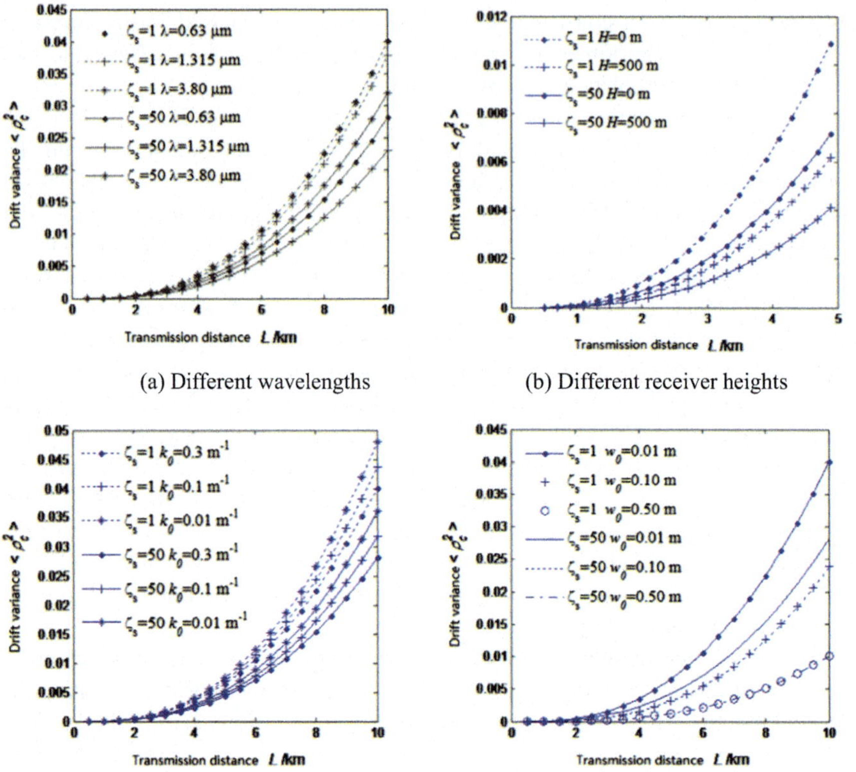

(a) Different wavelengths (b) Different receiver heights

(c) Different turbulent large scale (d) different initial radius

Fig. 3.10 Variation of drift variance of partially coherent light and fully coherent light with transmission distance

little effect on the drift. Therefore, it is not necessary to strictly use a single-mode laser when studying beam-drift characteristics.

(4) In contrast to horizontal propagation, beam propagation of partially coherent light is faster than that of fully coherent light, and the beam propagation effect is related to the propagation distance, initial beam radius, and receiver height. When the propagation distance is constant, the larger the initial radius, the higher the receiver height, and the smaller the beam-spreading effect.

(5) for the same slant propagation distance, the higher the receiver height, the smaller the drift effect of the beam and the influence of turbulence. The lower the coherence, the smaller the drift variance. The larger the outer scale of turbulence, the more pronounced the drift, and the drift effect of partially coherent light is smaller than that of fully coherent light.

(6) In oblique propagation, the variance of the beam centroid drift decreases with an increase in the initial radius of the beam, and the beam drift of fully coherent

light is less affected by the wavelength, whereas the longer the wavelength of the partially coherent light, the more obvious the drift.

References

1. Leader JC (1978) Atmospheric propagation of partially irradiation. Opt Soc Am 68:175–185
2. Ouyang CF. Irradiance scintillation of a partially coherent source in extremely strongly turbulence. Appl Opt 18:1133–1135
3. Wu J (1990) Propagation of a Gaussian-Schell beam through turbu lent media. J Mod Opt 37(4):671–684
4. Wu J, Boardman AD (1991) Coherence length of a Gaussian-Schell model beam and atmospheric turbulence. J Mod Opt 38(7):1355–1363
5. Ponomarenko AS, Greffet J, Wolf E (2002) The diffusion of partially coherent beams in turbulent media. Opt Commun 208(1):1–8
6. Gbur G, Wolf E (2002) Spreading of partially coherent bean in random media. J Opt Soc Am A 19:1592–1598
7. Ricklin JC, Davidson FM (2002) Atmospheric turbulence effects on a partially coherent Gaussian beam: implications for free-space laser communication. J Opt Soc Am A 19:1794–1802
8. Salem M, Shirai T, Sogariu A et al (2003) Long-distance propagation of partially coherent beams through atmospheric turbulence. Opt Commun 216:216–265
9. Dogariu A, Arnarande S (2003) Propagation of partially coherent beams:turbulence induced degradation. Opt Lett 28(1):10–12
10. Shirai T, Dogariu A, Wolf E (2003) Mode analysis of spreading of partially coherent beams propagation through atmospheric turbulence. J Opt Soc Am A 20:1094–1102
11. Borghi R, Guattari G (2003) Evaluation of the spatial coherence of a light beam through transverse intensity measurements. J Opt Soc Am A 20(9):1763–1770
12. Chernov LA (1960) Wave propagation in a random medium. McGraw-Hill, New York
13. Bechmann P (1965) Signal degeneration in laser beams propagated through a turbulent atmosphere radio. Science D69(4):629–640
14. Cook RJ (1975) Beam wander in a turbulent medium: an application of Ehrenfest's theorem. J Opt Soc Am 65:942–948
15. Chunrside JH, Lataitis RJ (1990) Wander of an optical beam in the turbulent atmosphere. Appl Opt 29(7):926–930
16. Andrews LC, Phillips RL (2005) Laser beam propagation through random media. SPIE Press, 83–99
17. Andrews LC, Phillips RL, Sasiela RJ et al (2005) Beam wander effects on the scintillation index of a focused beam. Proc SPIE: 5793
18. Berman GP, Chumak A, Gorshkov VN (2007) Beam wandering in the atmosphere: the effect of partial coherence. Phys Rev E 76(5):056606
19. Qian XM, Zhu W, Rao R (2008) Numerical simulation and analysis of turbulent effect of atmospheric slant propagation of ground to air laser. Infrared Laser Eng 37(5):787–792
20. Xiao X, Voelz DG (2012) Beam wander analysis for focused partially coherent beams propagating in turbulence. Opt Eng 51(2):026001-1–026001-7
21. Zhang Y, Song Z, Gong Z (1986) Discussion on some problems of beam drift in turbulent atmosphere. Quant Electron 3(1):57–61
22. Zhang Y, Song Z, Gong Z et al (1986) Laser beam drift in turbulent atmosphere. Acta optica Sinica (6):373–381
23. Tanqipei HX (1987) Calculation of laser beam drift in turbulent atmosphere. J Wuyi Univ (Soc Sci Edn) 1:69–74
24. Rao R (2000) Experimental study on laser beam drift in turbulent atmosphere. China Laser 11:1011–1015

25. Yi X (2005) Study on propagation characteristics of Gaussian beam in turbulent atmosphere. Xi'an University of Electronic Science and Technology, Xi'an
26. Gao C, Yu S (2006) Beam drift in strong turbulent region. High Power Laser Particle Beam 18(10):1597–1601
27. Wang C, Li J (2007) Detection of laser spot drift. Appl Opt 28(2):205–208
28. Wei H, Wu Z (2008) Broadening and drift characteristics of laser beam oblique propagation in atmospheric turbulence. J Radio Propag 23(4):611–615
29. Wei H (2009) Laser beam propagation and target echo characteristics in slant turbulent atmosphere. Xi'an University of Electronic Science and Technology, Xi'an
30. Qian X (2010) Numerical simulation for coherent and partially coherent beam propagation through atmospheric turbulence. Chin Phys Lett 27(4):044214
31. Wu GH, Luo B, Yu S et al (2011) The propagation of electromagnetic Gaussian-Schell model beams through atmospheric turbulence in a slanted path. J Opt A Pure Appl Opt 13(3):0357061–0357067
32. Wu Z, Xiang J, Wang M (2013) Broadening and drift of partially coherent Gaussian Schell beams in atmospheric turbulence. Infrared Laser Eng 42(3):658–662
33. Li Y, Wu Z (2012) Characteristics of a partially coherent Gaussian Schell-model beam propagating in slanted atmospheric turbulence. Chin Phys B 21(5):054203
34. Gao M, Liu Y, Wang F (2014) Propagation and drift of polarized partially coherent laser beams in turbulent atmosphere. [J/OL] Acta PHOTONICA Sinica. http://www.cnki.net/kcms/detail/61.1235.O4.20140513.1611.052.html
35. Han M (2015) Beam drift of partially coherent light in atmospheric turbulence. Xi'an University of Technology, Xi'an
36. Rao R (2005) Light propagation in turbulent atmosphere. Anhui Science and Technology Press, Hefei
37. Ishimaur A (1978) Wave propagation and scattering in random media. Academic Press New York, 1978(A. Shimaru, propagation and scattering of waves in random media, translated by Huang Runheng et al., Zhou Xiuji School, Beijing: Science Press, 1986)
38. Yura HT (1972) Mutual coherence function of a finite cross section optical beam propagating in a turbulent medium. Appl Opt 11(6):1399–1406
39. Korotkova O, Andrews LC, Phillips RL. Model for a partially coherent Gaussian beam in atmospheric turbulence with application in LaserCom. Opt Eng 43(2):330–341
40. Ji X, Chen X, Lü B (2008) Spreading and directionality of partially coherent Hermite-Gaussian beams propagating through atmospheric turbulence. J Opt Soc Am A 25(1):21–28
41. Ke X, Han M, Wang M (2014) Broadening and drift of partially coherent light on the horizontal transmission path in atmospheric turbulence. Acta optica Sinica 34(11):11060031–11060035
42. Wang L, Ke X (2004) Study and simulation of turbulence effects on optical communication links. J Light Scatt 3(3):250–255
43. Ke X, Han M, Wang M (2015) Broadening and drift of partially coherent light on slant path in atmospheric turbulence. Acta PHOTONICA Sinica 44 (3): 0306001-1–0306001-5
44. Duan M, Li J, Wei J et al (2013) Propagation of partially coherent Hermite Gaussian beams in slant path atmospheric turbulence. High Power Lasers Particle Beams 25(9):2252–2256

Chapter 4
Light Intensity Flicker and Angle of Arrival Fluctuation

The flicker of light intensity induces photoelectric detector noise and reduces the performance of communication systems. Moreover, the angle of arrival fluctuation reduces the efficiency of coherent detection, and even invalids the system. Therefore, it is necessary to consider the influence of light intensity flicker and angle of arrival fluctuation on wireless optical communication.

4.1 Introduction

There is a long history of studying the transmission and scattering theory of light waves in atmospheric turbulence. Newton once carefully observed and analyzed the natural phenomenon of starlight scintillation, and provided a scientific explanation in his seminal publication of Opticks [1]. Until the middle of the twentieth century, Obukhov used Rytov's gentle perturbation method to solve the wave transmission equation in random media. Tatarskii [2] studied the relationship between light intensity flicker and light transmission distance. The results were very close to the observed experimental phenomena, thus establishing the classical theory of light wave transmission under the condition of weak turbulence. In 1967, Fried [3, 4] described the average signal-to-noise ratio of a spatially coherent optical communication system using the light wave structure function based on statistical theory. However, since the probability density distribution of the average signal-to-noise ratio was not considered, the impact of atmospheric turbulence on the performance of the wireless optical communication system could not be fully described. In 1978, Churchside et al. [5, 6] and others attempted to overcome the aforementioned limitations to describe the statistical characteristics of coherent optical communication systems in detail. Although they determined the probability density distribution function of the average signal-to-noise ratio, they adopted a highly simplified atmospheric turbulence effect model. Given that it is based on a saturation phenomenon, Tatarskii's theoretical

X. Ke, *Partially Coherent Optical Transmission Theory in Optical Wireless Communication*, Optical Wireless Communication Theory and Technology,
https://doi.org/10.1007/978-981-97-7158-5_4

model encounters difficulty when attempting to explain several new phenomena. Using Markov approximation to solve the statistical moment equation of the light field, the asymptotic solution of scintillation intensity can be successfully derived, and the phenomenon of scintillation saturation under strong turbulence is explained. Although there are corresponding solutions for light wave transmission under weak turbulence and strong turbulence, a reasonable and effective solution has not been found under the critical conditions of the transition between weak turbulence and strong turbulence, and the development of a universal theory is an ongoing challenge. Since the 1970s, researchers have used numerical simulation methods to study the propagation characteristics of light waves in turbulence. In 1986, Winick [7] studied the effect of atmospheric turbulence on the performance of a spatially coherent optical communication system using numerical simulation. In 2005, Andrews and Phillips [8] studied the i–k distribution model of light wave transmission in a turbulent atmosphere. Miller et al. studied the logarithmic intensity fluctuation variance and structure function of a Gaussian beam [8, 9]. In 2008, Belmonte et al. [10] determined the probability density distribution function of the average signal-to-noise ratio. This model can evaluate the impact of the light intensity scintillation index and phase fluctuation variance on system performance and can describe the impact of atmospheric turbulence on a spatially coherent optical communication system.

Since the 1960s, different theoretical methods (perturbation theory, generalized Huygens-Fresnel integral, Rytov approximation, etc.), experimental methods, and numerical simulation approaches have been used to study the second-order statistics and fourth-order statistics of a light field for Gaussian beams propagating through random media [11–13]. The theory of fully coherent beam propagation in weak turbulence has been improved over the years, whereas the theory of beam propagation in strong turbulence needs to be investigated further. In the study of the coherence of beams, most of the initial beams are completely coherent or completely incoherent, and the study of partially coherent states has not been actively investigated. In fact, strictly coherent or incoherent light is a mathematical ideal. Even the sharpest spectral line has a certain spectral line width, therefore, it cannot be a point light source. Moreover, a beam is composed of a large number of radiation oscillators, which introduces a degree of ductility. Therefore, an actual light source is always partially coherent. The results show that compared to fully coherent light, partially coherent light is less affected by turbulence, and is suitable for suppressing the impact of atmospheric turbulence on laser communication. In 1975, Wolf and Carterl proposed the concept of a quasi-monochromatic partially coherent Gaussian light source [14]. Since this light source has the same degree of coherence as a Gaussian beam, it is much more convenient to use this model for research. Previously, it was considered that the spatial coherence of a laser guaranteed good directivity. In 1978, Wolf revealed the correlation between coherence and directivity in his work "Coherence and Geometry" [15]: (1) even if the coherence of beams is different, the same far-field light intensity distribution can be generated; (2) incompletely coherent light sources can also produce highly directional beams. It is generally believed that turbulence causes the fluctuation of the phase of a laser. Therefore, the wavefront will become very disordered after the laser is transmitted over a certain distance in the non-uniform medium. Cliford and Yura used this conclusion to explain the phenomenon of starlight scintillation [16]. However, the ability of a partially coherent light source to suppress the fluctuation of atmospheric light intensity and improve the bit error rate performance of a communication system needs to be investigated further.

In the 1990s, Wu et al. discovered that "partially coherent light is less sensitive to turbulence than fully coherent light" [17]. In 2002, Ponomarenko et al. [18]. studied the beam width and the diffusion angle of partially coherent light, and determined the physical mechanism that explains why partially coherent light is less affected by turbulence than fully coherent light. Salem et al. also theoretically explained the advantages of applying partially coherent light transmission in the atmosphere [19]. Subsequently, Dogariu and Amarande experimentally proved that "the influence of turbulence on partially coherent light is less than that on fully coherent light" [20]. Ricklin et al. derived the average intensity of partially coherent Gaussian beams propagating in the atmosphere using the cross-spectral intensity function [21]. In 2003, Tomohiro and Shirai studied the atmospheric turbulent propagation of the Hermite Gaussian mode of partially coherent Gaussian light [22], Riccardo, Borghi et al. [23] studied the coherent characteristics of the beam, and the results are consistent with the previous conclusions.

In the mid-1980s, a research project was undertaken at the Lincoln Laboratory of the Massachusetts Institute of technology called the "inter-satellite laser transmission experiment." In this research, semiconductor lasers and heterodyne detection methods were used to perform experiments related to high-speed satellite communication, and the acquisition and tracking functions were verified [24, 25]. In 1991, the California Institute of Technology built a 100 Mbps PSK inter-satellite coherent optical communication demonstration system. In 1998, the University of Florida conducted theoretical analysis and experimental verification involving the multi-aperture and equal gain coherent receiver. The results show that the multi-aperture receiving technology can effectively suppress the signal fading caused by atmospheric turbulence and improve the signal-to-noise ratio of the system [26]. In 1999, researchers at the University of Florida proposed a new coherent detection system based on polarization diversity, which was used to suppress the phase fluctuation and Doppler frequency shift caused by atmospheric turbulence [27]. In 1999, the signal-to-noise ratio and a bit error rate of different coherent optical communication schemes under the vibration condition of a low orbit satellite platform were investigated at the Lincoln Laboratory [28]. Researchers at Caltech have investigated the quantum limit of different modulation/demodulation schemes and an approach for expanding the channel capacity of inter-satellite optical communication links [29].

In 1990, the European Space Agency (ESA) conducted a 145 km free-space optical transmission experiment between two Canary Islands [30, 31]. The experimental results revealed that when the atmospheric coherence length is large, the aperture average effect is not pronounced. Moreover, is little difference between the measured value and the theoretical value of the variance of the angle of arrival fluctuation.

Japan's main research agencies for space optical communication are the Japan Space Agency (JAXA) and the National Institute of information and communication technology (NICT), which were merged from the National Space Development Agency (NASDA) and other smaller research agencies. Since 1998, NASA has conducted substantial research on satellite coherent optical communication. Different modulation and detection methods such as ASK (Amplitude shift keying), FSK (Frequency shift keying), PSK (Phase shift keying), and DPSK (Binary phase shift keying), have been compared. Heterodyne and homodyne schemes have been compared in detection systems. Distributed feedback lasers with a linewidth less than 50 kHz and a wavelength of 1550 nm have been applied to satellite coherent optical communication systems. The heterodyne PSK satellite analog link experiment with detection sensitivity better than 50 photons/bit was realized [32–34]. In 2007, NICT developed a 3 Gbps BPSK modulated real-time digital coherent receiver. In 2008, a 6 Gbps BPSK modulated real-time digital coherent receiver was realized [35].

In 2004, Liu and Wu determined the scintillation index of partially coherent light propagating through strong turbulence and conducted several numerical simulations based on their model [36]. In 2005, Rao introduced the concept of angle of arrival fluctuation when plane waves and spherical waves are transmitted in the atmosphere [37]. Subsequently, Zhang and Tao derived expressions to calculate the variance of the arrival angle fluctuation of partially coherent Gaussian-Schell beams in atmospheric turbulence [38].

Wireless laser communication has the potential to become the mainstream communication method in the future. In this regard, the study of the impact of atmospheric turbulence on laser transmission is important, especially for practical application.

4.2 Wavefront Distortion and Wavefront Correction in Coherent Optical Communication

Given that at least part of wireless optical communication links will be in the atmosphere, atmospheric absorption or scattering will lead to significant attenuation of laser intensity, and atmospheric turbulence will lead to beam drift, light intensity flicker, angle of arrival fluctuation, and wavefront distortion, which will greatly reduce the sensitivity of coherent detection. There are also technical bottlenecks in the field of coherent detection: (1) signal light and local oscillator light must have the same polarization so that their electric vector directions are consistent; (2) the mode structure of the signal light and local oscillator light must be the same; (3) the wavefront curvature of the signal light and local oscillator light must match; (4) the energy flow vector of the signal light and the local oscillator reference light should have the same direction as much as possible, i.e., the two beams should maintain their angular alignment in space, (5) the signal light and the local oscillator light must coincide on the optical mixing surface, and the spot diameters of the two beams should be equal to maximize the signal-to-noise ratio. If some parts of the signal spot cannot completely cover the local oscillator spot, the power in these regions will be lost; if some parts of the local oscillator beam are outside the signal beam, they will produce noise, in particular, atmospheric turbulence will cause spot jitter. Thus, it is difficult for the local oscillator beam to completely coincide with the signal spot. (6) During the transmission of signal light, the wavefront distortion caused by atmospheric turbulence will reduce the heterodyne mixing efficiency. Therefore, the influence of the atmospheric channel on the heterodyne receiver must be considered.

In the 1950s, Babcock, a famous American astronomer, proposed the idea of using a wavefront sensor to measure the wavefront error in real-time, and then using an arbitrarily deformable optical element to generate a controllable optical phase shift to correct it in real-time to address the dynamic disturbance caused by atmospheric turbulence and other factors. If the process from detection to correction is sufficient fast, the influence of the dynamic disturbance error caused by turbulence can be overcome, so that the optical system can automatically adjust to changes in the environment. This is the main idea of adaptive optics (AO) [39]. In 1972, Zedovich discovered a peculiar phenomenon in the process of stimulated Brillouin backscattering generated by optical waveguides filled with nonlinear media [40]. If an element that can cause distortion is placed in front of the CS_2 box, the distortion disappears. This is due to the phase conjugation phenomenon of nonlinear optics. Using this phenomenon, the wavefront distortion of light waves can be automatically corrected. Subsequently, it was determined that phase conjugation also exists in Raman scattering and Rayleigh scattering. Thus, another type of adaptive optical system has emerged, i.e., a nonlinear optical adaptive optical system. However, the practical nonlinear medium time constant is large, which limits the application range of this system.

In 1982, the U.S. Air Force installed the world's first adaptive optics telescope for space observation on Maui Island near Hawaii [41]. In 1984, Lockheed successfully developed an adaptive optics telescope for observing the sun [42]. In 1985 and 1988, researchers at the Massachusetts Institute of Technology (MIT) conducted two adaptive optics experiments using laser guide stars, with a measurement and correction period of < 500 μs. In 1990, Kaman airlines also conducted correction experiments involving laser beams [43], and successfully realized distortion compensation by correcting low-order Zernike aberration.

From 1987 to 1990, the Come-On adaptive optics system was jointly developed by the French Space Research Institute, Laserdot, and the European Southern Observatory (ESO) [44]. The system used a 19-element deformable mirror as the wavefront corrector, and the bandwidth was only 9 Hz. This system has been successfully applied in the field of astronomical observation and is considered to be a milestone in the development of astronomical observation technology. The ongoing Come-On plan has resulted in the expansion of the number of deformable mirrors to 52 and the frame rate is currently 400 Hz.

In 2005, Li et al. [45] investigated the adaptive compensation method for atmospheric turbulence in a space laser communication system. In 2008, Yang et al. [46] studied adaptive optics technology without a wavefront sensor that can suppress the wavefront distortion caused by atmospheric turbulence in atmospheric optical communication. In 2010, Chen [47] proposed an adaptive optics technology based on the random parallel gradient descent algorithm, which can realize closed-loop compensation for wavefront phase distortion without wavefront sensors. However, the slow convergence speed of the algorithm restricts its application in systems with high-speed real-time requirements. In 2010, Xia [48] et al. reconstructed distorted wavefronts, then corrected them using reflection deformable mirrors. In 2010, Qiao [49] studied the working principle and performance index of deformable mirrors in adaptive optics, and determined that the wavefront compensation of an adaptive optics system composed of detectors, controllers, and deformable mirrors can reach 90%. In 2010, Li and Juan [50] studied a closed-loop adaptive control optical system based on a genetic algorithm to compensate for the distortion of a laser beam. In 2011, Han et al. [51] analyzed the relationship between the coupling efficiency and turbulence intensity, the transmission distance, and the zenith angle. A random parallel gradient optimization algorithm was used to control deformable mirrors in real-time to optimize the coupling efficiency of a single-mode fiber. The adaptive optics system can sense the wavefront distortion caused by atmospheric disturbance, ambient temperature fluctuation, optical axis jitter, and other factors, and achieve real-time distortion compensation by adjusting the optical system.

4.3 Light Intensity Flicker and Angle of Arrival Fluctuation

4.3.1 Variance of Light Intensity Fluctuation

The logarithmic intensity fluctuation variance is defined as [36]

$$\sigma_{InI}^2 = \left\langle [In(\mathrm{I/I_0}) - In(\mathrm{I/I_0})]^2 \right\rangle \tag{4.1}$$

where I_0 is the incident light intensity and I is the light intensity at the receiving end.

Using the effective beam parameter method and Rytov approximation, when the refractive index spectrum satisfies the Kolmogorov model, the expression of the variance of the axial intensity fluctuation of partially coherent Gaussian Schell beams in weak turbulence is given by [36]

$$\begin{aligned}\sigma_I^2(0, L) &= \sigma_B^2 \\ &= 3.86\sigma_I^2 \times 0.4[(1 + 2\Theta_e)^2 + 4\Lambda_e^2]^{5/12} \\ &\quad \times \cos\left[\frac{5}{6}\arctan\left(\frac{1 + 2\Theta_e}{\Lambda_e}\right)\right] - 3.86\sigma_I^2 \times \frac{11}{16}\Lambda_e^{5/6}\end{aligned} \tag{4.2}$$

where, σ_1^2 is the Rytov exponent which is expressed as $\sigma_1^2 = 1.23C_n^2 k^{7/6} L^{11/6}$, L is the beam transmission distance, C_n^2 is the atmospheric refractive index constant,

k is wave number, $k = \frac{2\pi}{\lambda}$, λ is the wavelength, and Θ_e and Λ_e are the effective parameters after the propagation distance L of a partially coherent Gaussian Schell light in a vacuum. They can be expressed as

$$\Theta_e = \frac{\Theta_0}{\Theta_0^2 + \Lambda_0^2\left(1 + 4\frac{q_c}{\Lambda_0}\right)} \tag{4.3}$$

$$\Lambda_e = \frac{\Lambda_0\left(1 + 4\frac{q_c}{\Lambda_0}\right)}{\Theta_0^2 + \Lambda_0^2\left(1 + 4\frac{q_c}{\Lambda_0}\right)} \tag{4.4}$$

Θ_0 characterizes the degree of divergence of the beam, i.e., $\Theta_0 < 1$ represents a convergent beam, $\Theta_0 = 1$ represents a collimated beam, and $\Theta_0 > 1$ is a divergent beam. The parameter q_c represents the degree of coherence of partially coherent GSM beams. These three parameters can be expressed as follows:

$$\Theta_0 = 1 - \frac{L}{F_0} \tag{4.5}$$

$$\Lambda_0 = 2L/kw_0^2 \tag{4.6}$$

$$q_c = L/kl_c^2 \tag{4.7}$$

where F_0 is the beam wavefront curvature radius, w_0 is the initial beam radius, and l_c is the coherence length of the beam. The larger l_c, the better the coherence of the beam.

According to the modified Rytov theory, the variance of the axial logarithmic intensity fluctuation of partially coherent Gauss-Schell light for turbulent conditions can be represented as [51]

$$\sigma_1^2(0, \mathrm{L}) = \exp\left[\frac{0.49\sigma_B^2}{(1 + 0.56\sigma_B^{12/5})^{7/6}} + \frac{0.51\sigma_B^2}{(1 + 0.69\sigma_B^{12/5})^{5/6}}\right] - 1 \tag{4.8}$$

The variance of the off-axis logarithmic intensity fluctuation is given by

$$\sigma_2^2(\mathrm{r,L}) = 4.42\sigma_1^2\left(\frac{\Lambda_e}{1 + 4\Lambda_e q_c + 1.63\sigma_1^{12/5}\Lambda_e}\right)^{5/6} \times \frac{r^2}{w_e^2(1 + 4\Lambda_e q_c + 1.63\sigma_1^{12/5}\Lambda_e)} \tag{4.9}$$

where r is the vector diameter, i.e., the distance from any point on the receiving plane to the center of the spot. The total logarithmic intensity fluctuation variance is given by

$$\begin{aligned}\sigma_I^2 &= \sigma_1^2 + \sigma_2^2 \\ &= \exp\left[\frac{0.49\sigma_B^2}{(1+0.56\sigma_B^{12/5})^{7/6}} + \frac{0.51\sigma_B^2}{(1+0.69\sigma_B^{12/5})^{5/6}}\right] - 1 \\ &+ 4.42\sigma_1^2\left(\frac{\Lambda_e}{1+4\Lambda_e q_c + 1.63\sigma_1^{12/5}\Lambda_e}\right)^{5/6} \times \frac{r^2}{w_e^2(1+4\Lambda_e q_c + 1.63\sigma_1^{12/5}\Lambda_e)}\end{aligned} \tag{4.10}$$

When the beam propagates over a long distance, considering that the inhomogeneous medium of the atmospheric channel is primarily in the far-field of the transmitting end, a spherical wave can be used to simulate the radiation field. In addition, since a divergent beam is less affected by turbulence [52] and the variance of a divergent beam is similar to that of a spherical wave [36], a divergent GSM beam is used for simulation analysis, and the variance of the intensity fluctuation of a spherical wave is used to derive the approximate result for a partially coherent GSM beam.

Under weak turbulence, the normalized intensity fluctuation difference of a spherical wave is given by [53]

$$\sigma_I^2 = 0.496C_n^2 k^{7/6} L^{11/6} \tag{4.11}$$

Under strong turbulence, the normalized light intensity fluctuation variance of a spherical wave can be represented as [51]

$$\sigma_I^2 = 0.015\sigma_I^2\left(\frac{\eta Q}{\eta+Q}\right)^{7/6}\left[1+1.753\left(\frac{\eta}{\eta+Q}\right)^{1/2} - 0.252\left(\frac{\eta}{\eta+Q}\right)^{7/12}\right] \tag{4.12}$$

where $\eta = \frac{8}{1+0.069\sigma_I^2 Q^{1/6}}$, $Q = 10.89L/kl_0$, and l_0 is the internal scale of the turbulence.

Owing to its cascade nature [37], the internal scale of turbulence is greater than zero. Based on Davis's method, we can divide strong and weak turbulence according to the value of C_n^2, i.e., $C_n^2 < 6.4 \times 10^{-17}\text{m}^{-2/3}$ represents weak turbulence whereas $C_n^2 < 2.5 \times 10^{-13}\text{m}^{-2/3}$ is indicative of strong turbulence.

From Eqs. (4.10) and (4.11), it can be deduced that the expression for the variance of light intensity fluctuation when a partially coherent GSM beam passes through weak atmospheric turbulence is given by

$$\sigma_I^2 = 0.496C_n^2k^{7/6}L^{11/6}\left\{\exp\left[\frac{0.49\sigma_B^2}{(1+0.56\sigma_B^{12/5})^{7/6}}\right.\right.$$

$$\left.+\frac{0.51\sigma_B^2}{(1+0.69\sigma_B^{12/5})^{5/6}}\right] - 1 + 4.42\sigma_1^2\left(\frac{\Lambda_e}{1+4\Lambda_e q_c + 1.63\sigma_1^{12/5}\Lambda_e}\right)^{5/6}$$

$$\left.\times\frac{r^2}{w_e^2\left(1+4\Lambda_e q_c+1.63\sigma_1^{12/5}\Lambda_e\right)}\right\} \quad (4.13)$$

From Eqs. (4.10) and (4.12), the expression for the variance of intensity fluctuation of a partially coherent GSM beam propagating through strong atmospheric turbulence can be derived as follows:

$$\sigma_I^2 = 0.015\sigma_1^2\left(\frac{\eta Q}{\eta+Q}\right)^{7/6}\left[1+1.753\left(\frac{\eta}{\eta+Q}\right)^{1/2}\right.$$

$$\left.-0.252\left(\frac{\eta}{\eta+Q}\right)^{7/2}\right]\left[\exp\left[\frac{0.49\sigma_B^2}{(1+0.56\sigma_B^{12/5})^{7/6}}\right.\right.$$

$$\left.+\frac{0.51\sigma_B^2}{(1+0.69\sigma_B^{12/5})^{5/6}}\right] - 1 + 4.42\sigma_1^2\left(\frac{\Lambda_e}{1+4\Lambda_e q_c + 1.63\sigma_1^{12/5}\Lambda_e}\right)^{5/6}$$

$$\left.\times\frac{r^2}{w_e^2\left(1+4\Lambda_e q_c+1.63\sigma_1^{12/5}\Lambda_e\right)}\right] \quad (4.14)$$

Equations (4.13) and (4.14) are the results of the partial coherence of the beam re modulating the light intensity fluctuation variance of the fully coherent beam after passing through the weak and strong turbulence.

To better appreciate the improvement of the intensity fluctuation variance more intuitively, we define an improvement ratio of the intensity fluctuation variance α, which is the ratio of the intensity fluctuation variance of partially coherent light to the intensity fluctuation variance of fully coherent light under the same conditions. The smaller the value of α, the better the improvement in the light intensity fluctuation; the larger the value of α, the worse the improvement of the light intensity fluctuation.

4.3.2 *Variance of Arrival Angle Fluctuation*

For a quasi-monochromatic field $U(\vec{r})$ with frequency ω propagating from the plane $z = 0$ to the half-space $z > 0$ in a turbulent atmosphere, a random function of the position of the refractive index $n(\vec{r})$ in half space is assumed, where $n(\vec{r}) = 1 + n_1(\vec{r})$ and $n_1 << 1$. The field can then be described using the scalar wave equation [54]:

$$\{\nabla^2 + k^2[1 + 2n_1(\vec{r})]\}U(\vec{r}) = 0 \tag{4.15}$$

Assuming that the incident field $U(\vec{r}, 0)$ is incident close to the z-axis, the solution to Eq. (4.15) is given as

$$U(\vec{\rho}, 0) = \iint U(\vec{r}, 0)G_0(\vec{r}, \vec{r}, z)d^2\vec{r} \tag{4.16}$$

where $G_0(\vec{r}, \vec{\rho}, z) = \frac{-ik}{2\pi z}\exp\left[ikz + \frac{ik}{2z}|\vec{\rho} - \vec{r}|^2 + \psi(\vec{r}, \vec{\rho})\right]$ is the Green's function, and $\psi(\vec{r}, \vec{\rho})$ is the random perturbation of the complex phase of a spherical wave in a turbulent atmosphere, which depends on the characteristics of the medium. The preceding equation can then be expressed as

$$U(\vec{\rho}, z) = \frac{-ik}{2\pi z}\exp(ikz)\iint d^2rU(\vec{r}, 0)\exp\left[\frac{ik}{2z}|\vec{\rho} - \vec{r}|^2 + \psi(\vec{r}, \vec{\rho})\right] \tag{4.17}$$

The Huygens-Fresnel principle states that every point of a wavefront is a source of spherical waves, and the envelope of these spherical waves constitutes a new wavefront. Equation (4.17) describes the influence of atmospheric turbulence on transmission through $\psi(\vec{r}, \vec{\rho})$, which is called the generalized Huygens-Fresnel principle, and applies to regions of weak and strong turbulence fluctuation.

The transmission characteristics of light waves in a turbulent atmosphere are reflected in the complex cross-correlation function $\Gamma(\vec{\rho}_1, \vec{\rho}_2; \tau) = \langle E(\vec{\rho}_1; \tau)E^*(\vec{\rho}_2; t + \tau)\rangle$, where $\langle\cdot\rangle$ represents the overall average and E is the optical frequency of the electric field. In recent studies, the cross-spectral density function $W(\vec{\rho}_1, \vec{\rho}_2; \omega) = \langle U(\vec{\rho}_1; \omega)U^*(\vec{\rho}_2; \omega)\rangle$ has been used to investigate the transmission of partially coherent fields in a turbulent atmosphere [52], which is the time Fourier transform of $\Gamma(\vec{\rho}_1, \vec{\rho}_2; \tau)$ and ω is the frequency of the light waves. The cross-spectral density function obeys the Helmholtz equation and is a measure of the correlation between the fluctuations of two field components at the same frequency [55]. If the field is strictly monochromatic or the band is sufficiently narrow, we have

$$\frac{|\vec{\rho}_2 - \vec{\rho}_1|}{c} \ll \frac{1}{\Delta\omega} \tag{4.18}$$

The two characteristic quantities yield the same result. In laser communication systems, if the typical value $|\vec{\rho}_2 - \vec{\rho}_1|$ is on the order of 10 cm, the condition $\Delta\omega <$ 1 GHz must be achieved.

We consider the variation of $W(\vec{\rho}_1, \vec{\rho}_2; \omega = \omega_0)$, where ω_0 is the center frequency of the monochromatic Gaussian beam. The cross-spectral density function on the receiving surface is obtained using the generalized Huygens-Fresnel principle [56]:

$$W(\vec{\rho}_1, \vec{\rho}_2; \omega) = \left\langle U(\vec{\rho}_1; \omega)U^*(\vec{\rho}_2; \omega)\right\rangle$$
$$= \frac{1}{(\lambda z)^2}\iiint\int d^2\vec{r}_1 d^2\vec{r}_2 W(\vec{r}_1, \vec{r}_2)\langle\exp[\psi(\vec{r}_1, \vec{\rho}_1) + \psi^*(\vec{r}_2, \vec{\rho}_2)]\rangle$$

$$\times \exp\left\{\frac{ik}{2z}\left[(\vec{\rho}_1-\vec{r}_1)^2-(\vec{\rho}_2-\vec{r}_2)^2\right]\right\} \tag{4.19}$$

where $W_0(\vec{r}_1, \vec{r}_2)$ is the cross-spectral density function at the transmitter. For partially coherent beams, the transmitted light field can be corrected as

$$\widetilde{U}(\vec{r}, 0) = U(\vec{r}, 0)\exp[i\varphi_d(\vec{r})] \tag{4.20}$$

where $U(\vec{r}, 0) = \exp\left[-\left(\frac{1}{w_0^2}+\frac{ik}{2F_0}\right)r^2\right]$ and $\exp[i\varphi_d(\vec{r})]$ are small random perturbation terms caused by phase diffusion.

Assuming that the overall average value of some independent random phases caused by diffusion is Gaussian and only depends on the transmission distance of the beam rather than the actual diffusion path, the cross-spectral density function of the transmitter can be expressed as

$$\begin{aligned} W_0(\vec{r}_1, \vec{r}_2) &= \left\langle \widetilde{U}(\vec{r}_1; z)\widetilde{U}^*(\vec{r}_2; z)\right\rangle \\ &= U(\vec{r}_1, 0)U^*(\vec{r}_2, 0) \times \langle \exp[i\varphi_{d1}(\vec{r}_1)]\exp[i\varphi_{d2}(\vec{r}_2)]\rangle \\ &= U(\vec{r}_1, 0)U^*(\vec{r}_2, 0) \times \exp[-(\vec{r}_1-\vec{r}_2)^2/2l_c^2] \end{aligned} \tag{4.21}$$

where l_c is the partial coherence length, which describes the partial coherence characteristics of the transmitter. The expression for the cross-spectral density function $W(\vec{\rho}_1, \vec{\rho}_2; z)$ on the receiving surface is

$$\begin{aligned} W(\vec{\rho}_1, \vec{\rho}_2; z) = \frac{1}{(\lambda z)^2} \iiiint d^2\vec{r}_1 d^2\vec{r}_2 W_0(\vec{r}_1, \vec{r}_2)\left\langle \exp[\psi(\vec{r}_1, \vec{\rho}_1)+\psi^*(\vec{r}_2, \vec{\rho}_2)]\right\rangle \\ \times \exp\left\{\frac{ik}{2z}\left[(\vec{\rho}_1-\vec{r}_1)^2-(\vec{\rho}_2-\vec{r}_2)^2\right]\right\} \end{aligned} \tag{4.22}$$

The cross-correlation function of the complex phase of a spherical wave is expressed as

$$\left\langle \exp[\psi(\vec{r}_1, \vec{\rho}_1)+\psi^*(\vec{r}_2, \vec{\rho}_2)\right\rangle = \exp[-D_\psi/2] \tag{4.23}$$

where D_ψ is the phase structure function, which represents the mean square deviation of the phase between two points [36]. The expression is given as

$$\begin{aligned} D_\psi(\vec{r}_d, \vec{\rho}_d) \cong 8\pi^2k^2L \int_0^1\int_0^\infty \kappa\Phi_n(\kappa, L_0, l_0) \\ \times [1-J_0(|(1-\xi)(\vec{r}_1-\vec{r}_2)+\xi(\vec{\rho}_1-\vec{\rho}_2)|\kappa)]d\kappa d\xi \end{aligned} \tag{4.24}$$

where $\Phi_n(\kappa, L_0, l_0)$ is the refractive index turbulence spectrum.

In the Kolmogorov refractive index spectrum model, the structure constant is expressed as

$$
\begin{aligned}
D_\psi &= 2.92k^2 z \int_0^1 dt C_n^2(tz) |(\vec{r}_1 - \vec{r}_2)t + (1-t)(\vec{\rho}_1 - \vec{\rho}_2)|^{5/3} \\
&\cong 2\{[(\vec{r}_1 - \vec{r}_2)^2 + (\vec{r}_1 - \vec{r}_2)(\vec{\rho}_1 - \vec{\rho}_2) + (\vec{\rho}_1 - \vec{\rho}_2)^2]/\rho_T^2\}
\end{aligned} \tag{4.25}
$$

From Eqs. (4.23) and (4.25), we have

$$
\begin{aligned}
&\langle \exp[\psi(\vec{r}_1, \vec{\rho}_1) + \psi^*(\vec{r}_2, \vec{\rho}_2)\rangle \\
&\cong \exp\{-[(\vec{r}_1 - \vec{r}_2)^2 + (\vec{r}_1 - \vec{r}_2)(\vec{\rho}_1 - \vec{\rho}_2) + (\vec{\rho}_1 - \vec{\rho}_2)^2]/\rho_T^2\}
\end{aligned} \tag{4.26}
$$

where $\rho_T^2 = [1.46k^2 z \int_0^1 C_n^2(\xi z)(1-\xi)^{5/3} d\xi]^{-3/5}$ is the coherence length of a spherical wave with a slant propagation in a turbulent atmosphere. Equation (4.26) can then be written as

$$
\begin{aligned}
&W(\vec{\rho}_1, \vec{\rho}_2; z) \\
&= \frac{1}{(\lambda z)^2} \iiint\int d^2\vec{r}_1 d^2\vec{r}_2 W_0(\vec{r}_1, \vec{r}_2) \exp\left\{\frac{ik}{2z}[(\vec{\rho}_1 - \vec{r}_1)^2 - (\vec{\rho}_2 - \vec{r}_2)^2]\right\} \\
&\quad \times \exp\{-[(\vec{r}_1 - \vec{r}_2)^2 + (\vec{r}_1 - \vec{r}_2)(\vec{\rho}_1 - \vec{\rho}_2) + (\vec{\rho}_1 - \vec{\rho}_2)^2]/\rho_T^2\}
\end{aligned} \tag{4.27}
$$

Equation (4.27) is the cross-spectral density function at the receiving surface after the partially coherent beam is transmitted with a slant propagation in a turbulent atmosphere. For different types of partially coherent beams at the receiving surface, the cross-spectral density function depends on $W_0(\vec{r}_1, \vec{r}_2)$ at the transmitter.

For a narrow-band light source, the cross-correlation function of the beam can be approximately replaced by the cross-spectral density $W(\vec{\rho}_1, \vec{\rho}_2, z)$ [54], i.e.,

$$
\Gamma(\vec{\rho}_1, \vec{\rho}_2; z) = \langle U(\vec{\rho}_1, z) U^*(\vec{\rho}_2, z)\rangle \approx W(\vec{\rho}_1, \vec{\rho}_2, z) \tag{4.28}
$$

The relationship between the wave structure function and the cross-correlation function is then given as [52]

$$
\exp\left[-\frac{1}{2} D(\vec{\rho}_1, \vec{\rho}_2, z)\right] = \frac{|\Gamma(\vec{\rho}_1, \vec{\rho}_2, z)|}{[\Gamma(\vec{\rho}_1, \vec{\rho}_2, z)\Gamma(\vec{\rho}_1, \vec{\rho}_2, z)]^{1/2}} \tag{4.29}
$$

The cross-correlation function at the receiving surface of partially coherent GSM beams in a turbulent atmosphere is expressed as [55]

$$
\begin{aligned}
W(\vec{\rho}_c, \vec{\rho}_d; z) &= \langle I(\vec{\rho}_c, \vec{\rho}_d; z)\rangle \\
&= \frac{w_0^2}{W_\zeta^2(z)} \exp\left\{-\rho_d^2\left(\frac{1}{\rho_T^2} + \frac{1}{2w_0^2\Lambda_0^2}\right)\right.
\end{aligned}
$$

$$-\frac{4\vec{\rho}_c^2 - \varphi^2\vec{\rho}_d}{2W_\zeta^2(z)} - \frac{ik\vec{\rho}_c \cdot \vec{\rho}_d}{F_\zeta^2(z)}\Bigg\} \tag{4.30}$$

where $\varphi = \Theta_0/\Lambda_0 - \Lambda_0 w_0^2/\rho_T^2$ $W_\zeta(z)$ and $F_\zeta(z)$ are the effective beam radius and the phase wavefront curvature radius of the GSM beam in a turbulent atmosphere, respectively. These parameters are defined as follows:

$$W_\zeta(z) = w_0\left(\Theta_0^2 + \zeta\Lambda_0^2\right)^{1/2}, \quad \zeta = \zeta_s + \frac{2w_0^2}{\rho_T^2} \tag{4.31}$$

$$F_\zeta(z) = \frac{z(\Theta_0^2\zeta\Lambda_0^2)}{\varphi\Lambda_0 - \zeta\Lambda_0^2 - \Theta_0^2}, \quad \Lambda_0 = 2z/kw_0^2 \tag{4.32}$$

For a Gaussian beam at transmitters Θ_0 and Λ_0, when $\Theta_0 = 1$, the beam is collimated, when $\Theta_0 < 1$, the beam is convergent, and when $\Theta_0 > 1$, the beam is divergent [55]. ζ is the global coherence parameter, and $\zeta_s = 1 + w_0^2/l_c^2$ is the source coherence parameter of the beam at the transmitter. If $\zeta_s = 1$, the beam is completely coherent; if $\zeta_s > 1$, the beam is partially coherent.

When the beam propagates along a horizontal path, the atmospheric refractive index structure constant C_n^2 is constant as the height changes. ρ_T degenerates into $\rho_0 = [0.545C_n^2k^2L]^{-3/5}$. Combining (4.29) and (4.30), we can obtain an expression for the phase structure function of a partially coherent GSM beam propagating along a horizontal path in a turbulent atmosphere as follows:

$$D(\vec{\rho}_1, \vec{\rho}_2, z) = 2\rho^2\left\{\left(\frac{1}{\rho_0^2} + \frac{1}{2w_0^2\Lambda_0^2}\right) - \frac{\Theta_0^2}{2w_\zeta^2(z)\Lambda_0^2} + \frac{\Lambda_0 w_0^2}{w_\zeta^2(z)\rho_0^2} - \frac{\Lambda_0^2 w_0^4}{2w_\zeta^2(z)\rho_0^4}\right\} \tag{4.33}$$

The phase structure function represents the mean square deviation of the phase between two points [37]. Therefore Eq. (4.33) can be used to measure the phase fluctuation of partially coherent GSM beams. The wave structure function $D = D_\chi + D_s$ can be approximated as $D \approx D_s$ because the phase structure function D_s plays a critical role, so the influence of the logarithmic amplitude structure–function D_χ can be ignored [55]. From Eq. (4.33), the angle of arrival fluctuation variance of a partially coherent Gaussian-Schell beam at the receiving surface for horizontal propagation in a turbulent atmosphere is given by

$$\langle\alpha^2\rangle = \frac{D(\vec{\rho}_1, \vec{\rho}_2, L)}{(k\rho)^2} = \frac{2}{k^2}\left\{\left(\frac{1}{\rho_0^2} + \frac{1}{2w_0^2\Lambda_0^2}\right) - \frac{\varphi^2}{2W_\zeta^2}\right\} \tag{4.34}$$

4.3.3 Analysis of Light Intensity Flicker Characteristics

Figure 4.1 shows the logarithmic intensity fluctuation variance of partially coherent light as a function of the transmission distance for different atmospheric refractive index constants under weak turbulence. For a wavelength $\lambda = 1.55\ \mu\text{m}$, the initial beam radius $\omega_0 = 1$ cm, the internal scale $l_0 = 1$ cm, the coherence length $l_c = 0.001$ m, and the vector diameter $r = 1$ m. Figure 4.1 shows that the logarithmic intensity fluctuation variance of partially coherent light increases with the increase of the atmospheric refractive index constant for transmission over a given distance under weakly turbulent conditions. The larger the atmospheric refractive index constant, the faster the increases of the logarithmic intensity fluctuation variance with the propagation distance. This is because, with the increase of atmospheric refractive index, the detected light intensity caused by turbulence increases with the random change of time and space.

Figure 4.2 shows the variation of the logarithmic intensity fluctuation variance of fully coherent light ($l_c \to \infty$) and partially coherent light ($l_c = 0.01$ m, $l_c = 0.001$ m) with transmission distance in weak turbulence. In this case, the wavelength $\lambda = 1.55\ \mu\text{m}$, the initial beam radius $\omega_0 = 1$ cm, the internal scale $l_0 = 1$ cm, the atmospheric refractive index constant $C_n^2 = 3 \times 10^{-17}\ \text{m}^{-2/3}$, and the vector diameter $r = 1$ m. Based on this figure, it is evident that under the condition of weak turbulence, the logarithmic intensity fluctuation variance of the beam increases with the increase of the transmission distance. For a given distance, the logarithmic intensity fluctuation variance of the partially coherent light is noticeably smaller

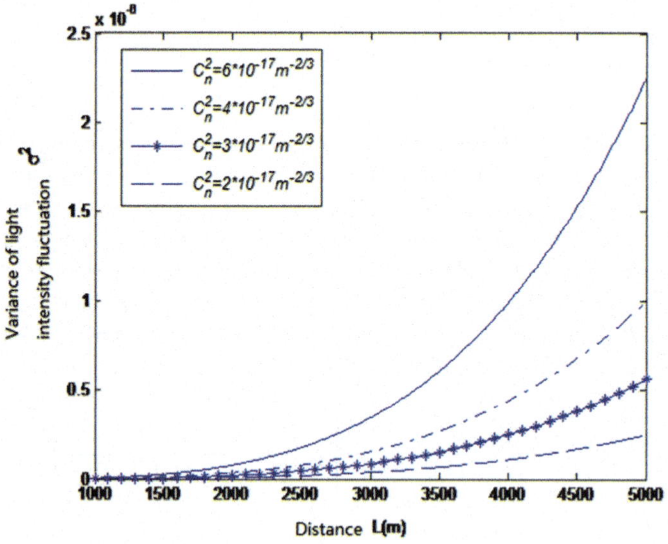

Fig. 4.1 Variation of intensity fluctuation of partially coherent light with distance for weak turbulence

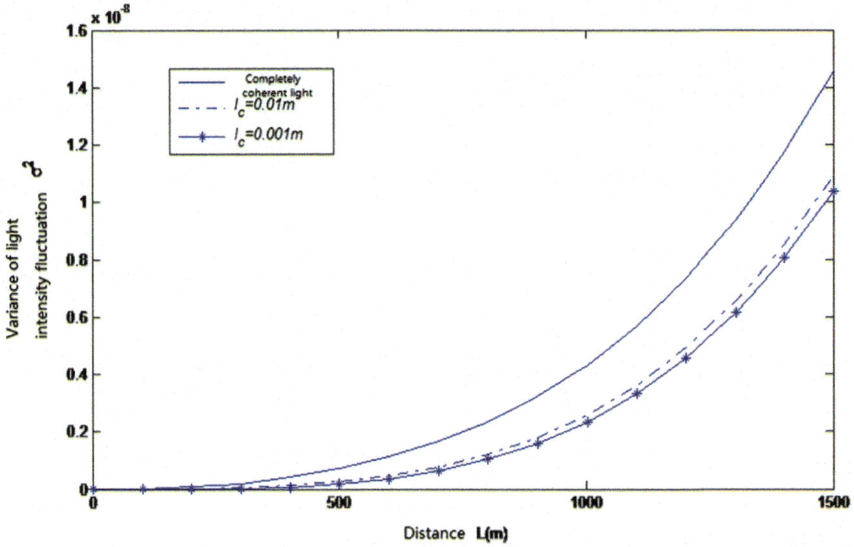

Fig. 4.2 Variation of light intensity fluctuation variance with distance for weak turbulence

than that of fully coherent light. As the coherence length decreases, the logarithmic intensity fluctuation variance of partially coherent light also decreases. Thus, under the condition of weak turbulence, partially coherent light is less affected than fully coherent light.

Figure 4.3 shows the variation of the logarithmic intensity fluctuation variance of fully coherent light ($l_c \to \infty$) and partially coherent light ($l_c = 0.01$ m, $l_c = 0.001$ m) with vector diameter in weak turbulence. In this case, the propagation distance $L = 5$ km, the wavelength $\lambda = 1.55$ μm, the initial beam radius $\omega_0 = 1$ cm, the internal scale $l_0 = 1$ cm, and the atmospheric refractive index constant $C_n^2 = 4 \times 10^{-17}\ \text{m}^{-2/3}$. The figure shows that under weak turbulence conditions, for a given transmission distance, the logarithmic intensity fluctuation variance of partially coherent light on a receiving surface is smaller than that of fully coherent light. Moreover, the worse the coherence, the better the improvement of intensity fluctuation. The curve of the logarithmic intensity fluctuation variance of partially coherent light with a vector diameter is relatively flat. When the coherence of the beam is reduced, the logarithmic intensity fluctuation variance on the receiving surface does not change significantly with the vector diameter. This indicates that reducing the coherence of the light source can effectively suppress the intensity fluctuation in weak turbulence.

Figure 4.4 shows the logarithmic intensity fluctuation variance of partially coherent light with transmission distance under different atmospheric refractive index constants under strong turbulence. In this case, the wavelength $\lambda = 1.55$ μm, the initial beam radius $\omega_0 = 1$ cm, the internal scale $l_0 = 1$ cm, the coherence length $l_c = 0.001$ m, and the vector diameter $r = 1$ m. Based on Fig. 4.4, similar to the case

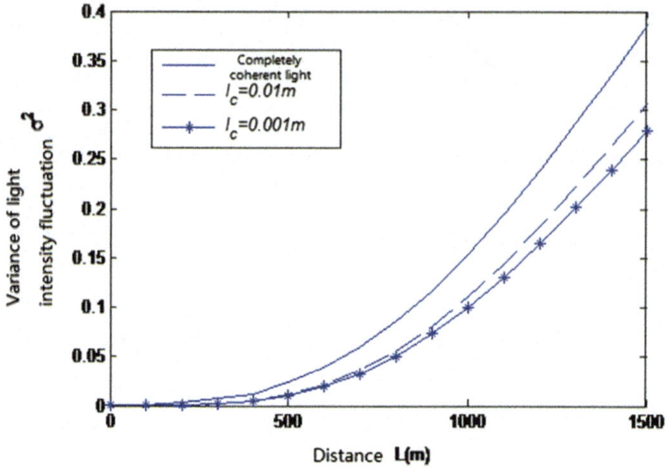

Fig. 4.3 Variation of light intensity fluctuation variance with vector diameter for weak turbulence

of weak turbulence condition, under strong turbulence, the logarithmic intensity fluctuation variance of partially coherent light also increases for a given distance with the increase of atmospheric refractive index constant. The larger the atmospheric refractive index constant, the faster the increase of the logarithmic intensity fluctuation variance with propagation distance. Moreover, the increase of the logarithmic intensity fluctuation variance in strong turbulence is significantly greater than that in weak turbulence.

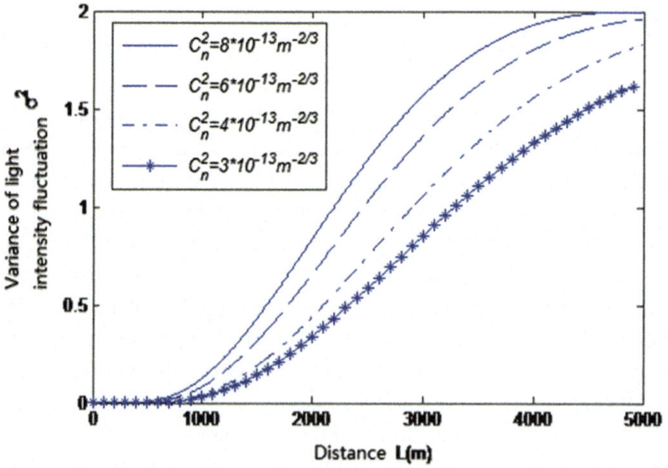

Fig. 4.4 Variation of intensity fluctuation of partially coherent light with distance for strong turbulence

Figure 4.5 shows the variation of the logarithmic intensity fluctuation variance of fully coherent light ($l_c \to \infty$) and partially coherent light ($l_c = 0.01$ m, $l_c = 0.001$ m) as a function of transmission distance under strong turbulence. The simulation parameters are wavelength $\lambda = 1.55\,\mu$m, initial beam radius $\omega_0 = 1$ cm, internal scale $l_0 = 1$ cm, atmospheric refractive index constant $C_n^2 = 3 \times 10^{-13}\,\text{m}^{-2/3}$, v and vector diameter $r = 1\,m$. In Fig. 4.5 under the condition of strong turbulence, the logarithmic intensity fluctuation variance of the beam increases with the increase of the transmission distance, which is similar to the case of weak turbulence. At a given distance, the logarithmic intensity fluctuation variance of partially coherent light is noticeably smaller than that of fully coherent light. As the coherence length decreases, the logarithmic intensity fluctuation variance of the partially coherent light also decreases, and the improvement is better than that of the case of weak turbulence. The influence of turbulence on partially coherent light is less than that of fully coherent light in strong turbulence.

Figure 4.6 shows the variation of the logarithmic intensity fluctuation variance of fully coherent light ($l_c \to \infty$) and a partially coherent GSM beam ($l_c = 0.01$ m, $l_c = 0.001$ m) with vector diameter under strong turbulence. In this case, propagation distance is $L = 5$ km, the wavelength is $\lambda = 1.55\,\mu$m, the initial beam radius is $\omega_0 = 1$ cm, the internal scale is $l_0 = 1$ cm, and the atmospheric refractive index constant is $C_n^2 = 3 \times 10^{-13}\,\text{m}^{-2/3}$. Form this figure, under the condition of strong turbulence, the logarithmic intensity fluctuation variance of partially coherent light on a receiving surface is smaller than that of fully coherent light at the same distance. Moreover, the worse the coherence, the greater the improvement in the light intensity fluctuation and the better the improvement compared to weak turbulence. The logarithmic intensity fluctuation variance of partially coherent light varies gradually with the vector diameter. When the coherence of the beam decreases, the intensity fluctuation variance on the receiving surface does not change significantly with the

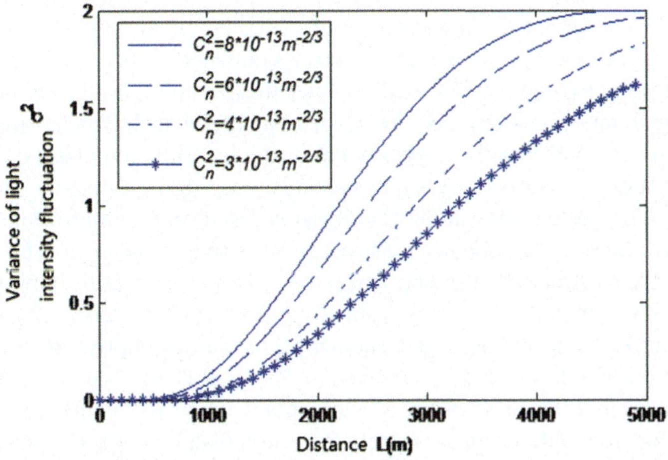

Fig. 4.5 Variation of light intensity fluctuation variance with distance for strong turbulence

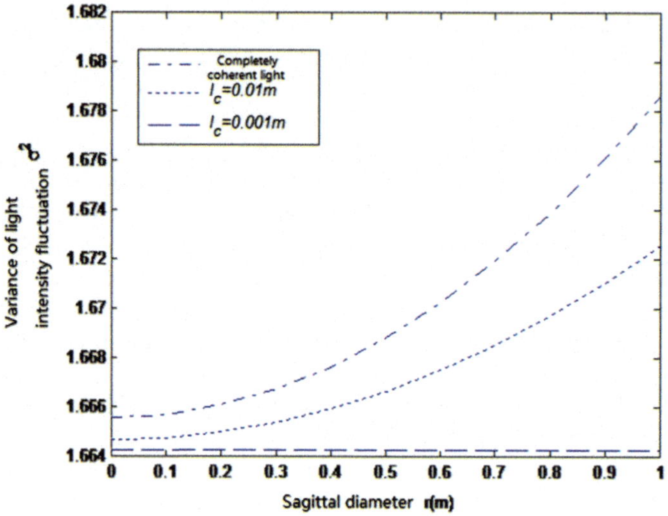

Fig. 4.6 Variation of intensity fluctuation variance of beam with vector diameter for strong turbulence

vector diameter. This shows that reducing the coherence of the light source can also effectively suppress the fluctuation of the light intensity under strong turbulence.

Figure 4.7 shows the ratio of the logarithmic intensity fluctuation variance of partially coherent light ($l_c = 0.001$ m) and fully coherent light ($l_c \to \infty$) in atmospheric turbulence, i.e., the change of the intensity fluctuation variance improvement ratio with transmission distance. In this case, the wavelength is $\lambda = 1.55\,\mu$m, the initial beam radius is $\omega_0 = 1$ cm, the internal scale is $l_0 = 1$ cm, the atmospheric refractive index constant is $C_n^2 = 3 \times 10^{-17}\,\mathrm{m}^{-2/3}$ for weak turbulence, $C_n^2 = 3 \times 10^{-13}\,\mathrm{m}^{-2/3}$ for strong turbulence, and the vector diameter is $r = 1$ m. In Fig. 4.7, the farther the transmission distance, the greater the improvement of the logarithmic intensity fluctuation variance compared to α, i.e., the smaller the improvement of the intensity fluctuation. When the transmission distance is short, the improvement ratio α under weak turbulence is smaller than that under strong turbulence. This indicates that the improvement of the logarithmic intensity fluctuation variance under weak turbulence is greater compared to the case of strong turbulence. However, with the increase in the transmission distance, the improvement ratio α under weak turbulence is higher than that under strong turbulence. This indicates that when the transmission distance increases to a certain extent, the improvement of the logarithmic intensity fluctuation variance under weak turbulence is less than that under strong turbulence. Based on the numerical results, the improvement ratio of the logarithmic intensity fluctuation variance under weak turbulence is 0.91116 when the transmission distance is 2.5 km. The logarithmic intensity fluctuation variance of partially coherent light is approximately 8.9% less than that of fully coherent light. For strong turbulence, the improvement ratio of the logarithmic intensity fluctuation variance is 0.75281. In the case of partially coherent light, the logarithmic intensity

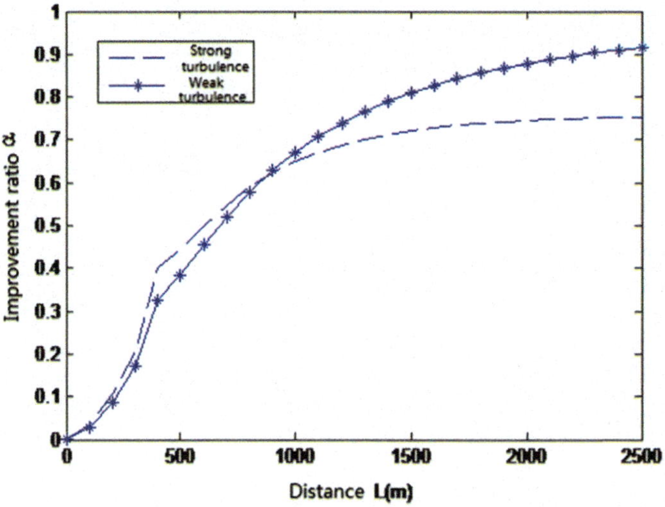

Fig. 4.7 Variance improvement ratio of light intensity fluctuation with distance

fluctuation variance is approximately 24.7% less than that of fully coherent light. Thus, the use of partially coherent light transmission in atmospheric turbulence can effectively reduce the intensity fluctuation.

4.3.4 Analysis of Arrival Angle Fluctuation Characteristics

Figure 4.8 shows a plot of the variance of the angle of arrival fluctuation of converged partially coherent Gaussian-Schell beams as a function of propagation distance for different coherence parameters. In this case, the initial beam radius is $w_0 = 1$ cm, the wavelength is $\lambda = 1.06\,\mu$m, the effective beam parameter is $\Theta_0 = 0.1$, and the atmospheric refractive index constant is $C_n^2 = 1 \times 10^{-14}\text{m}^{-2/3}$ In Fig. 4.8, within a transmission distance of 120 m, the variance of the arrival angle fluctuation of partially coherent light is greater than that of fully coherent light. Moreover, the worse the coherence, the greater the variance of the arrival angle fluctuation. When the propagation distance is greater than 120 m, the variance of the angle of arrival fluctuation of both the partially coherent and fully coherent light is essentially the same and does not changes with the propagation distance, i.e., the influence of beam coherence on the angle of arrival fluctuation is very small.

Figure 4.9 shows a plot of the variance of the arrival angle fluctuation of converged partially coherent Gaussian-Schell beams with waist radius for different coherence parameters. In this case, the propagation distance is $l = 1000$ m, the wavelength is $\lambda = 1.06$ μm, the effective beam parameter is $\Theta_0 = 0.1$ and the atmospheric refractive index constant is $C_n^2 = 1 \times 10^{-14}\text{m}^{-2/3}$. In Fig. 4.9, the variance of the

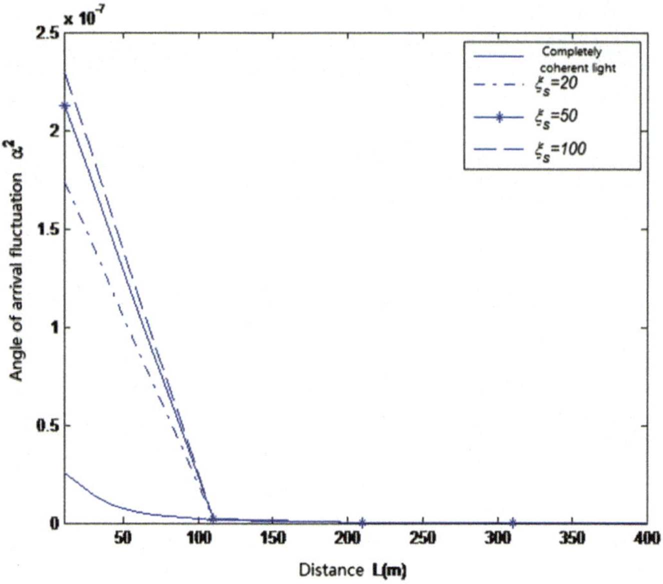

Fig. 4.8 Variance of arrival angle fluctuation of converged partially coherent Gaussian-Schell beams with propagation distance for different source coherence parameters

angle of arrival fluctuation of the beam increases initially, then decreases with the increase of the waist radius until the value becomes constant. When the waist radius is approximately 0.25 m, the variance of the arrival angle fluctuation is a maximum, and then decreases with the increase of the waist radius. Finally, the variance of arrival angle fluctuation is the same whether the light is fully coherent or partially coherent. When the waist radius is < 1 m, the variance of the angle of arrival fluctuation of the partially coherent light is greater than that of the fully coherent light. Moreover, the worse the coherence, the greater the variance of the angle of arrival fluctuation. Therefore, in practice, a beam with a waist radius less than 0.1 m should be selected.

Figure 4.10 shows a plot of the variance of the angle of arrival fluctuation of a partially coherent light beam under different atmospheric refractive index structure constants. In this case, the propagation distance is $l = 1000\,\text{m}$, the wavelength is $\lambda = 1.06\ \mu\text{m}$, the effective beam parameter is $\Theta_0 = 0.1$, and the initial beam radius is $w_0 = 1\,\text{cm}$. In this figure, when the beam is transmitted over 1000 m, the fluctuation of the arrival angle beam is essentially constant with a change in the coherence length. This shows that the beam coherence has little effect on the angle of arrival fluctuation. For a given distance, the greater the atmospheric refractive index constant, the greater the variance of the arrival angle fluctuation.

Figure 4.11 shows the variation of the phase mean square deviation of divergent partially coherent Gaussian-Schell beams with propagation distance for different coherence parameters. The simulation parameters are an initial beam radius $w_0 = 1\,\text{cm}$, wavelength $\lambda = 1.06\,\mu\text{m}$, effective beam parameter $\Theta_0 = 3$, and atmospheric

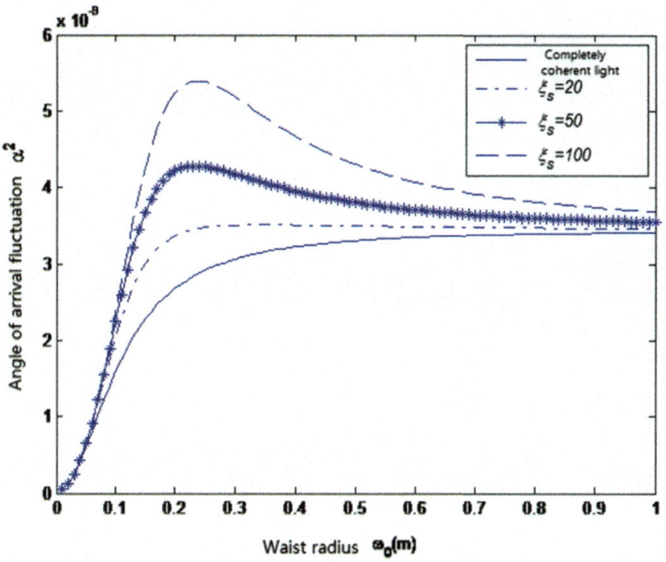

Fig. 4.9 Variance of the arrival angle fluctuation of converged partially coherent Gaussian-Schell beams with waist radius for different source coherence parameters

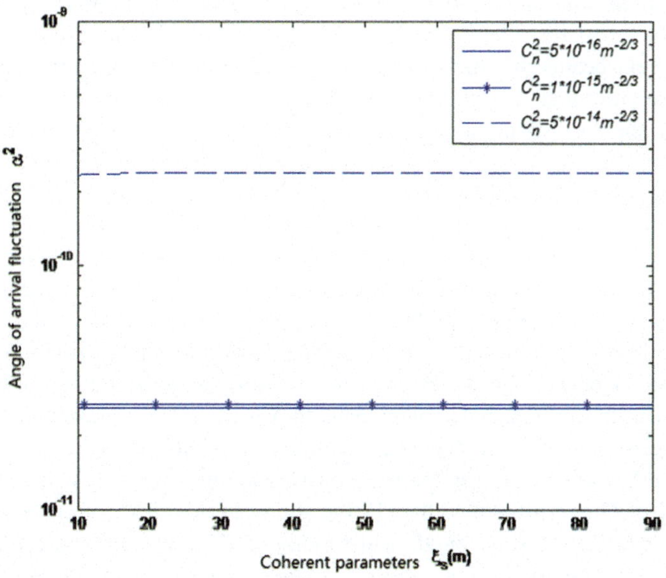

Fig. 4.10 Variance of arrival angle fluctuation with partial coherent parameters for different atmospheric refractive index structure constants

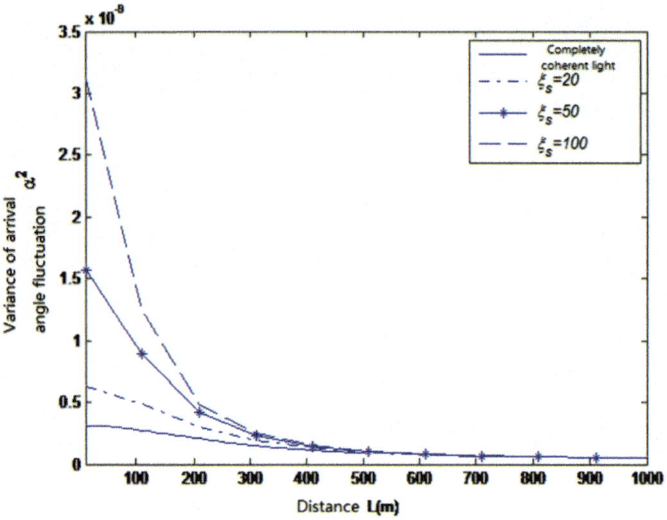

Fig. 4.11 Variance of arrival angle fluctuation of divergent partially coherent Gaussian-Schell beams as a function of propagation distance for different source coherence parameters

refractive index constant $C_n^2 = 1 \times 10^{-14}\text{m}^{-2/3}$. This figure shows that the variation of divergent partially coherent beams and fully coherent beams is the same as that of convergent beams, but the variance of the arrival angle fluctuation of divergent beams is smaller than that of convergent beams for the same transmission distance. Moreover, the corresponding transmission distance is longer when the influence of beam coherence on the arrival angle fluctuation is undefined and it tends to a fixed value at approximately 500 m.

Using the partially coherent GSM beam model, based on the modified Rytov method and the generalized Huygens-Fresnel principle in combination with Andrew's phenomenological scintillation model, this section systematically investigates intensity scintillation and the angle of arrival fluctuation of partially coherent light. In addition, the bit error rate associated with intensity scintillation was also investigated. Considering the different modulation methods and partially coherent light sources, the influence of light intensity flicker caused by atmospheric turbulence on the bit error rate of a partially coherent optical communication system is analyzed, and the method of using multi-aperture emission to improve the system performance is discussed. The results show that (1), the atmospheric refractive index structure is constant, and that the wavelength, waist radius, and other factors affect the performance of the communication system. (2) Under various turbulent conditions, the intensity fluctuation of partially coherent light is smaller than that of fully coherent light under the same conditions. (3) Using a multi-aperture to transmit partially coherent light can further reduce the fluctuation of light intensity and improve the performance of a communication system. (4) The bit error rate of the communication system using partially coherent light as the light source is significantly lower than that of the communication system that uses fully coherent light as the light source. (5) For specific distances, the worse the coherence of the beam, the greater the variance of the angle of arrival fluctuation. When the propagation distance increases, the angle of arrival fluctuation of both partially coherent and fully coherent light is fixed and does not change with the propagation distance, i.e., the influence of beam coherence on the angle of arrival fluctuation is very small. (7) The variation of divergent partially coherent beams and fully coherent beams is the same as that of the convergent beams, but the variance of the arrival angle fluctuation of the former is smaller than that of convergent beams at the same distance. Moreover, the corresponding propagation distance is greater when the influence of beam coherence on arrival angle fluctuation is not evident.

4.4 Relationship Between Angle of Arrival Fluctuation and Turbulence Scale and Other Parameters

4.4.1 Variance of Angle of Arrival Fluctuation of Beam Propagating in Turbulent Atmosphere

The expression of the generalized Huygens-Fresnel principle is given by [57]

$$U(\vec{\rho}, z) = \frac{-ik}{2\pi z}\exp(ikz)\iint d^2 r U(\vec{r}, 0)\exp\left[\frac{ik}{2z}|\vec{\rho} - \vec{r}|^2 + \psi(\vec{r}, \vec{\rho})\right] \tag{4.35}$$

The cross-spectral density function on the receiving surface is obtained using the generalized Huygens-Fresnel principle [58]:

$$\begin{aligned} W(\vec{\rho_1}, \vec{\rho_2}; z) &= \langle U(\vec{\rho_1}; z) U^*(\vec{\rho_2}; z)\rangle \\ &= \frac{1}{(\lambda z)^2}\iiint\int d^2\vec{r_1} d^2\vec{r_2} W(\vec{r_1}, \vec{r_2})\langle \exp[\psi(\vec{r_1}, \vec{\rho_1}) + \psi^*(\vec{r_2}, \vec{\rho_2})]\rangle \\ &\times \exp\left\{\frac{ik}{2z}\left[(\vec{\rho_1} - \vec{r_1})^2 - (\vec{\rho_2} - \vec{r_2})^2\right]\right\} \end{aligned} \tag{4.36}$$

where $W_0(\vec{r}_1, \vec{r}_2)$ is the cross-spectral density function at the transmitter. For partially coherent beams, a phase scatterer is placed at the aperture of the laser transmitter, and the transmitted light field can be corrected as

$$\widetilde{U}(\vec{r}, 0) = U(\vec{r}, 0)\exp[i\varphi_d(\vec{r})] \tag{4.37}$$

where $U(\vec{r}, 0) = \exp\left[-\left(\frac{1}{w_0^2} + \frac{ik}{2F_0}\right)r^2\right]$ and $\exp[i\varphi_d(\vec{r})]$ are small random perturbation terms caused by phase diffusion.

Assuming that the overall average value of some independent random phases caused by diffusion is Gaussian and only depends on the transmission distance of the beam rather than the actual diffusion path, the cross-spectral density function of the transmitter can be expressed as

$$
\begin{aligned}
W_0(\vec{r}_1, \vec{r}_2) &= \left\langle \widetilde{U}(\vec{r}_1; z\widetilde{U}^*(\vec{r}_2; z)) \right\rangle \\
&= U(\vec{r}_1, 0)U^*(\vec{r}_2, 0) \times \langle \exp[i\varphi_{d1}(\vec{r}_1)] \exp[i\varphi_{d2}(\vec{r}_2)] \rangle \\
&= U(\vec{r}_1, 0)U^*(\vec{r}_2, 0) \exp[-(\vec{r}_1 - \vec{r}_2)^2/2l_c^2]
\end{aligned}
\tag{4.38}
$$

where l_c is the partial coherence length, which describes the partial coherence characteristics of the transmitting source. The expression for the cross-spectral density function $W(\vec{\rho}_1, \vec{\rho}_2; z)$ on the receiving surface is

$$
\begin{aligned}
&W(\vec{\rho}_1, \vec{\rho}_2; z) \\
&= \frac{1}{(\lambda z)^2} \iiint \int d^2\vec{r_1} d^2\vec{r_2} W_0(\vec{r_1}, \vec{r_2}) \left\langle \exp[\psi(\vec{r_1}, \vec{\rho_1}) + \psi^*(\vec{r_2}, \vec{\rho_2})] \right\rangle \\
&\times \exp\left\{ \frac{ik}{2z} [(\vec{\rho_1} - \vec{r_1})^2 - (\vec{\rho_2} - \vec{r_2})^2] \right\}
\end{aligned}
\tag{4.39}
$$

The cross-correlation function of the complex phase of a spherical wave is expressed as

$$
\left\langle \exp[\psi(\vec{r_1}, \vec{\rho_1}) + \psi^*(\vec{r_2}, \vec{\rho_2})] \right\rangle = \exp[-D_\psi/2] \tag{4.40}
$$

where, D_ψ is the phase structure function, which represents the mean square deviation of the phase between two points [59], and its expression is

$$
\begin{aligned}
D_\psi(\vec{r}_d, \vec{\rho}_d) \cong 8\pi^2 k^2 L \int_0^1 \int_0^\infty \kappa \Phi_n(\kappa, L_0, l_0) \\
\times \left[1 - J_0(|(1-\xi)(\vec{r}_1 - \vec{r}_2) + \xi(\vec{\rho}_1 - \vec{\rho}_2)|\kappa)\right] d\kappa d\xi
\end{aligned}
\tag{4.41}
$$

where $\Phi_n(\kappa, L_0, l_0)$ is the refractive index turbulence spectrum. We utilize the modified von Karman spectrum [60], which takes both internal and external scales into account, and is expressed as follows:

$$
\phi_n(\kappa, h) = 0.033 C_n^2(h) \exp(-\kappa^2/\kappa_m^2) \times (\kappa^2 + \kappa_0^2)^{-11/6} \tag{4.42}
$$

where, $\kappa_m = 5.92/l_0$, $\kappa_0 \approx 2\pi/L_0$, and l_0 and L_0 represent the inner and outer scales of turbulence, respectively. From Eqs. (4.41) and (4.42), we have

$$
\left\langle \exp[\psi(\vec{r_1}, \vec{\rho_1}) + \psi^*(\vec{r_2}, \vec{\rho_2})] \right\rangle
$$

$$\cong \exp\left\{-\left[(\vec{r_1} - \vec{r_2})^2 + (\vec{r_1} - \vec{r_2})(\vec{\rho}_1 - \vec{\rho}_2) + (\vec{\rho}_1 - \vec{\rho}_2)^2\right] / \overline{\rho_0}^{-2}\right\} \tag{4.43}$$

where,

$$\overline{\rho_0}^{-2} = \rho_0^2 (1 - 0.715\kappa 0^{1/3})^{-1}$$

$$\rho_0^2 = \left(\frac{1.46k^2 z}{\cos\theta} \int_0^1 Cn^2(\xi z)(1-\xi)^{5/3} d\xi)^{-6/5} \times (1 - 0.103(3/8)^{-5/6}\kappa \mathrm{m}^{1/3} \right)$$

Equation (4.39) can then be simplified as

$$\begin{aligned} &W(\vec{\rho}_1, \vec{\rho}_2; z) \\ &= \frac{1}{(\lambda z)^2} \iiint\int d^2\vec{r_1} d^2\vec{r_2} W_0(\vec{r_1}, \vec{r_2}) \exp\left\{\frac{ik}{2z}[(\vec{\rho_1} - \vec{r_1})^2 - (\vec{\rho_2} - \vec{r_2})^2]\right\} \\ &\times \exp\left\{-\left[(\vec{r_1} - \vec{r_2})^2 + (\vec{r_1} - \vec{r_2})(\vec{\rho}_1 - \vec{\rho}_2) + (\vec{\rho}_1 - \vec{\rho}_2)^2\right] / \overline{\rho_0}^{-2}\right\} \end{aligned} \tag{4.44}$$

Equation (4.40) is the cross-spectral density function at the receiving surface when the partially coherent beam is transmitted with a slant propagation in a turbulent atmosphere.

For narrow-band light sources, the cross-correlation function of the beam can be approximated using the cross-spectral density $W(\vec{\rho}_1, \vec{\rho}_2, z)$, i.e.,

$$\Gamma(\vec{\rho}_1, \vec{\rho}_2; z) = \langle U(\vec{\rho}_1, z) U^*(\vec{\rho}_2, z)\rangle \approx W(\vec{\rho}_1, \vec{\rho}_2, z) \tag{4.45}$$

The relationship between the wave structure function and the cross-correlation function is then given by

$$\exp\left[-\frac{1}{2} D(\vec{\rho}_1, \vec{\rho}_2, z)\right] = \frac{|\Gamma(\vec{\rho}_1, \vec{\rho}_2, z)|}{[\Gamma(\vec{\rho}_1, \vec{\rho}_1, z)\Gamma(\vec{\rho}_2, \vec{\rho}_2, z)]^{1/2}} \tag{4.46}$$

The cross-correlation function of partially coherent GSM beams arriving at the receiving surface in a turbulent atmosphere is expressed as

$$\begin{aligned} W(\vec{\rho}_c, \vec{\rho}_d; z) &= \langle I(\vec{\rho}_c, \vec{\rho}_d; z)\rangle \\ &= \frac{w_0^2}{W_\zeta^2} \exp\left\{-\rho_d^2\left(\frac{1}{\rho_T^2} + \frac{1}{2w_0^2\Lambda_0^2}\right) - \frac{4\vec{\rho}_c^2 \varphi^2 \vec{\rho}_d}{2W_\zeta^2(z)} - \frac{ik\vec{\rho}_c \cdot \vec{\rho}_d}{F_\zeta^2(z)}\right\} \end{aligned} \tag{4.47}$$

where $\varphi = \Theta_0/\Lambda_0 - \Lambda_0 w_0^2/\rho_T^2$ $W_\zeta^2(z)$ $F_\zeta^2(z)$ is the effective beam radius and phase wavefront curvature radius of the GSM beam in a turbulent atmosphere, and can be expressed as follows:

$$W_\zeta(z) = w_0(\Theta_0^2 + \zeta\Lambda_0^2)^{1/2}, \zeta = \zeta_s + \frac{2w_0^2}{\rho_T^2} \tag{4.48}$$

$$F_\zeta(z) = \frac{z(\Theta_0^2 + \zeta\Lambda_0^2)}{\varphi\Lambda_0 - \zeta\Lambda_0^2 - \Theta_0^2}, \Lambda_0 = 2z/kw_0^2 \tag{4.49}$$

where Θ_0 and Λ_0 represent the effective parameters of the Gaussian beam at the transmitter. When $\Theta_0 = 1$, the beam is collimated, when $\Theta_0 < 1$, the beam is convergent, and when $\Theta_0 > 1$, the beam is divergent. ζ is the global coherence parameter, and $\zeta_s = 1 + w_0^2/l_c^2$ is the source coherence parameter of the beam at the transmitter. If $\zeta_s = 1$, the beam is completely coherent, and if $\zeta_s > 1$, the beam is partially coherent.

According to Eqs. (4.42) and (4.43), the phase structure function of a partially coherent GSM beam propagating along a horizontal path in a turbulent atmosphere is

$$D(\vec{\rho}_1, \vec{\rho}_2, z) = 2\rho^2\left\{\left(\frac{1}{\rho_0^2} + \frac{1}{2w_0^2\Lambda_0^2}\right) - \frac{\Theta_0^2}{2W_\zeta^2(z)\Lambda_0^2} + \frac{\Lambda_0 w_0^2}{w_\zeta^2(z)\rho_0^2} - \frac{\Lambda_0^2 w_0^4}{2w_\zeta^2(z)\rho_0^4}\right\} \tag{4.50}$$

The phase structure function represents the mean square deviation of the phase between two points. Thus, the phase fluctuation of partially coherent GSM beams can be determined using Eq. (4.46). The wave structure function $D = D_\chi + D_s$ can be approximated as $D \approx D_s$, because the phase structure function D_s plays a critical role. Therefore, the influence of the logarithmic amplitude structure function can be ignored. According to Eq. (4.46) for the variance of the angle of arrival fluctuation, the variance of the angle of arrival fluctuation at the receiving surface when the partially coherent GSM beam propagates horizontally in a turbulent atmosphere can be represented as follows:

$$\langle\alpha^2\rangle = \frac{D(\vec{\rho}_1, \vec{\rho}_2, L)}{(k\rho)^2} = \frac{2}{k^2}\left\{\left(\frac{1}{\rho_0^2} + \frac{1}{2w_0^2\Lambda_0^2}\right) - \frac{\varphi^2}{2W_\zeta^2}\right\} \tag{4.51}$$

4.4.2 Turbulent Inner and Outer Scales and Beam Propagation Distance

Figure 4.12 shows the influence of the internal and external scales of atmospheric turbulence on the angle of arrival fluctuation when partially coherent GSM beams are transmitted through atmospheric turbulence. The selected parameters are the initial waist radius $\omega_0 = 0.01$ m, the near-ground atmospheric refractive index structure

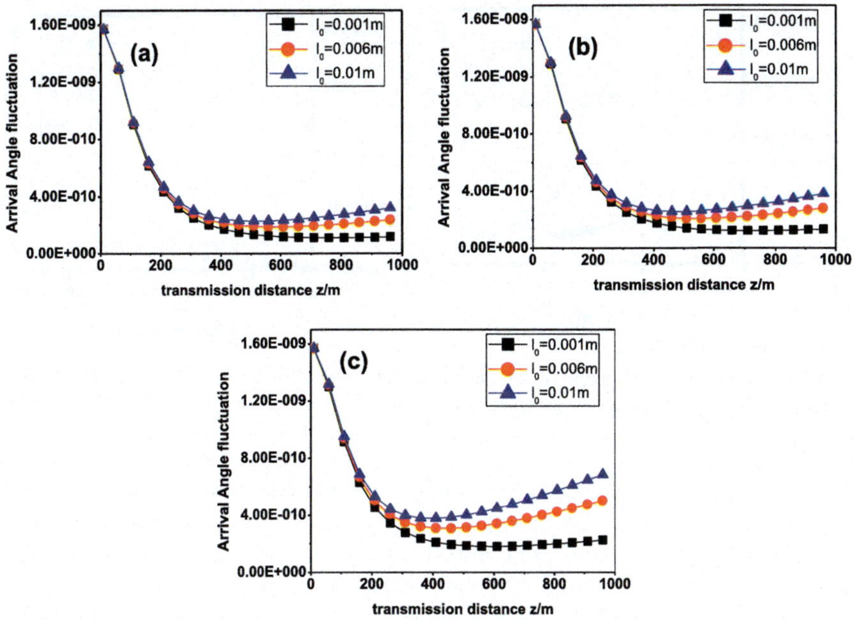

Fig. 4.12 Variation of the arrival angle fluctuation with transmission distance at different internal and external scales **a** $L_0 = 1$ m; **b** $L_0 = 10$ m; **c** $L_0 = 100$ m

constant $C_0 = 1.7 \times 10^{-14}$, wavelength $\lambda = 1.06 \times 10^{-6}$ m, the effective beam parameter $\Theta_0 = 0.1$, and other parameters that are shown in the figure.

As can be seen from Fig. 4.12a–c, the variance of the arrival angle fluctuation first decreases and then increases with the increase of transmission distance. When the propagation distance is less than 200 m, the variance of the angle of arrival fluctuation of beams with different turbulence internal and external scales is the same. The change in the propagation distance is not pronounced, which indicates that when the propagation distance is less than 200 m, the influence of atmospheric turbulence on the internal and external scales of the angle of arrival fluctuation is very small. When the internal scale l_0 increases and the propagation distance $z > 400$ m, the variance of the arrival angle fluctuation of partially coherent GSM beams also increases gradually. When the external scale L_0 increases and the transmission distance $z > 300$ m, the variance of the arrival angle fluctuation also increases.

4.4.3 Different Source Parameters and Angle of Arrival Fluctuation

Figure 4.13 shows the plot of the variance of the angle of arrival fluctuation with beam waist radius and the transmission distance for different coherent parameters.

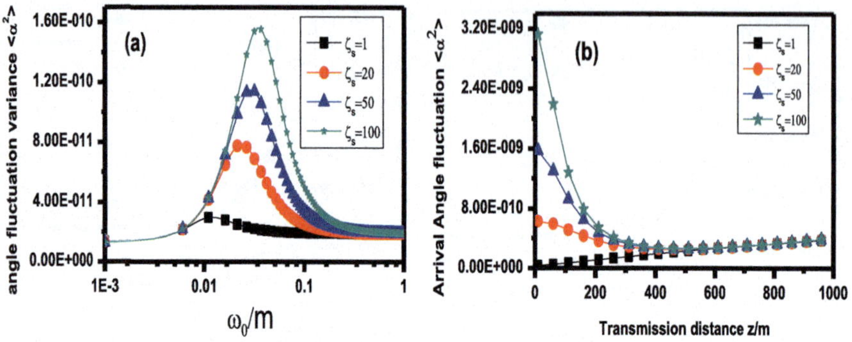

Fig. 4.13 Variation of the angle of arrival fluctuation with **a** beam waist radius, **b** transmission distance under different source coherence parameters

The selected parameters are the near-surface atmospheric refractive index structure constant $c_0 = 1.7 \times 10^{-14}$, wavelength $\lambda = 1.06 \times 10^{-6}$ m, effective beam parameter $\Theta_0 = 0.1$, inner scale $l_0 = 0.01$ m, and outer scale $L_0 = 10$ m. The source coherence parameters ζ_s are taken as 1, 20, 50, and 100 respectively.

As shown in Fig. 4.13a for the plot of the variance of the arrival angle fluctuation of beams with different source coherence parameters, a maximum value is achieved for a specific beam waist radius. When the beam waist radius $\omega_0 > 0.01$ m, the variance of the arrival angle fluctuation changes noticeably, and the greater the beam coherence, the more pronounced this change. As shown in Fig. 4.13b, the variance of the angle of arrival fluctuation of the fully coherent light ($\zeta_s = 1$) does not change significantly with the transmission distance, i.e., the angle of arrival fluctuation of the fully coherent light is not significantly affected by turbulence. However, the angle of arrival fluctuation variance of the partially coherent GSM beam decreases with the increase of the transmission distance, which is similar to the observation for the angle of arrival fluctuation variance of the fully coherent light. When the propagation distance is approximately 1 km, the variance of the angle of arrival fluctuation of beams with different coherence parameters is a minimum, and does not change appreciably with the change of propagation distance. In addition, the worse the coherence of the beam, the greater the variance of the angle of arrival fluctuation. This is because the phase of partially coherent light changes rapidly and randomly, so under the same conditions, the angle of arrival fluctuation of partially coherent light is greater than that of fully coherent light. Therefore, when the beam waist radius is within 1 cm, the influence of the source coherence parameters and the waist radius on the angle of arrival fluctuation is very weak.

4.4.4 Atmospheric Refractive Index Structure Constant and Angle of Arrival Fluctuation

Figure 4.14 shows the variance of the angle of arrival fluctuation of a light beam with coherent parameters under different atmospheric refractive index structure constants. The simulation parameters are transmission distance of $l = 1000\,\text{m}$, wavelength of $\lambda = 1.06\,\mu\text{m}$, effective beam parameter of $\Theta_0 = 0.1$, and initial beam radius of $w_0 = 0.01\,\text{m}$. Inner scale $l_0 = 0.01\,\text{m}$, outer scale $L_0 = 10\,\text{m}$. The atmospheric refractive index structure constant c_0 is taken as 1.7×10^{-13}, 1.7×10^{-14}, and 1.7×10^{-15}.

In Fig. 4.14, when the source coherence parameter $\zeta_s \leq 10$, with the increase of the source coherence parameter, the variance of the arrival angle fluctuation of partially coherent GSM beams also increases. When the source coherence parameter is $\zeta_s > 10$, the variance of the arrival angle fluctuation is almost constant with the increase of the source coherence parameter, which indicates that the influence on the variance of the arrival angle fluctuation becomes increasingly smaller with the increase of the source coherence parameter In addition, as the turbulence intensity decreases, the influence of the angle of arrival fluctuation of the partially coherent GSM beams gradually becomes smaller. The larger the atmospheric refractive index constant, the larger the angle of arrival fluctuation of the partially coherent GSM beams. This is because, with the increase of the atmospheric refractive index, the random variation of the beam arrival angle caused by turbulence increases with time and space.

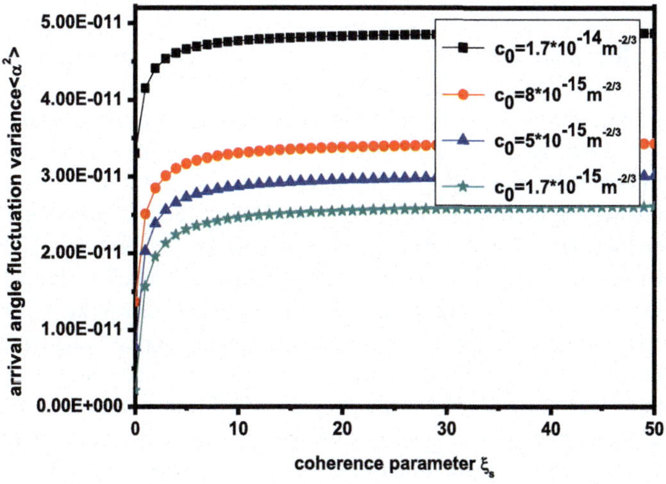

Fig. 4.14 Arrival angle fluctuation variance with source coherence parameters under different atmospheric refractive index structure constants

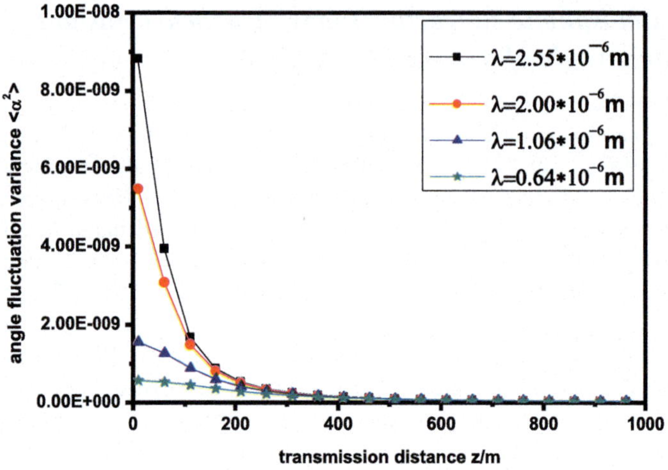

Fig. 4.15 Arrival angle fluctuation variance with transmission distance at different wavelengths

4.4.5 Relationship Between Angle of Arrival Fluctuation and Wavelength

Figure 4.15 shows the variance of the angle of arrival fluctuation with coherent parameters at different wavelengths. The selected parameters are an initial waist radius of $\omega_0 = 0.01$ m, a near-surface atmospheric refractive index structure constant of $c_0 = 1.7 \times 10^{-14}$, and an effective beam parameter of $\Theta_0 = 0.1$. The internal scale is $l_0 = 0.01$ m, the external scale is $L_0 = 10$ m, the source coherence parameter ζ_s is 50, and the values for λ are 2.55×10^{-6} m, 1.06×10^{-6} m, and 0.64×10^{-6} m.

In Fig. 4.15, with the increase of the transmission distance, the fluctuation of the arrival angle of the partially coherent GSM beams becomes gradually smaller. The larger the wavelength, the greater the fluctuation of the arrival angle. As the wavelength becomes gradually smaller, the plots become closer, indicating that the shorter the wavelength, the smaller the impact of multi-arrival angle fluctuation.

In this section, based on the generalized Huygens-Fresnel principle and the cross-spectral density function of partially coherent Gaussian-Schell beams, the expressions for the phase structure function and the variance of the angle of arrival fluctuation of partially coherent Gaussian-Schell beams are derived using the modified von Karman atmospheric turbulence refraction spectrum. The results of the numerical analysis are as follows: (1) as the propagation distance increases, the angle of arrival fluctuation becomes gradually smaller; (2) as the inner scale of atmospheric turbulence increases, the fluctuation of the arrival angle becomes larger, and with the increase of the outer scale of atmospheric turbulence, the fluctuation of the arrival angle also increases; (3) as the source coherence parameters increase, the fluctuation of the arrival angle becomes gradually larger; (4) the influence of turbulence on the angle of arrival fluctuation of fully coherent light is very small; (5) as the wavelength increases, the fluctuation of the arrival angle becomes larger; (6) as the turbulence intensity increases, the fluctuation of the arrival angle becomes larger. The theoretical results show that in practical application, the effects of internal and external scales of atmospheric turbulence, beam wavelength, beam waist radius, transmission distance, coherence length, and other factors on beam propagation characteristics should be carefully considered, and appropriate parameters should be selected to minimize the impact of atmospheric turbulence on the beam propagation characteristics.

References

1. Newton. Optics. Zhou Yueming, et al. (2007) Trans. Beijing: Peking University Press
2. Tatarskii VI (1961) Wave propagation in a turbulent medium. McGraw-Hill, New York
3. Fried DL (1967) Optical heterodyne detection of an atmospherically distorted signal wave front. Proc IEEE 55(1):57–67
4. Fried DL (1967) Atmospheric modulation noise in an optical heterodyne receiver. IEEE Quant Electron QE-3:213–221
5. Churnside JH, McIntyre CM (1978) Signal current probability distribution for optical heterodyne receivers in the turbulent atmosphere. Theory Appl Opt 17(14):2141–2147
6. Churnside JH, McIntyre CM (1980) Heterodyne receivers for atmospheric optical communications. Appl Opt 19(4):582–590
7. Winick KA (1986) Atmospheric turbulence-induced signal fades on optical heterodyne communication links. Appl Opt 25(11):1817–1825
8. Andrews LC, Phillips RL (1985) I-K distribution as a universal propagation model of laser beams in atmospheric turbulence. J Opt Soc Am A 2(2):160–163
9. Lu W (1997) Overview of optical communication. Optic Commun Technol 21(3):171–180
10. Belmonte A, Kahn JM (2008) Performance of synchronous optical receivers using atmospheric compensation techniques. Opt Express 16(18):14151–14162
11. Okayama H, Li ZW (1999) Measurement of the spatial coherence of light influenced by turbulence. Appl Opt 38(12):2342–2345
12. Belmonte A (2000) Feasibility study for the simulation of beam propagation: consideration of coherent lidar performance. Appl Opt 39(30):5426–5445
13. Morozov AV (2002) Two-frequency mutual coherence function of electromagnetic waves in random media: a path-integral variational solution. J Opt Soc Am A 19(10):2074–2084
14. Cliford F, Yura HT (1974) Equivalence of two theories of strong optical scintillation. Opt Soc Am 64:1641–1644
15. Wolf E (1978) Coherence and radiometry. Opt Soc Am 68:6–17
16. Yura HT (1974) Physical model for strong optical-amplitude fluctuations in a turbulent medium. Opt Soc Am 64:59–67
17. Wu J, Boardman AD (1991) Coherence length of a Gaussian-Schell model beam and atmospheric turbulence. Mod Opt 38:1355–1368
18. Ponomarenko AS, Greffet JJ, Wolf E (2002) The diffusion of partially coherent beams in turbulent media. Opt Commun 208:1–8
19. Gbur G, Wolf E (2002) Spreading of partially coherent beams in random media. Opt Soc Am 19:1592–1598
20. Salem M, Shirai T, Sogariu A, Wolf E (2003) Long-distance propagation of partially coherent beams through atmospheric turbulence. Opt Commun 216:261–265
21. Dogariu A, Amarande S (2003) Propagation of partially coherent beams: turbulence-induced degradation. Opt Lett 28(1):10–12
22. Ricklin JC, Davidson FM (2002) Atmospheric turbulence effects on a partially coherent Gaussian beam: implications for free-space laser communication. Opt Soc Am A 19:1794–1802

23. Shirai T, Dogariu A, Wolf E (2003) Mode analysis of spreading of partially coherent beams propagation through atmospheric turbulence. Opt Soc Am A 20:1094–1102
24. Gagliardi R, Robbins J, Taylor H (1987) Acquisition sequences in PPM communication. IEEE Trans Inf Theory 33(5):739–745
25. Boroson DM (1993) LITE engineering model-I: operation and performance of the communications and beam-control subsystems. Proc SPIE 1866:73–82
26. Pillsbury AD (1993) LITE engineering model-II: the flight-qualified subsystems. Proc SPIE 1866:58–72
27. Weeks AR, Xu J, Philips RR et al (1998) Experiment verification and theory for an eight-element multiple-aperture equal-gain coherent laser receiver for laser communications. Appl Opt 37(21):4782–4788
28. Xu J, Philips RR, Andrews LC (1999) A new scheme of coherent array detection to cancel out phase fluctuations and Doppler frequency shift due to atmospheric turbulence and target movement for laser communications. Proc SPIE 3615:325–330
29. Arnon S (1999) Power versus stabilization for laser satellite communication. Appl Opt 38(15):3229–3233
30. Vilnrotter VA, Lau CW (2002) Quantum detection and channel capacity for communications applications. Pro. SPIE 4635:103–115
31. Biswas A, Wright MW (2002) Mountain-top-to-mountain-top optical link demonstration: part I. IPN Prog Rep 2002:1–27
32. Biswas A, Wright MW (2002) Mountain-top-to-mountain-top optical link demonstration: part II. IPN Prog Rep 2002:1–16
33. Yamakawa S, Aroki T, Hisada Y (1999) Trade-off between IM/DD and coherent system in high-data-rate optical interorbit links. Proc SPIE 3615:80–89
34. Yamakawa S, Araki T, Morikawa H (2000) Analysis of noises generated in high-power-operated erbium-doped fiber amplifiers and their influences on IM-DD/Coherent OIC system performance. Proc SPIE 3932:196–204
35. Yamakawa S, Takata N (2003) Coherent lightwave receivers with a laser diode local oscillator for interorbit optical communication. Proc SPIE 4975:69–79
36. Liu W, Wu J (2004) Influence of partially coherent light on light intensity scintillation through strong turbulence. Opt Optoelectron Technol 2(1)
37. Rao R (2005) Propagation of light in atmospheric turbulence. Anhui Science and Technology Press, Hefei
38. Zhang Y, Tao C (2005) Angle of arrival fluctuation of Gaussian Schell beam propagating in turbulent atmosphere. Acta Photon Sin 34:424–427
39. Babcock HW (1953) The possibility of astronomical seeing. Publ Astron Soc Pac 65:229
40. Zeldovich BY et al (1972) JETP Lett 15:109
41. Roddier F (1991) Status of astronomical adaptive optics developments. In: ESO conference
42. Acton DS et al (1991) Solar astronomy with a 19-segment adaptive mirror. SPIE 1542:159
43. Neal DR (1991) Astigmatic unstable resonator with an intracavity deformable mirror. SPIE 1542:449
44. Merkle F et al (1991) Adaptive optics for the European very large telescope. SPIE 1542:283
45. Li H, Zhang H, Yin F (2008) An adaptive compensation method for atmospheric turbulence in space laser communication system. J Changchun Univ Technol Nat Sci Ed 2:1–3
46. Yang H, Cai D, Chen B et al (2008) Wavefront free sensing adaptive optics technology and its application in atmospheric optical communication. China Laser 5:681–684
47. Chen B, Li X, Jiang W (2010) Optimization of adaptive optics stochastic parallel gradient descent algorithm for atmospheric turbulence. China Laser 4:959–964
48. Xia L, Li X (2010) Wavefront correction experiment of atmospheric optical communication based on adaptive optics. Inf Electron Eng 3:331–335
49. Qiao R, Chen M (2010) Research on deformable mirror in adaptive optics. Electron Design Eng 3:74–78
50. Li J, Xiao J (2010) Closed loop adaptive optics system based on genetic algorithm. J Wuhan Inst Technol 1:47–49

51. Han L, Wang Q, Xin Taike GUI et al (2010) Performance Research of free space optical communication system based on adaptive optics compensation. Appl Opt 2:301–303
52. Andrews LC, Phillips RL (2005) Laser beam propagation through random media. SPIE Press, Bellingham, pp 533–600
53. Ricklin JC, Davidson FM (2003) Atmospheric optical communication with a Gaussian Schell beam. J Opt Am A 20(5):856–866
54. Wu J (1988) Happy dawn theory of light propagation in random media. Chengdu Institute of Telecommunication Engineering Press, Chengdu
55. Wang SCH, Plonus MA (1979) Optical beam propagation for a partially coherent source in the turbulent atmosphere. J Opt Soc Am 69:1297–1304
56. Li Y, Propagation characteristics of partially coherent beams in slant turbulent atmosphere. Xi'an University of Electronic Science and Technology, Xi'an
57. Zhu X, Kahn JM (2002) Free-space optical communication through atmospheric turbulence channels. IEEE Trans Commun 50(8):1293–1300
58. Ricklin JC, Davidson FM (2002) Atmospheric turbulence effects on a partially coherent Gaussian beam: implications for free-space laser communication. J Opt Soc Am A 19:1794–1802
59. Griadi RM, Karber, Chen Z et al (1982) Optical communication. People's Posts and Telecommunications Press, Beijing
60. Wei H (2006) Study on the characteristic of laser beam in the slant path through the atmospheric turbulence. Xi-an, Xidain University, pp 5–12

Chapter 5
Angular Reflector Return Intensity Characteristics

Based on the generalized Huygens-Fresnel principle, the modified von Karman spectral model, and the atmospheric structure constant model promulgated by International Telecommunications Union-Radio (ITU-R), the average intensity analytic equation of partially coherent Gaussian-Schell model (GSM) beams for upstream transmission can be derived. The expression for the average intensity of a GSM beam transmitted downward to a receiver after returning from an angular reflector is obtained, and the change in the average intensity of the GSM beam during the upward–downward transmission is analyzed.

5.1 Introduction

5.1.1 Progress in Target Echo Research

Laser beam target echoes in atmospheric turbulence have been widely used in ranging, target tracking, lidar imaging, passive modulation, laser communication, and other fields [1]. In addition to the investigation of the propagation of partially coherent beams in atmospheric turbulence [2–5], the target echo characteristics of partially coherent beams in atmospheric turbulence are also of interest [6–9].

Research on the characteristics of target echoes dates to the 1970s. In 1982, Kravtsov [10] conducted a preliminary study on the backward enhancement of echo from planar mirrors and diffusive surfaces. In 1984, Aksenov et al. [11] summarized the application of the local Green's function in two-way transmission problems. In 1987, using the "square approximation" hypothesis, Feng and Song [12] obtained an expression for the expansion of reflected laser beams in a turbulent atmosphere and verified that the expansion of the beams reflected by a planar mirror had an amplification effect. In 1993, Banakh used the generalized Huygens-Fresnel principle to calculate the effect of a hard-edged aperture planar mirror on the backward echo

X. Ke, *Partially Coherent Optical Transmission Theory in Optical Wireless Communication*, Optical Wireless Communication Theory and Technology,
https://doi.org/10.1007/978-981-97-7158-5_5

enhancement factor of spherical waves. An oscillating interval [13] was identified for the backward enhancement factor with a change in the size of the planar mirror. In 1995, Andrews and Miller [14] analyzed the return cross-correlation function and backward enhancement coefficient of a Gaussian beam reflected by a finite large planar mirror, considering the influence of the scale of atmospheric turbulence. In 1996, Andrews and Phillips [15] investigated the cross-correlation function and backward amplification characteristics of the returning Gaussian beams that were reflected by an angular reflector, considering the internal and external scales of atmospheric turbulence. In 1997, Andrews et al. [16] derived the intensity covariance function and variance function of a Gauss-Schell beam after reflection by the reflector target and compared them to the case of a smooth infinite diffusive target. In 2001, Andrews and Phillips considered the backward enhancement effect and calculated the backward return flicker index [1] for point and rough targets using the two-scale method. In 2006, based on the generalized Huygens-Fresnel principle and using the atmospheric turbulence structure constant model promulgated by ITU-R, Guo et al. deduced the cross-correlation function and power spectral density function of a Gaussian beam reflected by a diffusive target for oblique transmission in a turbulent atmosphere at the receiver [17]. In 2009, Wei Hongyan determined the echo characteristics of a reflector on the slope path and analyzed the echo characteristics of plane mirrors and corner reflectors with the target height [7, 18]. In 2011, Wu studied the statistical characteristics of fields scattered from rough surfaces in two-way transmission. They derived expressions for the correlation function, average scattering intensity, and intensity covariance of speckle fields for focused and collimated beams [8]. In 2012, based on the generalized Huygens-Fresnel principle, Li studied the scattering characteristics of partially coherent Gaussian-Schell model (GSM) beams as they propagated from a transmitter to a target in a turbulent atmosphere. The influence of the scattering characteristics of the diffusive target on the GSM beams in the two-slope path from the target to the receiver has been investigated [19]. In 2013, Ling et al. [20] reported that the effect of the inner scale and target height on the flicker index is much greater than that of the wavelength and transmission distance, based on the theory of light wave slope transmission, considering the inner scale. In 2014, using the Rytov method and ABCD transfer matrix theory, Wang et al. deduced the strength covariance function and flicker index equation [9, 21] for a single angular reflector echo on the receiving aperture plane for a two-way ground-air-ground turbulence atmospheric path.

5.1.2 Reverse Modulation System

Reverse modulation is an optical communication method that uses a target echo. The reverse modulation system consists of two unequal terminals: active and passive, as shown in Fig. 5.1. The active end is the transmission/reception terminal of the traditional laser communication system, which includes four components: a laser transmission system, laser receiving system, signal processing system, and control system. The passive end is the reverse modulation terminal, which includes the

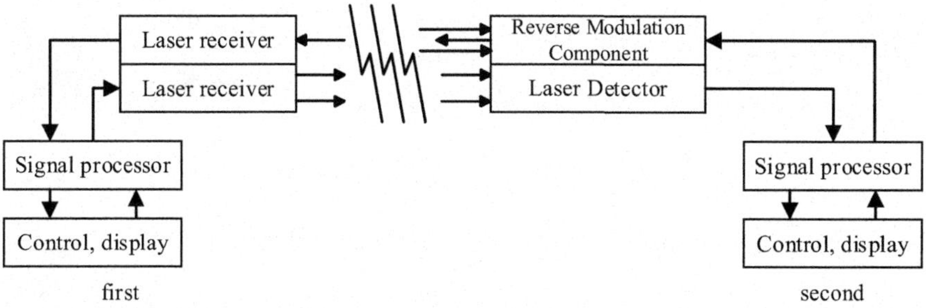

Fig. 5.1 Block diagram of the reverse modulation system

reverse modulator, information acquisition system, signal processing system, and control system. The reverse modulator includes a modulator and a reverse reflector.

The working principle of the MRR-FSO (Modulating Retro-reflector Free Space Optical) system is that the active end aligns with the passive end via the control system and emits a laser beam to the passive end. An incoming beam enters the passive-end optical system and is focused on the modulator. The passive end then modulates the incoming light by loading a signal onto the modulator to change its state. The dimmed beam is returned to the emitter along its original path through a reverse reflector. The active end converts the light signal of the dimmed beam into an electrical signal using the receiving system. Signal processing of the signal is then performed to obtain useful information embedded in the dimmed signal.

The reverse modulator loads electrical signals onto the modulator, which changes its structural characteristics. This causes changes in some of the properties of the laser to achieve modulation.

Reverse modulators can be divided into corner reflector-type reverse modulators and cat-eye-type reverse reflectors according to their structure. Two different reverse reflectors cause the reverse modulator to exhibit different modulation characteristics.

5.1.2.1 Corner Reflector Reverse Modulator

Angular reflectors consist of three high-reflectivity planes that are perpendicular to each other, and their geometric design ensures that the incoming beam has good reverse reflection properties. Reverse modulators based on angular reflectors can be divided into two categories according to different modulator principles involved. In the first case, a microelectromechanical system is used to change the characteristics of the angular reflector so that intensity modulation of the reflected beam can be achieved. In the second case, a transmitter light modulator is placed at the pupil of the corner reflector, which modulates the intensity of the reflected light signal base on its photoelectric properties.

As shown in Fig. 5.2, Ziph-Schatzberg et al. of Boston University developed an optical communication device based on a micro-electromechanical system (MEMS),

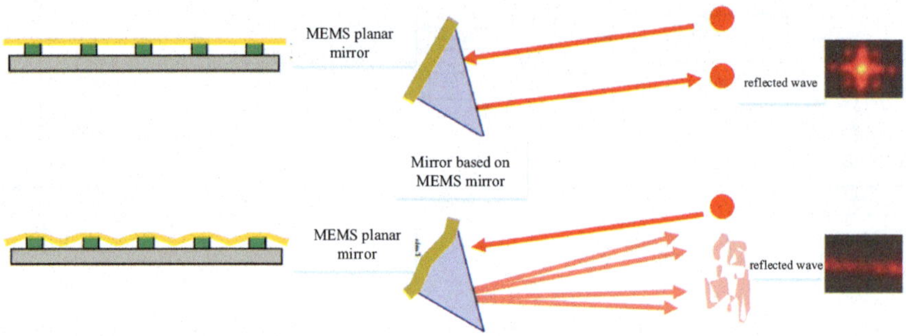

Fig. 5.2 MRR-FSO link and reverse modulator constructed using MEMS [22]

which electro-modulates a deformable micro-mechanical reflector on the reflector's surface to control deformations, thereby affecting the characteristics of the angular reflector to achieve signal modulation. Using this device, voice information transmission over a distance of 900 m was achieved for power consumption of 10 mW and a working frequency of 1 kbps.

In 2007, Zhang [23] coupled a giant magnetostrictive material with an organic silicon gel, which served as an angular reflector. By changing the magnetization state of the giant magnetostrictive bar to generate micro-displacement, the reflection characteristics of the angular reflector were adjusted, and intensity modulation of the returning light signal was detected at a rate of approximately 1 kbps.

5.1.2.2 Cat's Eye Structural Reverse Modulator

The caliber of the angular-reflector-type reverse modulator must be close to that of the angular reflector or the size of one of the reflectors of the angular reflector; i.e., the caliber of the modulator must be large enough to guarantee the performance. A large size implies that the maximum modulation rate of the modulator will decrease. There is a conflict between the caliber of the optical system and the modulation rate, which limits the application of the angular-reflector-type reverse modulator.

The cat-eye effect allows the incoming light to return to its original path; therefore, the cat-eye reverse modulator has the same reverse reflection characteristics as the corner reflector. A cat-eye reverse modulator can be constructed by combining the cat-eye structure with a modulator. The modulator was placed at the focal plane of the cat's eye reverse reflector, and the incoming light beam was focused on the surface of the modulator through the cat's eye optical system, which was affected by the modulator. This method can match a small modulator with a large-aperture optical system. A schematic of a typical cat's-eye reverse modulator is shown in Fig. 5.3.

In the Trident Warrior sea-level test, the U.S. Naval Research Laboratory (NRL) established a ship-to-ship reverse modulation link using a dual-mode optical emission system and a reverse modulator module, which facilitated MRR-FSO application on

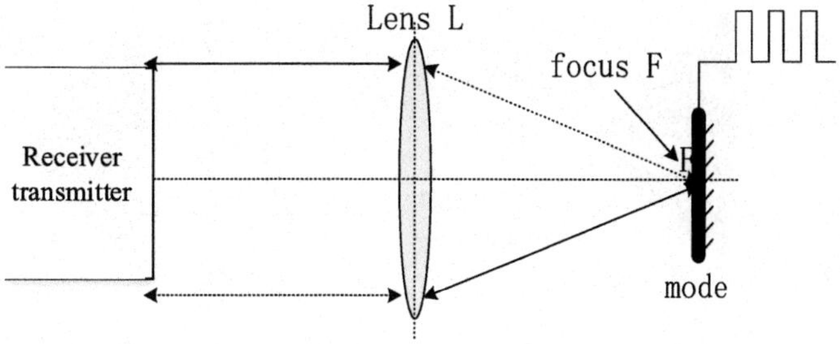

Fig. 5.3 Cat's eye MRR schematic diagram

naval vessels. In this test, the effective link length was approximately 4.5 km with a 2 Mbps bidirectional link, and the field-of-view angle of the reverse modulator was 60°. Coarse tracking is required to establish the link. Using the anti-electromagnetic interference characteristics of MRR-FSO, the NRL constructs a reverse modulator terminal using six reverse modulator units. The terminal was installed on an explosive ordinance removal robot. An 802.11 wireless link was replaced by an optical link, the link rate was 1.5 Mbps, and the robot was controlled using the optical link through a dual-mode optical transmission system beyond 30 m and 1 km. The results show that the MRR-FSO link works well and is not affected by the RF interference sources.

5.1.2.3 Optical Reverse Modulation Classification

As shown in Fig. 5.4, optical reverse modulators are divided into two main types according to their structure: pyramid prism reflective type and cat-eye structure echo modulation type.

Figure 5.4a, b show examples of the pyramidal prism reflection type; (c, d) belong to the "cat's eye" structure echo modulation type. In this chapter, the intensity distribution of partially coherent beams is analyzed when they are transmitted through atmospheric turbulence to a target angular reflector and reflected by an angular reflector when the returns are transmitted through atmospheric turbulence to the receiver, which provides a theoretical reference for reverse modulation.

5.2 Formula for Two-Way Slope Transmission of GSM Beam

The expression for partially coherent GSM beams as source fields is given as [24]:

$$u_0(\boldsymbol{r}) = \exp\left(-\frac{\boldsymbol{r}^2}{\omega_0^2} - \frac{ik \cdot \boldsymbol{r}^2}{2F}\right)\exp[i\varphi_d(\boldsymbol{r})] \tag{5.1}$$

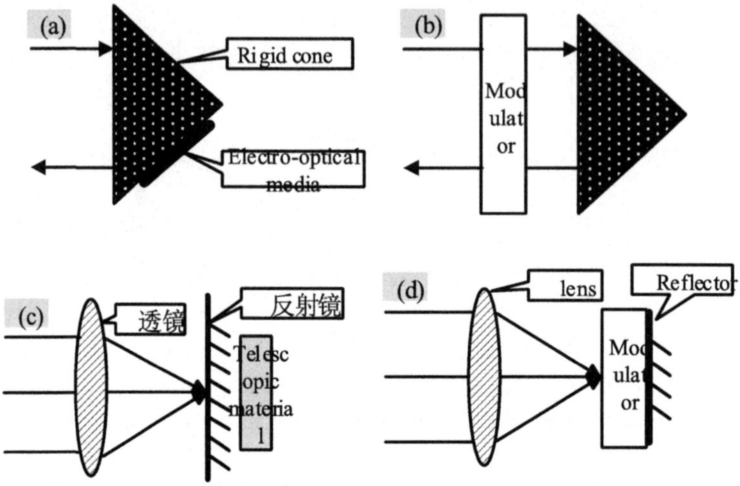

Fig. 5.4 Classification of reverse modulators: **a** Attenuated total internal reflection reverse modulator. **b** External modulation reverse modulator. **c** Off-focus modulation cat-eye structure reverse modulator. **d** Cat-eye structure reverse modulator [25] based on the modulator

where R is a vector perpendicular to the direction of beam transmission at the transmitter. W and f represent the waist and wavefront curvature radii, respectively. When $F \to \infty$, the source field beam is collimated. When $F < 0$, the source field beam is divergent. When $F > 0$, the source field beam is a converged beam, where $k = 2\pi / \lambda$ is the free-space wavenumber. $\exp[i\varphi_d(\boldsymbol{r})]$ represents small random fluctuations caused by phase scattering. The Cross-Spectral Density Function (CSDF) of a GSM beam at L = 0 can be expressed as follows [26]

$$W_0(\boldsymbol{r}_1, \boldsymbol{r}_2; 0) = \langle u_0(\boldsymbol{r}_1) u_0^*(\boldsymbol{r}_2) \rangle = \exp\left[-\frac{\boldsymbol{r}_1^2 + \boldsymbol{r}_2^2}{\omega_0^2} - \frac{ik \cdot \left(\boldsymbol{r}_1^2 - \boldsymbol{r}_2^2\right)}{2F} \right] \exp\left(-\frac{|\boldsymbol{r}_1 - \boldsymbol{r}_2|^2}{2\sigma_g^2} \right) \tag{5.2}$$

where σ_g is the coherence length of the source field. When increases, the coherence of the beam increases and vice versa. When $\sigma_g \to \infty$, the source field beam is a fully coherent Gauss–Schell beam. $\langle \cdot \rangle$ represents the statistical means.

The double-slope transmission of the partially coherent GSM beams under atmospheric turbulence is shown in Fig. 5.5. The figure shows the transmission distance, where H is the vertical distance from sea level. Laser beams are emitted from the emitter, incident on the target through atmospheric turbulence, and reflected. The echo returns to the receiver along the same path.

The target corner reflector, also known as the pyramid prism, is a total internal reflection prism. The corner reflector is a tetrahedral vertebral prism consisting of three perpendicular reflective surfaces and one incoming (outgoing) reflective surface. It is an optical element with directional reflection characteristics. When a

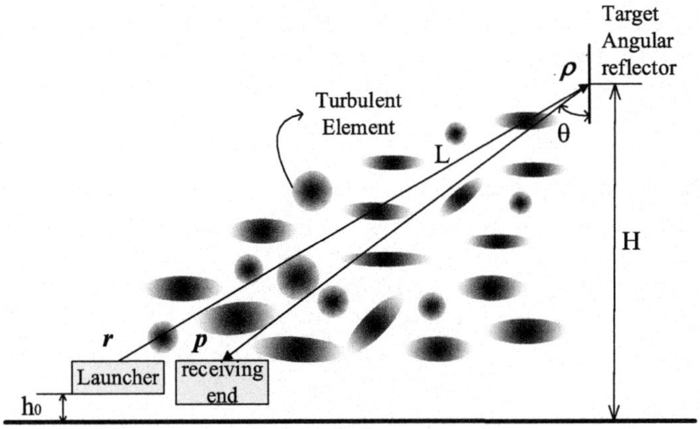

Fig. 5.5 Diagram of the double-slope transmission path in a turbulent atmosphere

parallel beam of light is incident perpendicular to the bottom and is reflected by three right-angled reflectors in turn, it is reflected in the opposite direction from the incident direction, forming a directional backward reflection. The working principle of the corner reflector is illustrated in Fig. 5.6.

Assuming that the source and target are relatively small relative to the transmission path, the field at any point on the target can be represented using the generalized Huygens-Fresnel principle [5, 24]:

$$u_i(\boldsymbol{\rho}; L) = \frac{k}{2\pi iL} \exp(ikL) \int d\boldsymbol{r} u_0(\boldsymbol{r}) \exp\left[\frac{ik|\boldsymbol{\rho} - \boldsymbol{r}|^2}{2L} + \psi_i(\boldsymbol{\rho}, \boldsymbol{r})\right] \tag{5.3}$$

where $\boldsymbol{\rho}$ is a two-dimensional vector on the target plane and L is the oblique beam transmission distance. In Eq. (5.3), $\psi_i(\boldsymbol{\rho}, \boldsymbol{r})$ represents the complex random perturbation of spherical waves in a turbulent medium, including the logarithmic amplitude

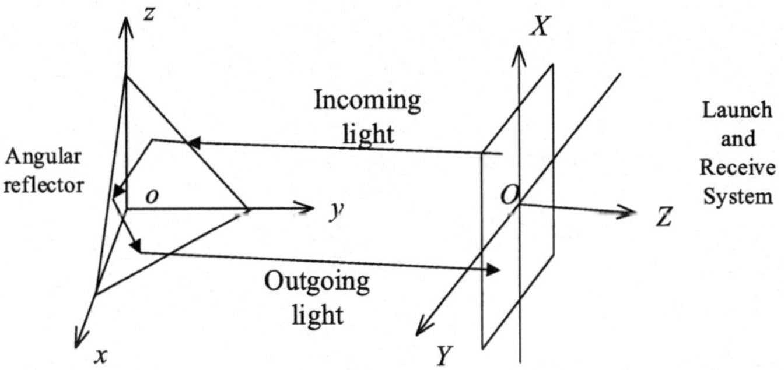

Fig. 5.6 Coordinate system of corner reflector [27]

and phase perturbations, transmitted from the emitted aperture ($\boldsymbol{r}$, 0) to a point on the target ($\boldsymbol{\rho}$; L). The expression is given as:

$$\psi_i(\rho, r) = \chi(\rho, r) + iS(\rho, r) \tag{5.4}$$

The CSDF of the GSM beam uplink transmitted to the field at the target can be expressed as:

$$\begin{aligned} W_i(\boldsymbol{\rho}_1, \boldsymbol{\rho}_2; L) &= \langle u_i(\boldsymbol{\rho}_1) u_i^*(\boldsymbol{\rho}_2) \rangle \\ &= \left(\frac{k}{2\pi L}\right)^2 \int d\boldsymbol{r}_1 \int d\boldsymbol{r}_2 \langle u_0(\boldsymbol{r}_1) u_0^*(\boldsymbol{r}_2) \rangle \\ &\exp\left[\frac{ik}{2L}\left(|\boldsymbol{\rho}_1 - \boldsymbol{r}_1|^2 - |\boldsymbol{\rho}_2 - \boldsymbol{r}_2|^2\right)\right] \\ &\times \langle \exp[\psi_i(\boldsymbol{\rho}_1, \boldsymbol{r}_1) + \psi_i^*(\boldsymbol{\rho}_2, \boldsymbol{r}_2)] \rangle \end{aligned} \tag{5.5}$$

Due to the directional reverse reflection characteristic of the Angle reflector, the beam reflected by the target Angle reflector is emitted from the symmetric position of the incident point about the center, and the field satisfies the formula [12]:

$$u_r(\boldsymbol{\rho}; L) = r(\boldsymbol{\rho}) u_i(-\boldsymbol{\rho}; L) \tag{5.6}$$

where $r(\boldsymbol{\rho})$ is the effective amplitude reflection coefficient of the angular reflector. Assuming that $r(\boldsymbol{\rho})$ obeys a Gaussian distribution, the expression for this term [15, 18] is given as $r(\boldsymbol{\rho}) = \exp\left(-2|\boldsymbol{\rho} - \boldsymbol{\rho}_0|^2 / r_0^2\right)$, and $u_i^*(\boldsymbol{\rho}; L)$ is the complex conjugate of the incident field to the target. The CSDF of the return field $u_r(\boldsymbol{\rho}; L)$ is calculated from (5.3) and (5.5):

$$\begin{aligned} W_r(\boldsymbol{\rho}_1, \boldsymbol{\rho}_2; L) &= \langle u_r(\boldsymbol{\rho}_1) u_r^*(\boldsymbol{\rho}_2) \rangle \\ &= \left(\frac{k}{2\pi L}\right)^2 \int d\boldsymbol{r}_1 \int d\boldsymbol{r}_2 \exp\left(-\frac{2|\boldsymbol{\rho}_1 - \boldsymbol{\rho}_0|^2 + 2|\boldsymbol{\rho}_2 - \boldsymbol{\rho}_0|^2}{r_0^2}\right) \\ &\langle u_0^*(\boldsymbol{r}_1) u_0(\boldsymbol{r}_2) \rangle \exp\left[\frac{ik}{2L}\left(|\boldsymbol{\rho}_1 - \boldsymbol{r}_1|^2 - |\boldsymbol{\rho}_2 - \boldsymbol{r}_2|^2\right)\right] \\ &\langle \exp[\psi_i(\boldsymbol{\rho} 1, \boldsymbol{r}_1) + \psi_i^*(\boldsymbol{\rho}_2, \boldsymbol{r}_2)] \rangle \end{aligned} \tag{5.7}$$

Considering the field reflected by the target as the initial field, the field at the receiving end can be obtained using the generalized Huygens-Fresnel principle [5, 16]:

$$u_{re}(\boldsymbol{p}) = \frac{k}{2\pi iL} \exp(ikL) \int d\boldsymbol{\rho} u_r(\boldsymbol{\rho}; L) \exp\left[\frac{ik|\boldsymbol{p} - \boldsymbol{\rho}|^2}{2L} + \psi_r(\boldsymbol{p}, \boldsymbol{\rho})\right] \tag{5.8}$$

where $\boldsymbol{p}$ is a two-dimensional transverse vector on the receiving plane, $\psi_r(\boldsymbol{p}, \boldsymbol{\rho})$ is the complex phase random disturbance caused by turbulence in the medium, which is transmitted as a spherical wave at the point above the target to the receiving aperture $(\boldsymbol{p}; 0)$. By substituting (5.3) and (5.5) into (5.8), we obtain,

$$
\begin{aligned}
u_{re}(\boldsymbol{p}) = & \left(\frac{k}{2\pi L}\right)^2 \int d\boldsymbol{r} \int d\boldsymbol{\rho} \exp\left(-\frac{2|\boldsymbol{\rho}-\boldsymbol{\rho}_0|^2}{r_0^2}\right) u_0^*(\boldsymbol{r}) \\
& \times \exp\left[\frac{ik}{2L}\left(|\boldsymbol{p}-\boldsymbol{\rho}|^2 - |\boldsymbol{\rho}-\boldsymbol{r}|^2\right)\right] \\
& \exp\left[\psi_r(\boldsymbol{p}, \boldsymbol{\rho}) + \psi_i^*(\boldsymbol{\rho}, \boldsymbol{r})\right]
\end{aligned} \tag{5.9}
$$

The Cross-Spectral Density Function of the beam on the receiving plane can be expressed as:

$$
\begin{aligned}
& W_{re}(\boldsymbol{p}_1, \boldsymbol{p}_2; 0) \\
& = \langle u_{re}(p_1) u_{re}^*(p_2) \rangle \\
& = \left(\frac{k}{2\pi L}\right)^4 \int\int\int\int d\boldsymbol{r}_1 d\boldsymbol{r}_2 d\boldsymbol{\rho}_1 d\boldsymbol{\rho}_2 \langle u_0(\boldsymbol{r}_1) u_0^*(\boldsymbol{r}_2) \rangle \\
& \times \exp\left(-\frac{2|\boldsymbol{\rho}_1-\boldsymbol{\rho}_0|^2 - 2|\boldsymbol{\rho}_2-\boldsymbol{\rho}_0|^2}{r_0^2}\right) \\
& \times \exp\left[\frac{ik}{2L}\left(|\boldsymbol{p}_1-\boldsymbol{\rho}_1|^2 + |\boldsymbol{\rho}_1+\boldsymbol{r}_1|^2 - |\boldsymbol{p}_2-\boldsymbol{\rho}_2|^2 - |\boldsymbol{\rho}_2+\boldsymbol{r}_2|^2\right)\right] \\
& \times \langle \exp\left[\psi_r(\boldsymbol{p}_1, \boldsymbol{\rho}_1) + \psi_i(-\boldsymbol{\rho}_1, \boldsymbol{r}_1) + \psi_r^*(\boldsymbol{p}_2, \boldsymbol{\rho}_2) + \psi_i^*(-\boldsymbol{\rho}_2, \boldsymbol{r}_2)\right]\rangle
\end{aligned} \tag{5.10}
$$

Based on Eq. (5.10), we have:

$$
M_L = \langle \exp\left[\psi_r(\boldsymbol{p}_1, \boldsymbol{\rho}_1) + \psi_i(-\boldsymbol{\rho}_1, \boldsymbol{r}_1) + \psi_r^*(\boldsymbol{p}_2, \boldsymbol{\rho}_2) + \psi_i^*(-\boldsymbol{\rho}_2, \boldsymbol{r}_2)\right]\rangle \tag{5.11}
$$

where $D_S(\boldsymbol{P}, \boldsymbol{Q})$ is a wave structure function, and $C_{\chi_{ir}}(\boldsymbol{\rho}; p)$ is a logarithmic amplitude covariance function on the backward and forward paths. $D_S(\boldsymbol{P}, \boldsymbol{Q})$ can be expressed as follows [24]:

$$
D_S(\boldsymbol{P}, \boldsymbol{Q}) = 8\pi^2 k^2 \sec\theta \int_{h_0}^{H} \int_0^{\infty} \kappa \phi_n(\kappa, h)\left\{1 - J_0\left[|(1-\xi)\boldsymbol{P} + \xi \boldsymbol{Q}|\kappa\right]\right\} d\kappa dh \tag{5.12}
$$

where J_0 is the zero-order Bessel function of the first kind and $\phi_n(\kappa, h)$ represents the refractive index power spectrum. Given that downward beam transmission and upward beam transmission are affected by turbulence, $\phi_n(\kappa, h)$ is smaller than upward beam transmission. The up and down transmissions are mainly determined by the following parameters:

$$\xi = \begin{cases} 1 - \frac{h-h_0}{H-h_0} & \begin{array}{l} H: \textit{Vertical Height of Receiver from Ground (Upward-transmission up)} \\ h_0: \textit{The vertical height of the transmitter from the ground} \end{array} \\ \frac{h-h_0}{H-h_0} & \begin{array}{l} H: \textit{The vertical height of the transmitter from the ground} \\ h_0: \textit{Vertical Height of Receiver from Ground (Down-load down)} \end{array} \end{cases} \tag{5.13}$$

The form of the $C_{\chi_{ir}}(\boldsymbol{\rho}; \boldsymbol{p})$ function is [21]:

$$C_{\chi_{ir}}(\boldsymbol{\rho}; \boldsymbol{p}) = 0.132\pi^2 k^2 L \int_0^1 \mathrm{d}t C_n^2(Ht) \int_0^\infty \mathrm{d}u \cdot u^{-\frac{8}{3}} \\ \times \sin^2\left[\frac{u^2 t(1-t)L}{2k}\right] J_0\left[u|\boldsymbol{p}t - \boldsymbol{\rho}(1-t)|\right] \tag{5.14}$$

The Bessel function in the (5.12) wave structure function can be used to approximate [3, 4] under strongly turbulent conditions.

$$J_0\left[|(1-\xi)\boldsymbol{P} + \xi\boldsymbol{Q}|\kappa\right] = 1 - \frac{1}{4} \times \left[(1-\xi)\boldsymbol{P} + \xi\boldsymbol{Q}\right]^2 \kappa^2 \tag{5.15}$$

Equation (5.15) was substituted into Eq. (5.12). After simplification using the integral, the following results can be obtained:

$$D_\phi(\boldsymbol{P}, \boldsymbol{Q}, z) = M_1 \boldsymbol{P}^2 + M_2 \boldsymbol{P} \cdot \boldsymbol{Q} + M_3 \boldsymbol{Q}^2 \tag{5.16}$$

$$\begin{cases} M_1 = 2\pi^2 k^2 \sec\theta \int_{h_0}^H \int_0^\infty \phi_n(\kappa, h)(1-\xi)^2 \kappa^3 d\kappa dh \\ M_2 = 4\pi^2 k^2 \sec\theta \int_{h_0}^H \int_0^\infty \phi_n(\kappa, h)(1-\xi)\xi \kappa^3 d\kappa dh \\ M_3 = 2\pi^2 k^2 \sec\theta \int_{h_0}^H \int_0^\infty \phi_n(\kappa, h)\xi^2 \kappa^3 d\kappa dh \end{cases} \tag{5.17}$$

When horizontally transmitted, $M_1 = M_2 = 0$, $M_3 = \frac{1}{3}\pi^2 k^2 z \int_0^\infty \kappa^3 \phi_n(\kappa) \mathrm{d}\kappa$ For propagation in free space, the turbulence term $M_j = 0\,(j = 1, 2, 3)$. For the power spectrum model of turbulence for $M_j\ (j = 1, 2, 3)$, we use the modified Von Karman spectrum [10, 16], which takes both internal and external scales into account, and the expression is given by:

$$\phi_n(\kappa, h) = 0.033 C_n^2(h) \frac{\exp(-\kappa^2/\kappa_m^2)}{(\kappa^2 + \kappa_0^2)^{11/6}} \tag{5.18}$$

where $\kappa_m = 5.92/l_0$, $\kappa_0 \approx 2\pi/L_0$, and l_0 and L_0 represent the internal and external scales of turbulence, respectively. $C_n^2(h)$ denotes the structural constant of the atmospheric refractive index for slope transmission. The structure constant model for the variation of the atmospheric refractive index with altitude, published by the International Telecommunications Union-Radio Communication Sector (ITU-R) in 2001, was used to characterize the intensity of turbulence [28]. It can be expressed as:

$$C_n^2(h) = 8.148 \times 10^{-56} v_{RMS}^2 h^{10} e^{-h/1000} + 2.7 \times 10^{-16} e^{-h/1500} + C_0 e^{-h/100} \tag{5.19}$$

where $v_{RMS}^2 = \sqrt{v_g^2 + 30.69 v_g + 348.91}$ is the vertical path wind speed, v_g represents the near ground wind speed, and C_0 represents the near ground atmospheric refractive index structure constant, i.e., $C_0 = 1.7 \times 10^{-14} \mathrm{m}^{-2/3}$.

Equation (5.10) can be simplified using Eqs. (5.12) and (5.16), and Eq. (5.2) can be substituted into Eq. (5.10) to yield the following

$$\begin{aligned}
&W_{re}(\boldsymbol{p}_1, \boldsymbol{p}_2; 0) \\
&= \left(\frac{k}{2\pi L}\right)^4 \int d\boldsymbol{r}_1 \int d\boldsymbol{r}_2 \int d\boldsymbol{\rho}_1 \int d\boldsymbol{\rho}_2 \exp\left[-\frac{\boldsymbol{r}_1^2 + \boldsymbol{r}_2^2}{\omega_0^2} - \frac{ik \cdot (\boldsymbol{r}_1^2 - \boldsymbol{r}_2^2)}{2F}\right] \\
&\times \exp\left(-\frac{|\boldsymbol{r}_1 - \boldsymbol{r}_2|^2}{2\sigma_g^2}\right) \exp\left(-\frac{2|\boldsymbol{\rho}_1 - \boldsymbol{\rho}_0|^2}{r_0^2} - \frac{2|\boldsymbol{\rho}_2 - \boldsymbol{\rho}_0|^2}{r_0^2}\right) \\
&\times \exp\left[\frac{ik}{2L}\left(|\boldsymbol{p}_1 - \boldsymbol{\rho}_1|^2 + |\boldsymbol{\rho}_1 + \boldsymbol{r}_1|^2 - |\boldsymbol{p}_2 - \boldsymbol{\rho}_2|^2 - |\boldsymbol{\rho}_2 + \boldsymbol{r}_2|^2\right)\right] \\
&\times \exp\left\{-\frac{1}{2}\left[M_{1down}(\boldsymbol{p}_1 - \boldsymbol{p}_2)^2 + M_{2down}(\boldsymbol{p}_1 - \boldsymbol{p}_2) \cdot (\boldsymbol{\rho}_1 - \boldsymbol{\rho}_2)\right.\right. \\
&\left.\left. + M_{3down}(\boldsymbol{\rho}_1 - \boldsymbol{\rho}_2)^2\right]\right\} \\
&\times \exp\left\{-\frac{1}{2}\left[M_{1up}(\boldsymbol{\rho}_1 - \boldsymbol{\rho}_2)^2 + M_{2up}(\boldsymbol{\rho}_1 - \boldsymbol{\rho}_2) \cdot (\boldsymbol{r}_1 - \boldsymbol{r}_2)\right.\right. \\
&\left.\left. + M_{3up}(\boldsymbol{r}_1 - \boldsymbol{r}_2)^2\right] + 4C_{\chi_{ir}}\right\}
\end{aligned} \tag{5.20}$$

where M_{idown} represents downlink transmission ($i = 1, 2, 3$), and M_{iup} indicates upstream transmission ($i = 1, 2, 3$).

5.3 Characteristics of Normalized Light Intensity Distribution

5.3.1 Light Intensity Distribution in Uplink

The average intensity of the emitter beam can be obtained using the CSDF of the emitter GSM beam, so that $\boldsymbol{r}_1 = \boldsymbol{r}_2 = \boldsymbol{r}$, and Eq. (5.2) can be expressed as:

$$\langle I(r) \rangle = W_0(r, r; 0) = \exp\left[-\frac{2r^2}{\omega_0^2}\right] \tag{5.21}$$

By substituting Eqs. (5.16) and (5.2) into the cross-spectral density function (CSDF) of the field using Eq. (5.5) and setting $\boldsymbol{\rho}_1 = \boldsymbol{\rho}_2 = \boldsymbol{\rho}$, the average intensity distribution at the incident target can be obtained using Eq. (5.5):

$$\begin{aligned}\langle I(\boldsymbol{\rho}; L)\rangle &= W_i(\boldsymbol{\rho}, \boldsymbol{\rho}\,; 0)\\ &= \left(\frac{k}{2\pi L}\right)^2 \int d\boldsymbol{r}_1 \int d\boldsymbol{r}_2 \exp\left[-\frac{\boldsymbol{r}_1^2 + \boldsymbol{r}_2^2}{\omega_0^2} - \frac{ik\cdot\left(\boldsymbol{r}_1^2 - \boldsymbol{r}_2^2\right)}{2F}\right]\\ &\times \exp\left(-\frac{|\boldsymbol{r}_1 - \boldsymbol{r}_2|^2}{2\sigma_g^2}\right)\exp\left[\frac{ik}{2L}\left(|\boldsymbol{\rho} - \boldsymbol{r}_1|^2 - |\boldsymbol{\rho} - \boldsymbol{r}_2|^2\right)\right]\\ &\exp\left[-\frac{1}{2}M_{3up}(\boldsymbol{r}_1 - \boldsymbol{r}_2)^2\right]\end{aligned} \tag{5.22}$$

To facilitate calculation, an integral transformation is performed as follows:

$$\begin{cases}\boldsymbol{r}_c = \frac{1}{2}(\boldsymbol{r}_1 + \boldsymbol{r}_2)\\ \boldsymbol{r}_d = \boldsymbol{r}_1 - \boldsymbol{r}_2\end{cases} \quad \begin{cases}\boldsymbol{r}_1 = \boldsymbol{r}_c + \frac{1}{2}\boldsymbol{r}_d\\ \boldsymbol{r}_2 = \boldsymbol{r}_c - \frac{1}{2}\boldsymbol{r}_d\end{cases} \tag{5.23}$$

$$\int d\boldsymbol{r}_1 \int d\boldsymbol{r}_2 = \int d\boldsymbol{r}_c \int d\boldsymbol{r}_d \tag{5.24}$$

By substituting Eqs. (5.23) and (5.24) into Eq. (5.22), we have:

$$\begin{aligned}\langle I(\boldsymbol{\rho}; L)\rangle &= \frac{k^2}{4\pi^2 L^2}\int d\boldsymbol{r}_c \int d\boldsymbol{r}_d \exp\left(-\frac{4\boldsymbol{r}_c^2 + \boldsymbol{r}_d^2}{2\omega_0^2} - \frac{ik\cdot \boldsymbol{r}_c\boldsymbol{r}_d}{F}\right)\exp\left(-\frac{\boldsymbol{r}_d^2}{2\sigma_g^2}\right)\\ &\times \exp\left[\frac{ik}{L}\left(\boldsymbol{r}_c\boldsymbol{r}_d - \boldsymbol{\rho}\,\boldsymbol{r}_d\right)\right] \times \exp\left(-\frac{1}{2}M_{3up}\boldsymbol{r}_d^2\right)\end{aligned} \tag{5.25}$$

The two-dimensional Fourier transformation corresponding to the two-dimensional Gaussian function $g(\boldsymbol{\rho}) = e^{-a\rho^2}$ is given by:

$$\begin{aligned}Ft\{g(\boldsymbol{\rho})\} &= \int_{-\infty}^{+\infty}\left[e^{-ax^2}\right]e^{-i2\pi w_x x}dx \cdot \int_{-\infty}^{+\infty}\left[e^{-ay^2}\right]e^{-i2\pi w_y y}dy\\ &= \frac{\pi}{a}\cdot e^{-\frac{\pi^2 w^2}{a}}\end{aligned} \tag{5.26}$$

Equation (5.26) can be used to solve the integral of the variables $\boldsymbol{r}_c$ and $\boldsymbol{r}_d$ in (5.25), and the expression for the average intensity at the incident target can be obtained as follows:

$$\langle I(\boldsymbol{\rho}; L)\rangle = \frac{k^2\omega_0^2}{8L^2 A}\exp\left(-\frac{k^2 p^2}{4L^2 A}\right) \tag{5.27}$$

$$A = \frac{1}{2\omega_0^2} + \frac{1}{2\sigma_g^2} - M_{3up} + \frac{\omega_0^2 k^2 T^2}{8} \tag{5.28}$$

Similarly, by simplifying the calculation for Eq. (5.7) as $\boldsymbol{\rho}_1 = \boldsymbol{\rho}_2 = \boldsymbol{\rho}$, the expression for the average intensity distribution of the angular reflector returns can be obtained:

$$\langle I_r(\boldsymbol{\rho}; L)\rangle = \frac{k^2\omega_0^2}{8L^2A}\exp\left(-\frac{4|\boldsymbol{\rho} - \boldsymbol{\rho}_0|^2}{r_0^2}\right)\exp\left(-\frac{k^2\boldsymbol{\rho}^2}{4L^2A}\right) \tag{5.29}$$

where A is shown in formula (5.28).

5.3.2 Light Intensity Distribution at Receiver

Using the cross-spectral density of the receiving beam represented by Eq. (5.20), the average light intensity on the receiving plane can be obtained; thus, $\boldsymbol{p}_1 = \boldsymbol{p}_2 = \boldsymbol{p}$, and Eq. (5.20) can be expressed:

$$\begin{aligned}
&\langle I(\boldsymbol{p})\rangle \\
&= W_{re}(\boldsymbol{p}, \boldsymbol{p}; 0) \\
&= \left(\frac{k}{2\pi L}\right)^4 \int d\boldsymbol{r}_1 \int d\boldsymbol{r}_2 \int d\boldsymbol{\rho}_1 \int d\boldsymbol{\rho}_2 \exp\left[-\frac{\boldsymbol{r}_1^2 + \boldsymbol{r}_2^2}{\omega_0^2} - \frac{ik\cdot\left(\boldsymbol{r}_1^2 - \boldsymbol{r}_2^2\right)}{2F}\right] \\
&\times \exp\left(-\frac{|\boldsymbol{r}_1 - \boldsymbol{r}_2|^2}{2\sigma_g^2}\right)\exp\left(-\frac{2|\boldsymbol{\rho}_1 - \boldsymbol{\rho}_0|^2}{r_0^2} - \frac{2|\boldsymbol{\rho}_2 - \boldsymbol{\rho}_0|^2}{r_0^2}\right) \\
&\times \exp\left[\frac{ik}{2L}\left(|\boldsymbol{p}_1 - \boldsymbol{\rho}_1|^2 + |\boldsymbol{\rho}_1 + \boldsymbol{r}_1|^2 - |\boldsymbol{p}_2 - \boldsymbol{\rho}_2|^2 - |\boldsymbol{\rho}_2 + \boldsymbol{r}_2|^2\right)\right] \\
&\times \exp\left[-\frac{1}{2}M_{3down}(\boldsymbol{\rho}_1 - \boldsymbol{\rho}_2)^2\right] \\
&\times \exp\left\{-\frac{1}{2}\left[M_{1up}(\boldsymbol{\rho}_1 - \boldsymbol{\rho}_2)^2 + M_{2up}(\boldsymbol{\rho}_1 - \boldsymbol{\rho}_2)\cdot(\boldsymbol{r}_1 - \boldsymbol{r}_2) + M_{3up}(\boldsymbol{r}_1 - \boldsymbol{r}_2)^2\right]\right. \\
&\left. +4C_{\chi_{ir}}\right\}
\end{aligned} \tag{5.30}$$

By substituting Eqs. (5.23) and (5.24) into Eq. (5.3), it is possible to obtain the following:

$$\langle I(\boldsymbol{p})\rangle = \left(\frac{k}{2\pi L}\right)^4 \int d\boldsymbol{r}_c \int d\boldsymbol{r}_d \int d\boldsymbol{\rho}_{\mathrm{c}} \int d\boldsymbol{\rho}_{\mathrm{d}} \exp\left(-\frac{2\boldsymbol{r}_c^2}{\omega_0^2}\right)$$
$$\exp\left(\frac{ik\cdot \boldsymbol{r}_c\boldsymbol{r}_d}{T}\right)\exp\left(-\frac{\boldsymbol{r}_d^2}{2\delta^2}\right)$$
$$\times \exp\left(-\frac{4\boldsymbol{\rho}_{\mathrm{c}}^2}{r_0^2} - \frac{\boldsymbol{\rho}_{\mathrm{d}}^2}{r_0^2} + \frac{8\boldsymbol{\rho}_{\mathrm{c}}\boldsymbol{\rho}_0}{r_0^2} + \frac{4\boldsymbol{\rho}_0^2}{r_0^2}\right)$$
$$\times \exp\left[\frac{ik}{L}(-\boldsymbol{p}\boldsymbol{\rho}_{\mathrm{d}} + \boldsymbol{\rho}_c\boldsymbol{r}_d + \boldsymbol{\rho}_d\boldsymbol{r}_c + \boldsymbol{\rho}_c\boldsymbol{\rho}_d)\right]$$
$$\times \exp\left[-\frac{1}{2}(M_{3down} + M_{1up})\boldsymbol{\rho}_{\mathrm{d}}^2 - \frac{1}{2}M_{2up}\boldsymbol{\rho}_{\mathrm{d}}\cdot\boldsymbol{r}_d\right.$$
$$\left.-\frac{1}{2}M_{3up}\boldsymbol{r}_d^2 + 4C_{\chi_{ir}}\right] \tag{5.31}$$

where,

$$\frac{1}{\delta} = \frac{1}{\omega_0^2} + \frac{1}{\sigma_g^2} \tag{5.32}$$

$$T = \frac{1}{F} - \frac{1}{L} \tag{5.33}$$

Equation (5.31) can be written as:

$$\langle I(\boldsymbol{p})\rangle = \left(\frac{k}{2\pi L}\right)^4 \exp\left(\frac{4\boldsymbol{\rho}_0^2}{r_0^2} + 4C_{\chi_{ir}}\right) \int d\boldsymbol{\rho}_{\mathrm{d}} \int d\boldsymbol{\rho}_{\mathrm{c}} \int d\boldsymbol{r}_d \exp\left(-\frac{\boldsymbol{r}_d^2}{2\delta^2}\right)$$
$$\times \exp\left(-\frac{4\boldsymbol{\rho}_{\mathrm{c}}^2}{r_0^2} - \frac{\boldsymbol{\rho}_{\mathrm{d}}^2}{r_0^2} + \frac{8\boldsymbol{\rho}_{\mathrm{c}}\boldsymbol{\rho}_0}{r_0^2}\right)\exp\left[\frac{ik}{L}(-\boldsymbol{p}\boldsymbol{\rho}_{\mathrm{d}} + \boldsymbol{\rho}_c\boldsymbol{r}_d + \boldsymbol{\rho}_c\boldsymbol{\rho}_d)\right]$$
$$\times \exp\left[-\frac{1}{2}(M_{3down} + M_{1up})\boldsymbol{\rho}_{\mathrm{d}}^2 - \frac{1}{2}M_{2up}\boldsymbol{\rho}_{\mathrm{d}}\cdot\boldsymbol{r}_d - \frac{1}{2}M_{3up}\boldsymbol{r}_d^2\right]$$
$$\times \int \exp\left(-\frac{2\boldsymbol{r}_c^2}{\omega_0^2}\right)\exp\left(\frac{ik}{L}\boldsymbol{\rho}_d\boldsymbol{r}_c\right)\exp\left(\frac{ik\cdot \boldsymbol{r}_d\boldsymbol{r}_c}{T}\right)d\boldsymbol{r}_c \tag{5.34}$$

Equation (5.26) can be used to solve the integral got the variable in Eq. (5.34) as follows:

$$\int \exp\left(-\frac{2\boldsymbol{r}_c^2}{\omega_0^2}\right)\exp\left(\frac{ik}{L}\boldsymbol{\rho}_d\boldsymbol{r}_c\right)\exp\left(\frac{ik\cdot \boldsymbol{r}_d\boldsymbol{r}_c}{T}\right)d\boldsymbol{r}_c$$
$$= \frac{\pi\omega_0^2}{2}\exp\left(-\frac{\omega_0^2k^2T^2}{8}\boldsymbol{r}_d^2 - \frac{\omega_0^2k^2T}{4L}\boldsymbol{\rho}_d\boldsymbol{r}_d - \frac{\omega_0^2k^2}{8L^2}\boldsymbol{\rho}_d^2\right) \tag{5.35}$$

By introducing (5.35) into (5.34), we obtain,

$$
\begin{aligned}
\langle I(\boldsymbol{p})\rangle = {} & \frac{k^4\omega_0^2}{32\pi^3L^4}\exp\left(\frac{4\boldsymbol{\rho}_0^2}{r_0^2}+4C_{\chi_{ir}}\right)\\
& \iint \exp\left(\frac{8\boldsymbol{\rho}_{\mathrm{c}}\boldsymbol{\rho}_0}{r_0^2}-\frac{4\boldsymbol{\rho}_{\mathrm{c}}^2}{r_0^2}-\frac{\boldsymbol{\rho}_{\mathrm{d}}^2}{r_0^2}\right)\times\exp\left[\frac{ik}{L}(\boldsymbol{\rho}_{c}\boldsymbol{\rho}_{d}-\boldsymbol{p}\boldsymbol{\rho}_{\mathrm{d}})\right]\\
& \exp\left[-\frac{1}{2}(M_{3down}+M_{1up})\boldsymbol{\rho}_{\mathrm{d}}^2-\frac{\omega_0^2k^2}{8L^2}\boldsymbol{\rho}_{d}^2\right]d\boldsymbol{\rho}_{\mathrm{d}}d\boldsymbol{\rho}_{\mathrm{c}}\\
& \times\int\exp\left(-\frac{\boldsymbol{r}_d^2}{2\delta^2}-\frac{1}{2}M_{3up}\boldsymbol{r}_d^2-\frac{\omega_0^2k^2T^2}{8}\boldsymbol{r}_d^2\right)\\
& \times\exp\left(\frac{ik}{L}\boldsymbol{\rho}_{c}\boldsymbol{r}_d-\frac{1}{2}M_{2up}\boldsymbol{\rho}_{\mathrm{d}}\cdot\boldsymbol{r}_d-\frac{\omega_0^2k^2T}{4L}\boldsymbol{\rho}_{d}\boldsymbol{r}_d\right)d\boldsymbol{r}_d
\end{aligned}
\tag{5.36}
$$

Similarly, Eq. (5.26) can be used to solve the integral of the variables in (5.36) as follows:

$$
\begin{aligned}
& \int\exp\left(\frac{ik}{L}\boldsymbol{\rho}_{c}\boldsymbol{r}_d-\frac{\boldsymbol{r}_d^2}{2\delta^2}-\frac{1}{2}M_{3up}\boldsymbol{r}_d^2-\frac{\omega_0^2k^2T^2}{8}\boldsymbol{r}_d^2-\frac{1}{2}M_{2up}\boldsymbol{\rho}_{\mathrm{d}}\cdot\boldsymbol{r}_d-\frac{\omega_0^2k^2T}{4L}\boldsymbol{\rho}_{d}\boldsymbol{r}_d\right)d\boldsymbol{r}_d\\
& =\frac{\pi}{A_1}\exp\left(-\frac{k^2}{4L^2A_1}\boldsymbol{\rho}_{c}^2-\frac{ikQ}{2LA_1}\boldsymbol{\rho}_{c}\boldsymbol{\rho}_{\mathrm{d}}+\frac{Q^2}{4A_1}\boldsymbol{\rho}_{\mathrm{d}}^2\right)
\end{aligned}
\tag{5.37}
$$

where,

$$
A_1=\frac{1}{2\delta^2}+\frac{M_{3up}}{2}+\frac{\omega_0^2k^2T^2}{8}
\tag{5.38}
$$

$$
Q=\frac{\omega_0^2k^2T}{4L}+\frac{M_{2up}}{2}
\tag{5.39}
$$

By introducing (5.37) into (5.36), we obtain the following:

$$
\begin{aligned}
\langle I(\boldsymbol{p})\rangle = {} & \frac{k^4\omega_0^2}{32\pi^2L^4A_1}\exp\left(\frac{4\boldsymbol{\rho}_0^2}{r_0^2}+4C_{\chi_{ir}}\right)\int\exp\left[-\frac{\boldsymbol{\rho}_{\mathrm{d}}^2}{r_0^2}-\frac{1}{2}(M_{3down}+M_{1up})\boldsymbol{\rho}_{\mathrm{d}}^2\right]\\
& \times\exp\left(-\frac{\omega_0^2k^2}{8L^2}\boldsymbol{\rho}_{d}^2+\frac{Q^2}{4A_1}\boldsymbol{\rho}_{\mathrm{d}}^2-\frac{ik}{L}\boldsymbol{p}\boldsymbol{\rho}_{\mathrm{d}}\right)d\boldsymbol{\rho}_{\mathrm{d}}\\
& \int\exp\left(\frac{ik}{L}\boldsymbol{\rho}_{c}\boldsymbol{\rho}_{d}-\frac{4\boldsymbol{\rho}_{\mathrm{c}}^2}{r_0^2}-\frac{k^2}{4L^2A_1}\boldsymbol{\rho}_{c}^2+\frac{8\boldsymbol{\rho}_{\mathrm{c}}\boldsymbol{\rho}_0}{r_0^2}-\frac{ikQ}{2LA_1}\boldsymbol{\rho}_{c}\boldsymbol{\rho}_{\mathrm{d}}\right)d\boldsymbol{\rho}_{\mathrm{c}}
\end{aligned}
\tag{5.40}
$$

Similarly, the integral of the variable $\boldsymbol{\rho}_{\mathrm{c}}$ in Eq. (5.40) can be solved using Eq. (5.26):

$$
\int\exp\left(\frac{ik}{L}\boldsymbol{\rho}_{c}\boldsymbol{\rho}_{d}-\frac{4\boldsymbol{\rho}_{\mathrm{c}}^2}{r_0^2}-\frac{k^2}{4L^2A_1}\boldsymbol{\rho}_{c}^2+\frac{8\boldsymbol{\rho}_{\mathrm{c}}\boldsymbol{\rho}_0}{r_0^2}-\frac{ikQ}{2LA_1}\boldsymbol{\rho}_{c}\boldsymbol{\rho}_{\mathrm{d}}\right)d\boldsymbol{\rho}_{\mathrm{c}}
$$

$$= \frac{\pi}{A_2} \exp\left(\frac{16\boldsymbol{\rho}_0^2}{r_0^4 A_2}\right) \exp\left[-\frac{k^2(Q-A_1)^2}{16L^2A_1^2A_2}\boldsymbol{\rho}_{\mathrm{d}}^2 - \frac{2ik(Q-A_1)}{r_0^2 L A_1 A_2}\boldsymbol{\rho}_0\boldsymbol{\rho}_{\mathrm{d}}\right] \tag{5.41}$$

where:

$$A_2 = \frac{4}{r_0^2} + \frac{k^2}{4L^2A_1} \tag{5.42}$$

By introducing (5.41) into (5.40), we obtain

$$\begin{aligned} \langle I(\boldsymbol{p})\rangle &= \frac{k^4\omega_0^2}{32\pi L^4 A_1 A_2} \exp\left[\left(\frac{4}{r_0^2} + \frac{16}{r_0^4 A_2}\right)\boldsymbol{\rho}_0^2 + 4C_{\chi_{ir}}\right] \\ &\times \int \exp\left[-\frac{ik}{L}\boldsymbol{p}\boldsymbol{\rho}_{\mathrm{d}} - \frac{2ik(Q-A_1)}{r_0^2 L A_1 A_2}\boldsymbol{\rho}_0\boldsymbol{\rho}_{\mathrm{d}} - \frac{\boldsymbol{\rho}_{\mathrm{d}}^2}{r_0^2} - \frac{1}{2}\left(M_{3down} + M_{1up}\right)\boldsymbol{\rho}_{\mathrm{d}}^2\right] \\ &\times \exp\left[-\frac{\omega_0^2 k^2}{8L^2}\boldsymbol{\rho}_d^2 + \frac{Q^2}{4A_1}\boldsymbol{\rho}_{\mathrm{d}}^2 - \frac{k^2(Q-A_1)^2}{16L^2A_1^2A_2}\boldsymbol{\rho}_{\mathrm{d}}^2\right] d\boldsymbol{\rho}_{\mathrm{d}} \end{aligned} \tag{5.43}$$

Similarly, (5.26) can be used to solve the integral of variable $\boldsymbol{\rho}_d$ in (5.43), that is,

$$\begin{aligned} &\int \exp\left[-\frac{\boldsymbol{\rho}_{\mathrm{d}}^2}{r_0^2} - \frac{1}{2}\left(M_{3down} + M_{1up}\right)\boldsymbol{\rho}_{\mathrm{d}}^2 - \frac{\omega_0^2 k^2}{8L^2}\boldsymbol{\rho}_d^2 + \frac{Q^2}{4A_1}\boldsymbol{\rho}_{\mathrm{d}}^2 - \frac{k^2(Q-A_1)^2}{16L^2A_1^2A_2}\boldsymbol{\rho}_{\mathrm{d}}^2\right] \\ &\times \exp\left[-\frac{ik}{L}\boldsymbol{p}\boldsymbol{\rho}_{\mathrm{d}} - \frac{2ik(Q-A_1)}{r_0^2 L A_1 A_2}\boldsymbol{\rho}_0\boldsymbol{\rho}_{\mathrm{d}}\right] d\boldsymbol{\rho}_{\mathrm{d}} \\ &= \frac{\pi}{A_3} \exp\left\{-\frac{k^2}{4A_3}\left[\frac{\boldsymbol{p}}{L} + \frac{2(Q-A_1)}{r_0^2 L A_1 A_2}\boldsymbol{\rho}_0\right]^2\right\} \end{aligned} \tag{5.44}$$

and we have:

$$\langle I(\boldsymbol{p})\rangle = \frac{k^4\omega_0^2}{32L^4A_1A_2A_3} \exp\left\{\left(\frac{4}{r_0^2} + \frac{16}{r_0^4 A_2}\right)\boldsymbol{\rho}_0^2 - \frac{k^2}{4A_3}\left[\frac{2(Q-A_1)}{r_0^2 L A_1 A_2}\boldsymbol{\rho}_0 + \frac{\boldsymbol{p}}{L}\right]^2 + 4C_{\chi_{ir}}\right\} \tag{5.45}$$

where,

$$T = \frac{1}{F} - \frac{1}{L};\ Q = \frac{\omega_0^2 k^2 T}{4L} + \frac{M_{2up}}{2}\ \frac{1}{\delta^2} = \frac{1}{\omega_0^2} + \frac{1}{\sigma_g^2};\ \frac{1}{\delta^2} = \frac{1}{\omega_0^2} + \frac{1}{\sigma_g^2} \tag{5.46}$$

$$A_1 = \frac{1}{2\delta^2} + \frac{M_{3up}}{2} + \frac{\omega_0^2 k^2 T^2}{8} \tag{5.47}$$

$$A_2 = \frac{4}{r_0^2} + \frac{k^2}{4L^2A_1} \tag{5.48}$$

$$A_3 = \frac{1}{r_0^2} + \frac{M_{3down}}{2} + \frac{M_{1up}}{2} + \frac{\omega_0^2 k^2}{8L^2} - \frac{Q^2}{4A_1} + \frac{k^2(Q - A_1)^2}{16L^2 A_1^2 A_2} \tag{5.49}$$

To analyze the effect of atmospheric turbulence on the beam intensity distribution, the normalized beam intensity is defined as:

$$I^N(p) = \frac{\langle I(p) \rangle}{\langle I_{fs}(p) \rangle |p =, \sigma_{g \to \infty, M=0}} \tag{5.50}$$

where $\langle I(\boldsymbol{p}) \rangle$ is the average light intensity of the beam on the receiving plane, and $\langle I_{fs(p)} \rangle |p = 0, \sigma_g \to \infty, M = 0$ is the light intensity value on the axis of the receiving plane when a fully coherent beam travels in free space with a two-way slope. When the average light intensity at the receiving end is affected by turbulence, this parameter refers to the light intensity distribution in free space. Therefore, this normalization is better suited for determining the effect of turbulence on the light intensity distribution. The expression for $\langle I_{fs(p)} \rangle |p = 0, \sigma_g \to \infty, M = 0$ is given as:

$$\langle I_{fs(p)} \rangle |p = 0, \sigma_g \to \infty, M = 0 = \frac{k^4 \omega_0^2}{32L^4 A_{fs1} A_{fs2} A_{fs3}} \frac{\exp\left[\left(\frac{4}{r_0^2} + \frac{16}{r_0^4 A_{fs2}}\right)\rho_0^2\right]}{\exp\left\{\frac{k^2}{4A_{fs3}}\left[\frac{2(Q - A_{fs1})}{r_0^2 L A_{fs1} A_{fs2}}\boldsymbol{\rho}_0 + \frac{p}{L}\right]^2\right\}} \tag{5.51}$$

$$A_{fs1} = \frac{1}{2\omega_0^2} + \frac{\omega_0^2 k^2}{8L^2} \tag{5.52}$$

$$A_{fs2} = \frac{4}{r_0^2} + \frac{k^2}{4L^2 A_{fs1}} \tag{5.53}$$

$$A_{fs3} = \frac{1}{r_0^2} + \frac{\omega_0^2 k^2}{8L^2} - \frac{\omega_0^4 k^4}{64L^4 A_{fs1}} + \frac{k^2\left(\omega_0^2 k^2 T - 4LA_{fs1}\right)^2}{16^2 L^4 A_{fs1}^2 A_{fs2}} \tag{5.54}$$

By substituting Eqs. (5.45) and (5.51) into Eq. (5.49), an expression for the normalized light intensity distribution on the receiving plane can be obtained.

5.4 Numerical Analysis

5.4.1 Uplink Average Light Intensity Characteristics

Based on Eqs. (5.21) and (5.27), the influence of each parameter on the distribution of light intensity at the emitter and target was calculated and analyzed. For incident wavelength $\lambda = 650$ nm, coherence length $\sigma_g = 0.02$ m, transmission distance

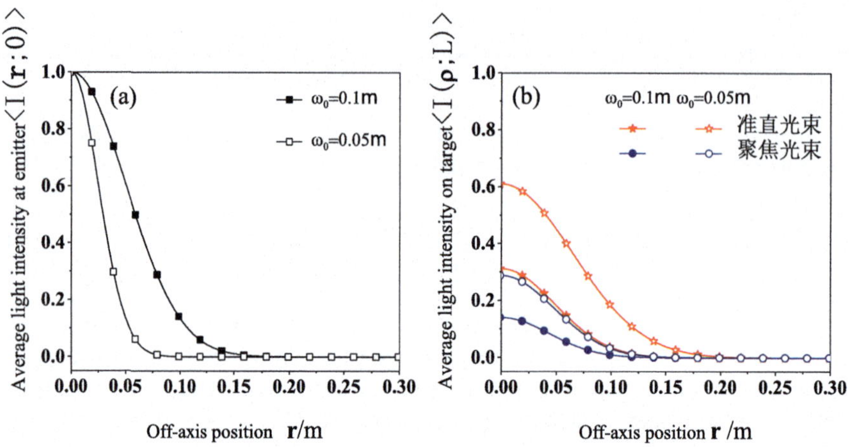

(a) Transmitter; (b) upstream transmission to the destination

Fig. 5.7 Average intensity distribution at emitter and target for different beam radii

$L = 5$ km, and inner scale and outer scales of turbulence of $l_0 = 0.01$ m and $L_0 = 10$ m, respectively, the average intensity distribution at the emitter and for upstream transmission to the target is shown in Fig. 5.7 for different beam waist radii.

Theoretically, when the beam radius is equal to 1 m [27], the effect on the propagation of the plane wave is small. By comparing (a) and (b) in Fig. 5.2, it is found that the larger the beam radius, the greater the intensity attenuation on the axis at the target, and the smaller the beam expansion. As the beam radius decreases, the average intensity of the focused beam decreases rapidly after it travels upward, whereas the opposite is true for a collimated beam. In general, the effect of turbulence on the focused beam is greater than that on a collimated beam, which is consistent with the results shown in Fig. 5.2 in [7]. As shown in Fig. 5.2b, the average intensity of the focused beam decreases rapidly with a decrease in the beam radius, whereas the opposite is true for the collimated beam. Overall, the focused beam was more affected by turbulence than the collimated beam.

For wavelength $\lambda = 650$ nm, coherence length $\sigma_g = 0.02$ m, waist radius $\omega_0 = 0.1$ m, and inner and outer scales of the turbulence of $l_0 = 0.01$ m and $L_0 = 10$ m, respectively, the average intensity distribution of the focused beam and the straight beam upward to the target is shown in Fig. 5.4 at different transmission distances.

From Fig. 5.8, it is evident that as the beam transmission distance increases, the intensity of the focused beam and the straight beam that is incident on the target axis decreases. Moreover, the intensity of the focused beam decreased faster compared to that of the collimated beam. In Eq. (5.1), when $F > 0$, the beam is focused, and when $F \to \infty$, the beam is collimated. From Eqs. 5.19 and 5.20, it is determined that the average light intensity on the axis of the target is relatively large when $F > 0$.

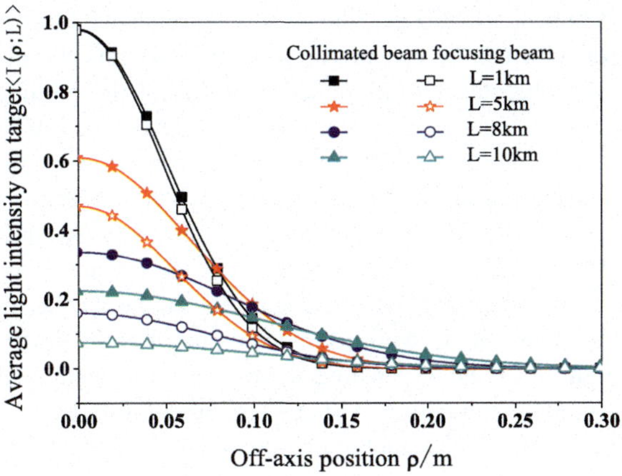

Fig. 5.8 Average intensity distribution on target at different transmission distances

For wavelength $\lambda = 650\,\text{nm}$, coherence length $\sigma_g = 0.02\,\text{m}$, beam waist radius $\omega_0 = 0.1\,\text{m}$, transmission distance $L = 10\,\text{km}$, and turbulent internal scale, and turbulent external scale of the incident beam of $l_0 = 0.01\,\text{m}$ and $L_0 = 10\,\text{m}$, respectively, the beam is transmitted upward to the target, and the average light intensity distribution after the target echo is shown in Fig. 5.9 for different corner reflector radii.

From Fig. 5.10, it can be observed that the radius of the corner reflector has a significant influence on the normalized light intensity along the axis. When the corner reflector radius increases, the normalized light intensity along the axis is initially constant, then increases rapidly near the corner reflector radius $r_0 = 5\,\text{cm}$, and finally stabilizes near the corner reflector radius $r_0 = 500\,\text{cm}$. When $r_0 < 0.5\,\text{cm}$, there is little change in the intensity of the axially normalized light, at which point the angular reflector can be considered as a point target. At this distance, both the focused and collimated beams have the same normalized intensity along the axis.

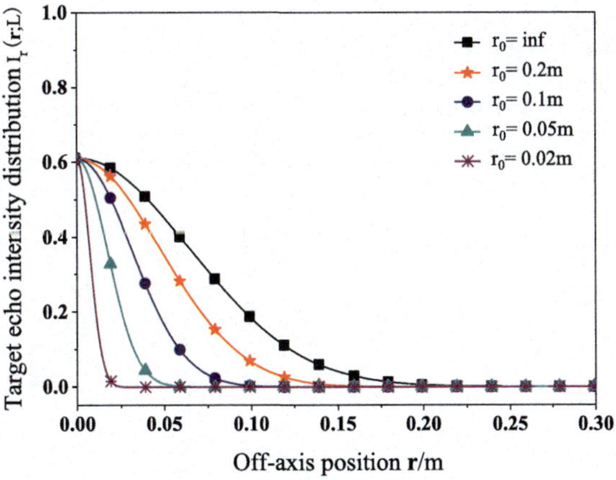

Fig. 5.9 Average light intensity distribution of target echo for different corner reflector radii

When $r_0 >$ 500 cm, the axially normalized light intensity is stable, the angular reflector can be considered to be infinite, and the intensity of the focused beam is more affected by turbulence than that of the collimated beam in free space. This result is in agreement with the main conclusions in [9].

The influence of different slope transmission distances on the normalized intensity distribution on the receiving plane when $\lambda = 650$ nm, $\sigma_g = 0.02$ m, $\omega_0 = 0.1$ m, corner reflector, $r_0 = 0.05$ m, and inner and the outer scales of turbulence of $l_0 = 0.01$ m and $L_0 = 10$ m, respectively, as shown in Fig. 5.11.

From Fig. 5.11, it is evident that the larger the transmission distance, the more dispersed the normalized intensity distribution on the receiving plane, and the more significant the expansion. Conversely, the smaller the transmission distance, the closer the transmission characteristics in turbulence are to those in free space. This indicates that the smaller the oblique transmission distance, the smaller the influence of atmospheric turbulence on the light intensity, and the normalized light with the

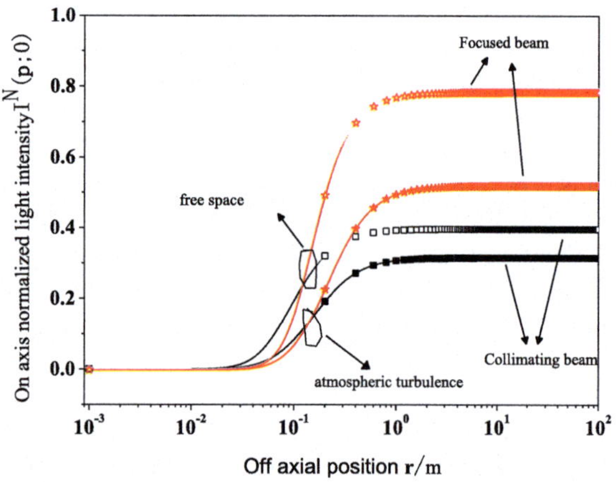

Fig. 5.10 Changes of light intensity with corner reflector radius on axis normalization

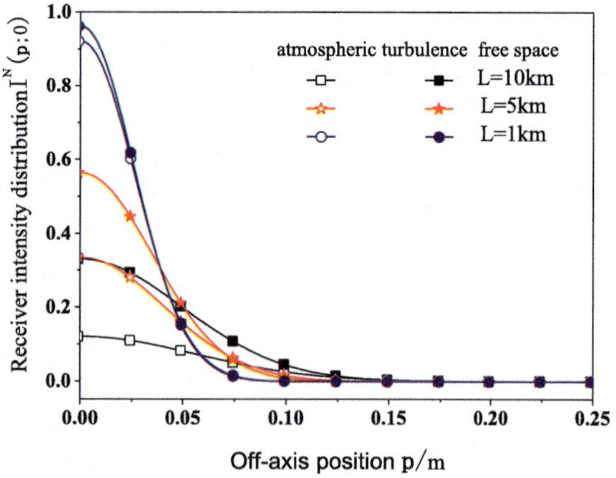

Fig. 5.11 Normalized intensity distribution on the receiving plane for different transmission distances

highest intensity is on the axis of the receiving plane. The physical interpretation of the propagating beam is that it is affected by the atmospheric refractive index and free-space diffraction when it travels in atmospheric turbulence. The phase distortion of the wavefront becomes significant with an increase in the transmission distance, thus, the beam is more affected by turbulence.

When the parameters are set to wavelength $\lambda = 650\,\text{nm}$, coherence length $\sigma_g = 0.05\,\text{m}$, waist radius $\omega_0 = 0.1\,\text{m}$, corner reflector radius $r_0 = 0.05\,\text{m}$, transmission distance $L = 10\,\text{km}$, and inner and outer scales of turbulence of $l_0 = 0.01\,\text{m}$, $L_0 = 10\,\text{m}$, respectively, the normalized intensity distribution on the receiving plane varies with the coherence length of the incoming wave, as shown in Fig. 5.9. When increases, the degree of coherence of the beam increases, whereas the coherence decreases. When $\sigma_g \to \infty$, the source field beam is a fully coherent Gauss–Schell beam.

Figure 5.12 shows that the normalized return intensity distribution variation with the off-axis distance of the receiving plane after collimated Gaussian-Schell beams of different coherence lengths are reflected by the angular reflector at the axis position in a turbulent atmosphere. The figure shows that the smaller the coherence length, the more dispersed the intensity distribution and the more significant the expansion in two-way oblique transmission. The smaller the beam's coherence length, the closer the transmission curve in turbulence relative to that in free space, i.e., the smaller the beam's coherence length, the less the effect of turbulence on the light intensity. At $\sigma_g \to \infty$, the emitted Gauss-Schell beam is fully coherent, and its intensity is most affected by turbulence. The intensity of light is dependent on free space diffraction and turbulence. The lower the coherence of the beam, the less the effect of turbulence on the light intensity. However, the light intensity is reduced because the beam is more affected by vacuum diffraction.

Owing to the different zenith angles, the normalized intensity distribution at the receiver side was different, as shown in Fig. 5.10. The parameters were set as wavelength of the incoming beam $\lambda = 650\,\text{nm}$, coherence length $\sigma_g = 0.02\,\text{m}$, waist

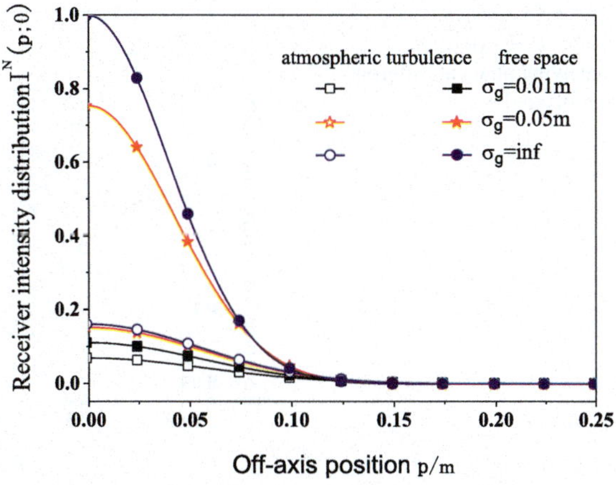

Fig. 5.12 Normalized intensity distribution on the receiving plane for different coherence lengths of the incident wave

radius $\omega_0 = 0.1$ m, corner reflector radius $r_0 = 0.2$ m, and inner and outer scales of the turbulence $l_0 = 0.01$ m and $L_0 = 10$ m, respectively.

The results in Fig. 5.13 show that the difference between the normalized intensity distribution of the returns transmitted in atmospheric turbulence and those transmitted in free space increases with the zenith angle. Moreover, the larger the zenith angle, the smaller the light intensity distribution. At the zenith angle $\theta = 0^{\circ}$, i.e., vertical transmission, the distribution of the light intensity is most concentrated, and the intensity is the highest on the receiving plane axis. At the zenith angle $\theta = 90^{\circ}$, i.e., horizontal transmission, the distribution of the light intensity is the most dispersed, and the light intensity on the receiving plane axis is the lowest. The main physical reason for this result is that the structure constant $C_n^2(h)$ of the atmospheric refractive index decreases with increasing altitude when the altitude is lower than 10 km. When the altitude is greater than 10 km, $C_n^2(h)$ increases slightly and decreases owing to the influence of wind speed at the high altitude. The closer the atmospheric channel is to the ground when the beam is transmitted, the greater the influence of turbulence; i.e., the larger the zenith angle, the more dispersed the distribution of the light intensity.

Considering that when the parameters of the incoming beam are set to a wavelength $\lambda = 650$ nm, coherence length $\sigma_g = 0.05$ m, radius of the corner reflector $r_0 = 0.2$ m, and inner and outer scales of the turbulence of $l_0 = 0.01$ m and $L_0 = 10$ m, respectively, the normalized intensity distribution on the receiving plane changes with the different structural constants of the atmospheric refractive index, as shown in Fig. 5.14.

This figure shows that under the same turbulence conditions, the normalized intensity distribution for slope transmission is more concentrated compared to horizontal transmission, and the light intensity on the axis is greater. This is because when the structure constant of the atmospheric refractive index is lower than the height, it decreases with the increase of the height, i.e., the closer the atmospheric channel to the ground during beam transmission, the greater the turbulence effect. It can also be seen that the greater the turbulence intensity, the more dispersed the light intensity

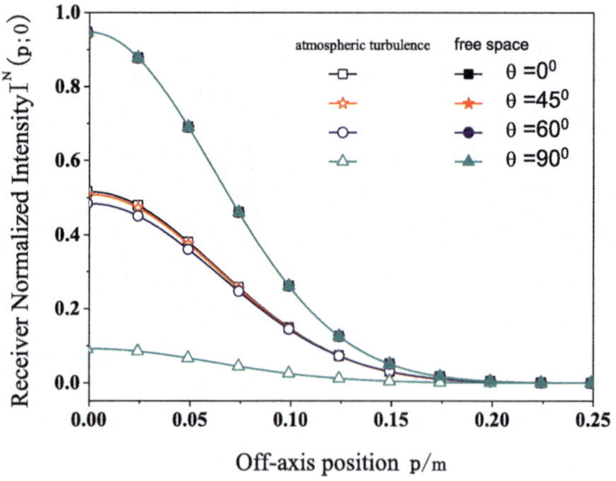

Fig. 5.13 Normalized intensity distribution on receiving plane at different zenith angles

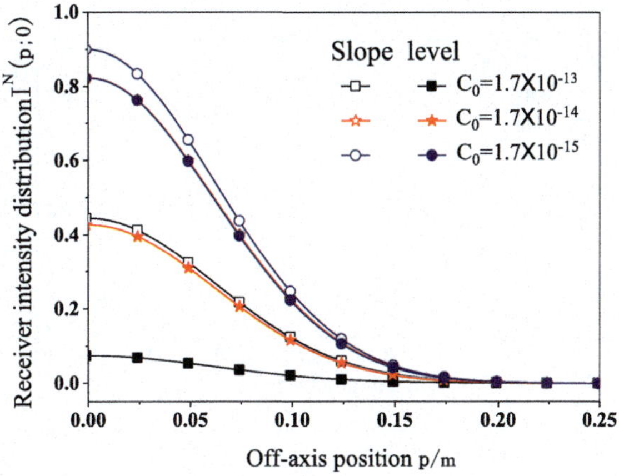

Fig. 5.14 Normalized intensity distribution on the receiving plane for different refractive index structure constants

distribution, and the smaller the light intensity on the axis. This is because when a beam travels in atmospheric turbulence, the random fluctuation of the atmospheric refractive index results in phase distortion of the beam, which causes the beam to expand after a certain transmission distance. As a result, the light distribution is more dispersed at the receiving end, and the light intensity on the axis is lower. The larger the random fluctuation, the weaker the light intensity.

The normalized intensity distribution on the receiving end plane is shown in Fig. 5.15, for different atmospheric turbulence scales. The parameters are set as wavelength of the incoming beam $\lambda = 650$ nm, coherence length $\sigma_g = 0.02$ m, waist radius $\omega_0 = 0.1$ m, corner reflector radius $r_0 = 0.2$ m, and inner and outer scales of turbulence of $l_0 = 0.01$ m and $L_0 = 10$ m, respectively.

Figure 5.15 shows that for the same outer scale, the larger the inner scale, the more concentrated the normalized intensity distribution on the receiving plane and the larger the intensity on the axis. At different external scales, the normalized intensity distribution curves are very close. Thus, the influence of the outer scale on the normalized intensity distribution of the oblique beam transmission echo can be ignored when analyzing the effect of atmospheric turbulence. The main physical reason for this result is that the smaller the scale of the turbulence, the smaller the turbulent swirls that are included in the beam section, and the more significant the diffraction of the light beam that is incident on these swirls, which causes the random distribution of the intensity of the light beam to be more severe in space and time, resulting in a more dispersed intensity. The larger the scale outside the turbulence, the more significant the beam drift on the receiving surface, but the less the influence on the distribution of light intensity.

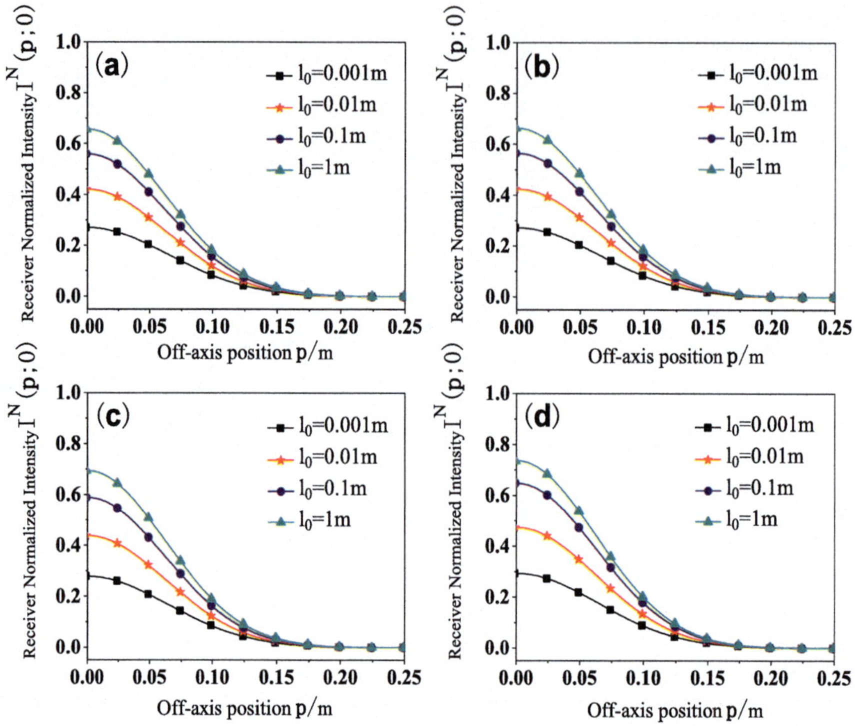

Fig. 5.15 Normalized intensity distribution on the receiving plane for different internal and external scales

5.5 Summary

The change in the average intensity of partially coherent GSM beams during upward-downward transmission through atmospheric turbulence was analyzed theoretically. The transmitter and receiver were at the same height. The influence of atmospheric turbulence on the average light intensity was analyzed numerically, and the main conclusions are as follows:

(1) For oblique transmission in a turbulent atmosphere, the return of partially coherent beams on the receiving plane changes significantly with the angular reflector radius. When the angular reflector is considered to be a point target, the axial light intensity on the receiving plane tends to the same value. When an angular reflector is considered as an infinite target, the axial light intensity on the receiving plane tends to the same value with the increase of the radius of the corner reflector.

(2) For oblique transmission in a turbulent atmosphere, whether the transmission is up to the target or down to the receiver, the intensity distribution is significantly influenced by the scale of the turbulent flow. However, and the outer scale of turbulence has almost no effect.

In practice, all parameters must be considered to select the optimal beam and the most appropriate target size. However, there are few experimental results on beam propagation in a double-slope turbulent atmosphere, and the advancement of theoretical research requires more experimental verification.

References

1. Andrews LC, Phillips RL (2001) Monostatic lidar in weak-to-strong turbulemce. Waves Random Media 11:233–245
2. Wei HY, Wu ZS (2008) Study on the effect of laser beam propagation on the slant path through atmospheric turbulence. J Electromagn Waves Appl 22(5–6):787–802
3. Wei L, Liu LR, Sun JF, Yang QG, Zhu YJ (2007) Change in degree of coherence of partially coherent electromagnetic beams propagating through atmospheric turbulence. Opt Commun 271:1–8
4. Li J, Lv B (2011) A comparative study of the upward and downward propagation of partially coherent eddy beams through atmospheric turbulence. J Phys 7(60):0742051–0742057
5. Wang SCH, Plonus MA (1979) Optical beam propagation for a partially coherent source in the turbulent atmosphere. J Opt Soc Am 69(9):1297–1304
6. Belmonte A (2005) Coherent return turbulent fluctuations in ground lidar systems profiling along slant paths. Opt Exp 13(23):9598–9604
7. Wei H, Wu Z, Peng H (2008) Scattering characteristics of diffuse objects in oblique atmospheric turbulence. J Phys 57(10):6666–6672
8. Wu Y, Wu Z (2011) Study on statistical properties of scattered field from rough surface in two-way transmission. Laser Technol 35(2):234–239
9. Wang L, Wu Z, Wang M (2013) Second-order statistical properties of reflected waves from planar mirrors with limited aperture in turbulent atmosphere. J Optics 33(11):11010021–11010026
10. Kravtsov YA, Saichev AI (1982) Effects of double passage of waves in randomly inhomogeneous media. Sov Phys Usp 25(7):494–508
11. Aksenov VP, Banakh VA, Mironov VL (1984) Fluctuations of retroreflected laser radiation in a turbulent atmosphere. J Opt Soc Am A 1(3):263–274
12. Feng Y, Song Z (1987) Expansion of reflected beams in a turbulent atmosphere. J Opt 7(9):844–850
13. Banakh VA (1993) Influence of reflector edges diffraction on manifestation of backscatter amplification effect. Proc SPIE 1993:312–318
14. Andrews LC, Miller WB (1996) The mutual coherence function and the backscatter amplification effect for a reflected Gaussian beam wave in atmospheric turbulence. Waves Random Med 5(2):167–182
15. Andrews LC, Phillips RL, Miller WB (1997) Mutual coherence function for a double-passage retroreflected optical wave in atmospheric turbulence. Appl Opt 36(3):698–708
16. Andrews LC, Phillips RL, Weeks AR (1997) Rytov approximation of the irradiance covariance and variance of a retroreflected optical beam in atmospheric turbulence. J Opt Soc Am A 14(8):1938–1948
17. Guo L, Wu W (2006) Study on the statistical properties of laser beam propagation through the turbulent atmosphere in the slant-path. In: Asia-Pacific Conference on Environmental Electromagnetics, pp 881–884

18. Wei H (2009) Laser beam transmission and target echo characteristics in oblique turbulent atmosphere. Xi'an University of Electronic Science and Technology, Xi'an
19. Li Y-Q, Wu Z-S (2012) Characteristics of a partially coherent Gaussian Scell-model beam propagating in slant atmospheric turbulence. Chin Phys B 21:0542031–0542038
20. Ling C, Hongyan W, Hongjian Z, Dongmei C (2013) Scintillation index of laser echo in oblique atmospheric turbulence. Intense Laser Particle Beam 25(9):2262–2267
21. Kingdom of Interest (2014) Target echo characteristics of laser beam in turbulent atmosphere. Xi'an University of Electronic Science and Technology, Xi'an
22. Leah Z-S, Thomas B, Steven C et al (2009) Secure optical communication system utilizing deformable MEMS mirrors. In: SPIE, 7209; 72090T
23. Zhang J (2007) Preliminary study on optical reverse modulator based on giant magnetostrictive materials. University of Electronic Science and Technology, Chengdu, pp 16–18
24. Andrews LC, Phillips RL (2005) Laser beam propagation through random media. SPIE Press, Bellingham, pp 671–680
25. Li Z, Peng R, Wu J, Zhang J (2007) Optical reverse modulation technology. Adv Laser Photoelectron 44(11):47–53
26. Xang NJ, Wu ZS, Hua XX, Wang MJ (2014) Statistical properties of Gaussian-Schell beam from diffuse target in turbulent atmosphere. High Power Laser Particle Beams 26(2):0210031–0210037
27. Li L, Sun H (2012) Structure design of airborne optical angular reflector array. Sichuan Military Eng J 33(11):117–120
28. Wei H, Wu Z (2008) Widening and drift characteristics of laser beam slope transmission in atmospheric turbulence. J Radio Sci 23(4):611–615

Chapter 6
Polarization Characteristics of Partially Coherent Beams in Atmospheric Turbulence

Based on theoretical descriptions of coherence and polarization, and considering the partially coherent Gaussian-Schell model (EGSM), the expressions for the degree of polarization (DoP) and the angle of polarization of a beam propagating in atmospheric turbulence are derived. The effects of turbulence on the DoP and the angle of polarization of upstream and downstream beams were analyzed. A comparative analysis of the polarization characteristics of light beams propagating upstream and downstream through atmospheric turbulence was performed. The effects of the beam waist radius, amplitude ratio, wavelength, and atmospheric refractive index structure constant on the polarization characteristics of the EGSM beams were analyzed.

6.1 Introduction

6.1.1 Changes in Polarization Characteristics of Light Beams Passing Through Turbulent Media

The study of the propagation characteristics of laser beams in atmospheric turbulence has wide applications in remote sensing, target tracking, and lidar imaging [1–3]. In recent years, there has been increasing attention directed to the polarization characteristics of light beams propagating through atmospheric turbulence [4–19]. Previous studies have shown that [19] the polarization characteristics of a beam in atmospheric turbulence change to varying degrees, which is closely related to the coherence of the light source components and the turbulence intensity.

In 1998, Gori proposed a beam coherence-polarization matrix to analyze the polarization characteristics of spatial points in the time domain [4]. The research on the polarization characteristics of light beams starts with the cross spectral density matrix (CSDM), which will make the results more simpler [6–10]. In 2003, Wolf presented a

X. Ke, *Partially Coherent Optical Transmission Theory in Optical Wireless Communication*, Optical Wireless Communication Theory and Technology,
https://doi.org/10.1007/978-981-97-7158-5_6

paper on the theory of coherence and polarization of random electromagnetic waves, which is of great significance in the study of the polarization characteristics of light beams [5]. In 2004, Wolf et al. analyzed the polarization characteristics of partially coherent beams in atmospheric turbulence and free space with transmission distance [6]. In 2005, Korotkova et al. deduced expression for the DoP and polarization angle of a light beam using coherent-polarization theory, and analyzed the change in the polarization state of a light beam during free-space transmission [7, 8]. In the same year, Shirai comparatively studied the far-field polarization characteristics of fully and partially coherent beams in free space [9]. In 2008, Tao studied the DoP [10] of a field when EGSM beams propagate in atmospheric turbulence. In 2009, Zhao reported that the polarization of a GSM beam does not change when it travels through free space [11]. In 2011, Chen analyzed and summarized research progress on light correlation and polarization [12]. In 2014, Yang et al. studied the polarization characteristics of partially coherent Airy, Hermite-Gauss beams, and pulsed beams in atmospheric turbulence [13–15]. In 2015, Ou et al. studied the polarization characteristics of elliptically polarized eddy beams, angularly polarized undifferentiated beams, and radially polarized beams [17–19]. The polarization characteristics of partially coherent beams when they are transmitted upstream and downstream and the mechanism by which atmospheric turbulence affects the polarization characteristics of beams are active areas of current interest.

6.1.2 Coherent Detection and Polarization Control

Line optical communication includes incoherent and coherent optical communication. Incorrect optical communication generally utilizes intensity modulation/direct detection (IM/DD) whereas coherent optical communication involves the optical communication mode that modulates the amplitude, frequency, or phase of light waves at the transmitter and demodulates information at the receiver using coherent detection (heterodyne or zero-difference) technology.

The basic principle of coherent detection is the addition of local oscillating light when an optical signal is detected, such that the signal light and the local oscillating light are coherent on the surface of a photoelectric detector, thereby forming an intermediate frequency signal. In a coherent light detection system, the local optical power is much greater than the signal optical power. Thus, the IF signal gain is improved after the two are mixed.

In a coherent optical communication system, the signal beam and local oscillator beam must have the same or similar polarization states. In theory, the signal beam is mixed with the local beam to project the electric vector of the signal light wave in the direction of the local light wave electric vector, and only the projection component contributes to the output intermediate-frequency current after mixing. If the angle between the two beams of light is greater than 60° [20], the receiver's sensitivity

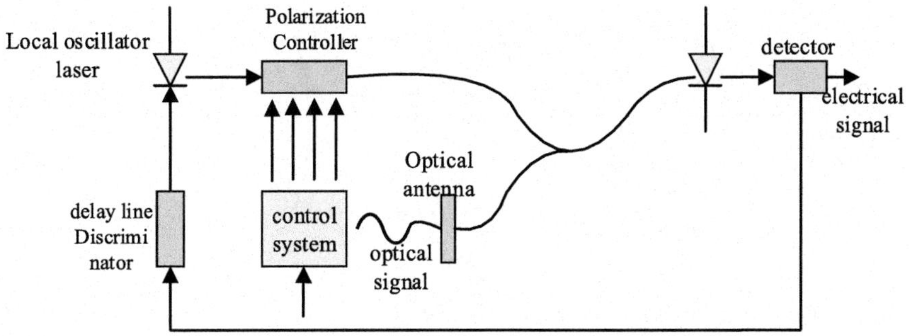

Fig. 6.1 Polarization control system for coherent optical communication

cannot be improved. Thus, the advantage of coherent light detection is lost. Polarization control has become an indispensable technology for coherent light detection to effectively exploit the advantages of this technology.

The polarization control system of coherent optical communication is shown in Fig. 6.1. An optical antenna is used to couple the optical signal to a multimode optical fiber at the receiving end. The polarization state of the local oscillator is changed using a polarization controller, which is coupled to a single-mode optical fiber through a fixed-direction polarizer. The resulting local and optical signals are transmitted along the same optical fiber using an optical fiber combiner and mixed on the surface of the detector, where the output signal of the detector is a medium-frequency current signal. The intermediate-frequency current signal is used as a feedback signal, and the polarization state of the local light is changed using the control system and the polarization controller. The polarization of the local light is continuously adjusted based on the detected changes in the feedback current. When the feedback current is a maximum, the polarization of the optical signal is consistent with that of the local light [21]. Small adjustments are then made within the allowable error range to ensure that the system is stable.

A polarization controller is a device that can change the polarization of light. Polarization controllers often use the birefringence effect of crystals to change the polarization state of polarized light. These crystals are known as phase retarders. Two main variables are involved: the principal axis azimuth of the refractive index and the phase delay. Any parameter can be changed to adjust the polarization state of light. There are three types of changes according to the following parameters:

6.1.2.1 Azimuth-Controlled Polarization Controller

Common azimuth-controlled polarization controllers are shown in Fig. 6.2, and consist of three rotatable waveforms with phase delays of $\lambda/4$, $\lambda/2$, and $\lambda/4$, each of which can rotate arbitrarily along the optical axis. First, the input polarized light is changed to linearly polarized light through a $\lambda/4$ wave plate. Then, the light output from the first wave plate is rotated to the desired polarization direction using the $\lambda/2$

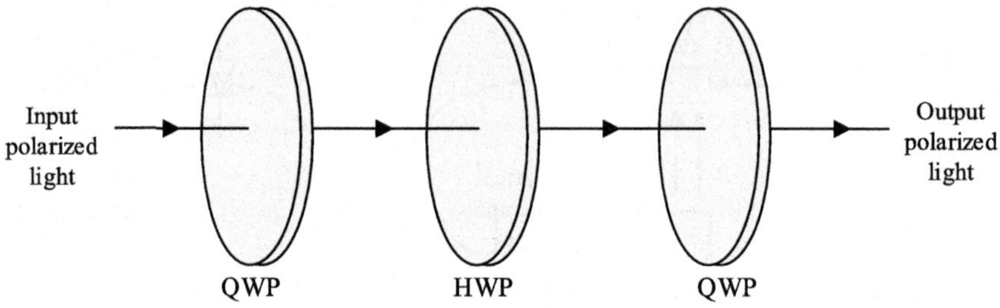

Fig. 6.2 Azimuth-controlled polarization controller

wave plate. Finally, the linearly polarized light output from the $\lambda/2$ wave plate is converted into circularly or elliptically polarized light using the second $\lambda/4$ wave plate. The phase delay of the waveband is fixed, and the azimuth of the waveband is rotatable.

6.1.2.2 Delay Controlled Polarization Controller

Delay-controlled polarization controllers generally utilize three or four 45°-bearing waveforms. The position of the waveband is fixed and the magnitude of the delay is varied using the load voltage, as shown in Fig. 6.3.

The two 45°-bearing fast axes of the wave correspond to the two orthogonal axes on the equatorial plane of a Poincare sphere, i.e., the rotational axis. The polarization state of the input light is equivalent to a point on the Poincare sphere and rotates around these two axes sequentially. For different loading voltages, the angles of rotation around the axes are different. When an appropriate voltage is applied, the desired polarization state can be obtained. An offset angle of 45°could effectively prevent the occurrence of dead deflection ends. In theory, when the delay range of a waveband is larger than 2π, the two wavebands can arbitrarily change the polarization state of the polarized light. However, owing to the limitation associated with the utilization of the delay of the waveband, the use of only two wavebands

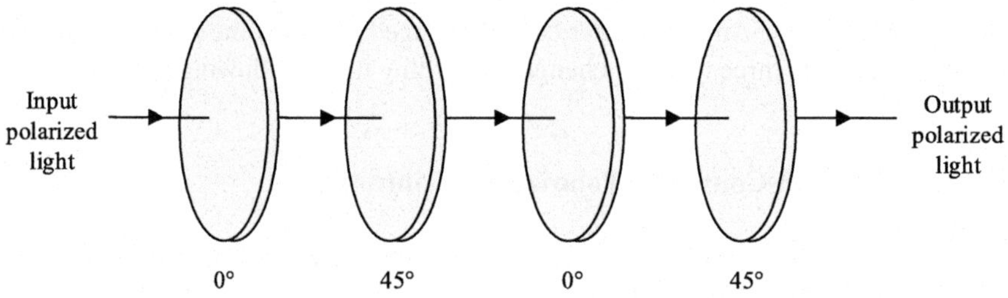

Fig. 6.3 Delay polarization controller

does not facilitate arbitrary control. Therefore, three or four waveforms are typically used to achieve arbitrary polarization control.

6.1.2.3 Delay-Azimuth Controlled Polarization Controller

The delay-azimuth polarization controller is a combination of azimuth-controlled polarization controller and delay controlled polarization controller, i.e., only one of the waveforms is needed to achieve arbitrary rotation of the polarization state of the light wave. This polarization controller can simultaneously change both parameters (refractive index principal axis and delay) to change the polarization state of polarized light.

6.1.2.4 Progress in Beam Polarization Control

The most widely used polarization control technologies in coherent optical communication are polarization diversity-receiving and polarization control technologies based on a polarization controller.

Walker and Walker [22] used polarization-maintaining optical sensors and integrated lithium niobate optical devices for polarization control. Heismann [23] analyzed the operation of a non-reset polarization controller consisting of three infinite rotating waveforms and found that for any angular offset between two waveforms, the controller can rotate continuously and non-reset to change any input polarization state to any arbitrary output polarization state. Hirabayashi et al. [24] introduced a delay polarization controller consisting of several wavebands that utilized feedback polarization control and exhibited a high control speed. Yu et al. [25] developed a hollow optical fiber polarization controller based on the acousto-optic effect. In the approach, radio-frequency signals are converted into sound waves of a certain frequency using piezoelectric ceramics, and transmitted through hollow optical fibers. The internal shape of the fibers was periodically changed, thereby introducing birefringence to achieve polarization control. Lyons et al. [26] developed an all-fiber polarization controller for an electronically controlled microheater based on the thermal-optical effect of polarization-maintaining optical fibers, which has a low insertion loss, compact structure, and low driving voltage. Terrel et al. [27] proposed a polarization controller consisting of three six cm long optical fibers fused. The working principle is that three hollow optical fibers twisted in different directions deflect their optical fiber spindles. Oh et al. [28] investigated the application of an electro-optical fiber polarization controller in which a hollow optical fiber was filled with a liquid crystal. Polarization control was accomplished by changing the delay of the controller by adjusting the control voltage loaded on the electrodes.

Li et al. [29–31] have investigated the algorithms and control circuits used in polarization controllers (mainly the ferroelectric ceramic polarization controller). Polarization control of coherent light was achieved using Digital Signal Processing

(DSP). Yao et al. [32, 33] studied the control algorithm of a phase delay polarization controller, and obtained an expression for the delay during the change of the polarization state, and analyzed the error introduced by the delay. Zhao [34] conducted an in-depth study on the principles and algorithms used in various polarization controllers. Zhang et al. [35, 36] investigated polarization controller theory. Wennan [37] implemented the control circuit of a crystal polarization controller using a field-programmable gate device. Guo et al. [38] developed a complete perturbation algorithm for a polarization controller and tested it using simulations. Yuan et al. [39] analyzed the reset principle of a four-channel squeezed optical polarization controller using the Jones matrix that is suitable for high-speed applications. In addition, they designed a reset scheme using two channels as the main delay and two others as the auxiliary reset delay. Chen et al. [40] used a variable step algorithm and prediction model to improve the searching and tracking speed, to achieve uninterrupted polarization control without resetting. Qi [41] studied the change and control of the polarization state of light in single-mode optical fibers. Yi et al. [42] studied polarization control technology in a single-mode optical fiber. An optical polarization controller was analyzed in detail using a Poincaré sphere representation. Eight methods were identified to achieve polarization control, and the performance of each approach was verified experimentally. Jin et al. [43] investigated the effect of the force acting on a single-mode optical fiber on the polarization state of light. It was determined that the polarization state of light changes periodically with an increase in the radial stress (the force acting along the radial direction). The axial stress (the force acting along the fiber's length) had little effect.

6.2 Theoretical Analysis

6.2.1 Cross Spectral Density Matrix

Based on the published work on the coherence and polarization of random electromagnetic waves [5], the second-order coherence polarization of a beam can be expressed using the CSDM of an electromagnetic beam as follows [5, 6]:

$$\overleftrightarrow{W}(\boldsymbol{r}_1, \boldsymbol{r}_2; 0) \equiv W_{ij}(\boldsymbol{r}_1, \boldsymbol{r}_2; 0) = \begin{bmatrix} W_{xx}(\boldsymbol{r}_1, \boldsymbol{r}_2; 0) & W_{xy}(\boldsymbol{r}_1, \boldsymbol{r}_2; 0) \\ W_{yx}(\boldsymbol{r}_1, \boldsymbol{r}_2; 0) & W_{yy}(\boldsymbol{r}_1, \boldsymbol{r}_2; 0) \end{bmatrix} \tag{6.1}$$

where $W_{ij}(\boldsymbol{r}_1, \boldsymbol{r}_2; 0) = \left\langle E_i(\boldsymbol{r}_1; 0) E_j^*(\boldsymbol{r}_2; 0) \right\rangle$, $(i, j = x, y)$. E_i and E_j represent two orthogonal fields that are perpendicular to the plane of the transmission direction and E_j^* is the complex conjugation of E_j. $\langle \cdot \rangle$ represents the statistical mean.

Assuming a partially coherent EGSM beam as the source field, the CSDM at the source plane is given by [11]:

$$W_{ij}(\boldsymbol{r}_1,\boldsymbol{r}_2;0)=A_iA_jB_{ij}\exp\left[-\left(\frac{r_1^2}{4\sigma_i^2}+\frac{r_2^2}{4\sigma_j^2}\right)\right]\exp\left(-\frac{|\boldsymbol{r}_1-\boldsymbol{r}_2|^2}{2\delta_{ij}^2}\right) \tag{6.2}$$

The parameters A_i, and A_j are the amplitudes of the field in the x, and y directions, respectively. σ_i and σ_j are the beam waist radii in the x and y directions, respectively. δ_{ij} is the coherence length of the beam at the plane $z=0$, which satisfies $\delta_{ij}=\delta_{ji}$. Parameter B_{ij} satisfies the relationship [44]:

$$B_{ij}=1 \quad i=j;\ |B_{ij}|\le 1 \quad i\ne j;\ B_{ij}=B_{ji}^* \tag{6.3}$$

In atmospheric turbulence, EGSM beams propagate from the source plane $z=0$ to the receiving plane $z>0$. Using the generalized Huygens-Fresnel principle [45], the CSDM of the receiving beam is given as:

$$\begin{aligned}W_{ij}(\boldsymbol{\rho}_1,\boldsymbol{\rho}_2;z)=&A_iA_jB_{ij}\left(\frac{k}{2\pi z}\right)^2\int d\boldsymbol{r}_1\int d\boldsymbol{r}_2\\&\exp\left[-\left(\frac{r_1^2}{4\sigma_i^2}+\frac{r_2^2}{4\sigma_j^2}\right)\right]\exp\left(-\frac{|\boldsymbol{r}_1-\boldsymbol{r}_2|^2}{2\delta_{ij}^2}\right)\\&\exp\left[\frac{ik}{2L}\left(|\boldsymbol{\rho}_1-\boldsymbol{r}_1|^2-|\boldsymbol{\rho}_2-\boldsymbol{r}_2|^2\right)\right]\\&\langle\exp[\psi(\boldsymbol{\rho}_1,\boldsymbol{r}_1)+\psi^*(\boldsymbol{\rho}_2,\boldsymbol{r}_2)]\rangle\end{aligned} \tag{6.4}$$

where z is the beam transmission distance, $k=\frac{2\pi}{\lambda}$ is the wave number, and λ is the wavelength. $\psi(\boldsymbol{\rho},\boldsymbol{r})$ is a complex random perturbation of a turbulent medium. The expression for $\langle\exp[\psi(\boldsymbol{\rho}_1,\boldsymbol{r}_1)+\psi^*(\boldsymbol{\rho}_2,\boldsymbol{r}_2)]\rangle$ is:

$$\langle\exp[\psi(\boldsymbol{\rho}_1,\boldsymbol{r}_1)+\psi^*(\boldsymbol{\rho}_2,\boldsymbol{r}_2)]\rangle=\exp\left[-\frac{1}{2}D_\psi(\boldsymbol{\rho}_1-\boldsymbol{\rho}_2,\boldsymbol{r}_1-\boldsymbol{r}_2)\right] \tag{6.5}$$

where $D_\psi(\boldsymbol{\rho}_1-\boldsymbol{\rho}_2,\boldsymbol{r}_1-\boldsymbol{r}_2)$ is a wave structure function that can be expressed as [46]:

$$D_\psi(\mathbf{p},\mathbf{Q},z)=8\pi^2k^2\sec\eta\int_{h_0}^{H}\int_0^\infty\kappa\phi_n(\kappa,h)\left\{1-J_0\left[|(1-\xi)\mathbf{p}+\xi\mathbf{Q}|\kappa\right]\right\}d\kappa dh \tag{6.6}$$

In the equation J_0 is the first zero-order Bessel function, η is the zenith angle, and $\phi_n(\kappa,h)$ represents the refractive index spectral density function of the turbulent medium. Equation 6.6 can be simplified as:

$$D_\phi(\boldsymbol{P},\mathbf{Q},z)=M_1\boldsymbol{P}^2+M_2\boldsymbol{P}\cdot\mathbf{Q}+M_3\mathbf{Q}^2 \tag{6.7}$$

$$\begin{cases} M_1 = 2\pi^2 k^2 \sec\eta \int_{h_0}^{H} \int_0^{\infty} \phi_n(\kappa, h)(1-\xi)^2 \kappa^3 \mathrm{d}\kappa \mathrm{d}h \\ M_2 = 4\pi^2 k^2 \sec\eta \int_{h_0}^{H} \int_0^{\infty} \phi_n(\kappa, h)(1-\xi)\xi \kappa^3 \mathrm{d}\kappa \mathrm{d}h \\ M_3 = 2\pi^2 k^2 \sec\eta \int_{h_0}^{H} \int_0^{\infty} \phi_n(\kappa, h)\xi^2 \kappa^3 \mathrm{d}\kappa \mathrm{d}h \end{cases} \tag{6.8}$$

The difference between upstream and downstream is that in (6.8), ξ is different. For the upstream case, $\xi = 1 - (h - h_0)/(H - h_0)$, H is the receiver height, h_0 is the light source height. For the downstream case, $\xi = (h - h_0)/(H - h_0)$, H is the height of the light source, and h_0 is the height of the receiver. For $\phi_n(\kappa, h)$, the modified von Karman spectrum is used, and it is expressed as [47]:

$$\phi_n(\kappa, h) = 0.033C_n^2(h)\exp(-\kappa^2/\kappa_m^2)(\kappa^2 + \kappa_0^2)^{-11/6} \tag{6.9}$$

Where $K_m = 5.92/l_0$, $K_0 \approx 2\pi/L_o$, l_o and L_0 respectively represent the inner and outer scales of turbulence; C_n^2stands the oblique atmospheric refractive index structure constant model, a model of atmospheric refractive index structure constant varying with height proposed by ITU-R in 2001 is adopted, and the expression is as follows:

$$C_n^2(h) = 8.148 \times 10^{-56} v_{RMS}^2 h^{10} e^{-h/1000} + 2.7 \times 10^{-16} e^{-h/1500} + C_o e^{-h/1000} \tag{6.10}$$

$C_n^2(h) = 8.148$ is the vertical path wind speed; v_g is the near-ground wind speed when the ground wind speed is unknown, v_g is approximately 2.8 m/s, and C_0 is the near-ground atmospheric structure constant, i.e., $C_0 = 1.7 \times 10^{-14}\mathrm{m}^{-2/3}$。

Substituting Eq. (6.7) into Eq. (6.4), and letting $\boldsymbol{\rho}_1 = \boldsymbol{\rho}_2 = \boldsymbol{\rho}$, (6.4) can be expressed as follows:

$$\begin{aligned} & W_{ij}(\boldsymbol{\rho}, \boldsymbol{\rho}; z) \\ & = A_i A_i B_{ij} \left(\frac{k}{2\pi z}\right)^2 \int d\boldsymbol{r}_c \int d\boldsymbol{r}_d \exp\left[-\left(\frac{1}{\alpha_{ij}^2}\boldsymbol{r}_c^2 + \frac{1}{\beta_{ij}}\boldsymbol{r}_c\boldsymbol{r}_d + \frac{1}{4\alpha_{ij}^2}\boldsymbol{r}_d^2\right)\right] \\ & \times \exp\left(-\frac{\boldsymbol{r}_d^2}{2\delta_{ij}^2}\right) \exp\left[\frac{ik}{L}(\boldsymbol{r}_c\boldsymbol{r}_d - \boldsymbol{\rho}\boldsymbol{r}_d)\right] \exp\left[-M_3\boldsymbol{r}_d^2\right] \end{aligned} \tag{6.11}$$

where $\boldsymbol{r}_c = \frac{1}{2}(\boldsymbol{r}_1 + \boldsymbol{r}_2)$, $\boldsymbol{r}_d = \boldsymbol{r}_1 - \boldsymbol{r}_2$, $\frac{1}{\alpha_{ij}^2} = \frac{1}{4\sigma_i^2} + \frac{1}{4\sigma_j^2}$, $\frac{1}{\beta_{ij}} = \frac{1}{4\sigma_i^2} - \frac{1}{4\sigma_j^2}$.

The two-dimensional Fourier transform of the two-dimensional Gaussian function $g(\boldsymbol{\rho}) = e^{-a\rho^2}$ is given as:

$$\begin{aligned} Ft\{g(\boldsymbol{\rho})\} & = \int_{-\infty}^{+\infty} \left[e^{-ax^2}\right] e^{-i2\pi w_x x} dx \cdot \int_{-\infty}^{+\infty} \left[e^{-ay^2}\right] e^{-i2\pi w_y y} dy \\ & = \frac{\pi}{a} \cdot e^{-\frac{\pi^2 w^2}{a}} \end{aligned} \tag{6.12}$$

Equation (6.11) can be used to solve for the variables $\boldsymbol{r}_c$ and $\boldsymbol{r}_d$ in Eq. (6.11), and the expression is given by:

$$W_{ij}(\boldsymbol{\rho}, \boldsymbol{\rho}; z) = \frac{A_i A_j B_{ij}}{\Delta_{ij}^2} \exp\left(-\frac{\boldsymbol{\rho}^2}{\Delta_{ij}^2 \cdot \alpha_{ij}^2}\right) \tag{6.13}$$

$$\text{where } \Delta_{ij}^2 = 1 - \frac{2i}{k\beta_{ij}} z + \frac{1}{k^2}\left[\left(\frac{1}{\chi_{ij}^2} + M_3\right)\frac{4}{\alpha_{ij}^2} - \frac{1}{\beta_{ij}^2}\right] z^2 \tag{6.14}$$

$$\frac{1}{\chi_{ij}^2} = \frac{1}{4\alpha_{ij}^2} + \frac{1}{2\delta_{ij}^2} \tag{6.15}$$

6.2.2 Polarization Characteristics of Light Beam

The DoP can be described using the generalized Stokes parameter, and the DoP expression of the end beam is [48]

$$P(\boldsymbol{\rho}; z) = \sqrt{S_1^2 + S_2^2 + S_3^2} \Big/ S_0 \tag{6.16}$$

where the Stokes parameters S_0, S_1, S_2, and S_3 can be obtained from the receiver CDSM.

$$\begin{aligned} S_0 &= W_{xx}(\boldsymbol{\rho}, \boldsymbol{\rho}; z) + W_{yy}(\boldsymbol{\rho}, \boldsymbol{\rho}; z)\ S_1 = W_{xx}(\boldsymbol{\rho}, \boldsymbol{\rho}; z) - W_{yy}(\boldsymbol{\rho}, \boldsymbol{\rho}; z) \\ S_2 &= W_{xy}(\boldsymbol{\rho}, \boldsymbol{\rho}; z) + W_{yx}(\boldsymbol{\rho}, \boldsymbol{\rho}; z)\ S_3 = i\left[W_{yx}(\boldsymbol{\rho}, \boldsymbol{\rho}; z) - W_{xy}(\boldsymbol{\rho}, \boldsymbol{\rho}; z)\right] \end{aligned} \tag{6.17}$$

Based on Eqs. (6.13) and (6.16), the Stokes parameters for the receiving plane $z > 0$ can be obtained as follows:

$$\begin{aligned} S_0 &= \frac{A_x^2}{\Delta_{xx}^2} \exp\left(-\frac{\boldsymbol{\rho}^2}{\Delta_{xx}^2 \cdot \alpha_{xx}^2}\right) + \frac{A_y^2}{\Delta_{yy}^2} \exp\left(-\frac{\boldsymbol{\rho}^2}{\Delta_{yy}^2 \cdot \alpha_{yy}^2}\right) \\ S_1 &= \frac{A_x^2}{\Delta_{xx}^2} \exp\left(-\frac{\boldsymbol{\rho}^2}{\Delta_{xx}^2 \cdot \alpha_{xx}^2}\right) - \frac{A_y^2}{\Delta_{yy}^2} \exp\left(-\frac{\boldsymbol{\rho}^2}{\Delta_{yy}^2 \cdot \alpha_{yy}^2}\right) \\ S_2 &= \frac{A_x A_y B_{xy}}{\Delta_{xy}^2} \exp\left(-\frac{\boldsymbol{\rho}^2}{\Delta_{xy}^2 \cdot \alpha_{xy}^2}\right) + \frac{A_y A_x B_{yx}}{\Delta_{yx}^2} \exp\left(-\frac{\boldsymbol{\rho}^2}{\Delta_{yx}^2 \cdot \alpha_{yx}^2}\right) \\ S_3 &= i\left[\frac{A_y A_x B_{yx}}{\Delta_{yx}^2} \exp\left(-\frac{\boldsymbol{\rho}^2}{\Delta_{yx}^2 \cdot \alpha_{yx}^2}\right) - \frac{A_x A_y B_{xy}}{\Delta_{xy}^2} \exp\left(-\frac{\boldsymbol{\rho}^2}{\Delta_{xy}^2 \cdot \alpha_{xy}^2}\right)\right] \end{aligned} \tag{6.18}$$

From Eqs. (6.17) and (6.18), the DoP of the beam axis on the receiving plane is given by:

$$P(\boldsymbol{\rho} = \mathbf{0}; z) = \frac{\sqrt{\left(\frac{A_x^2}{A_y^2}\frac{\Delta_{yy}^2}{\Delta_{xx}^2} - 1\right)^2 + 4|B_{xy}|^2 \frac{A_x^2}{A_y^2}\frac{\Delta_{yy}^4}{\Delta_{xy}^2 \Delta_{yx}^2}}}{\frac{A_x^2}{A_y^2}\frac{\Delta_{yy}^2}{\Delta_{xx}^2} + 1} \tag{6.19}$$

where $A_x^2 / A_y^2 = I_x / I_y$ is the ratio of source-polarized light intensity. From Eq. (6.19), the DOP of the point on the beam axis of the receiving plane is only related to the ratio of A_x^2 / A_y^2, but not directly to A_x and A_y.

6.2.3 *Polarization Direction Angle of Light Beam*

The direction angle $\theta_0(\boldsymbol{\rho}; z)$ refers to the angle between the principal axis of the elliptical polarization and the x direction of the light vibration. The direction of the light vibration of the two orthogonal beams rotates counterclockwise with respect to the direction of light transmission. The direction angle is given by [6]:

$$\theta_0(\boldsymbol{\rho}; z) = \frac{1}{2}\arctan\left(\frac{2\mathrm{Re}\left[W_{xy}(\boldsymbol{\rho}, \boldsymbol{\rho}; z)\right]}{W_{xx}(\boldsymbol{\rho}, \boldsymbol{\rho}; z) - W_{yy}(\boldsymbol{\rho}, \boldsymbol{\rho}; z)}\right) \tag{6.20}$$

By substituting Eq. (6.13) into Eq. (6.20), we obtain

$$\theta_0(\boldsymbol{\rho}; z) = \frac{1}{2}\arctan\left(\frac{2\mathrm{Re}\left[\frac{A_x A_y B_{xy}}{\Delta_{xy}^2}\exp\left(-\frac{\rho^2}{\Delta_{xy}^2 \cdot \alpha_{xy}^2}\right)\right]}{\frac{A_x^2}{\Delta_{xx}^2}\exp\left(-\frac{\rho^2}{\Delta_{xx}^2 \cdot \alpha_{xx}^2}\right) - \frac{A_y^2}{\Delta_{yy}^2}\exp\left(-\frac{\rho^2}{\Delta_{yy}^2 \cdot \alpha_{yy}^2}\right)}\right) \tag{6.21}$$

where $\theta_0(\boldsymbol{\rho}; z)$ ranges from $-\pi / 2$ to $\pi / 2$.

6.3 Numerical Analysis

The following is a comparative analysis of the changes in the DoP and the polarization angle between the upstream and downstream transmission of EGSM beams. For numerical calculation, the parameters $B_{ij} = 0.5\exp(i\pi / 4)$, $\delta_{xy} = \delta_{yx} = 0.025$ m, $\delta_{xx} = 0.02$ m, $\delta_{yy} = 0.01$ m, $\lambda = 650$ nm, $C_0 = 1.7 \times 10^{-14}$ $\mathrm{m}^{-2/3}$, $H = 0.5$ km, $z = 10$ km, $l = 0.01$ m, $L_0 = 10$ m, and zenith angle $\eta = \arccos(H / z)$. The correlation length δ_{xy} satisfies $\max\{\delta_{xx}, \delta_{yy}\} \le \delta_{xy} \le \min\left\{\delta_{xx} / |B_{xy}|^{0.5}, \delta_{yy} / |B_{xy}|^{0.5}\right\}$.

6.3.1 Comparison of Polarization Characteristics Between up and Down Transmission

6.3.1.1 Comparison of Polarization Degrees

Based on Eqs. (6.16) and (6.18), the distribution of the DoP when the GSM beam is transmitted up and down to the receiving end is compared and analyzed, as shown in Figs. 6.4 and 6.5.

Figures 6.4 and 6.5 show the distribution curves of the EGSM beams for upstream and downstream to the DoP at the receiving end at the transmission distance $z = 10$ km when σ_x, σ_y are the same and different, respectively. Figures 6.3 and 6.4 show that the change in the DoP of the field is more concentrated in the downward transmission of EGSM beams than in the upstream transmission. This is because the difference between the upstream and downstream transmissions is the turbulence term M_3, and the downstream transmission M_3 is smaller than the upstream transmission M_3. Therefore, the DoP of the receiver approaches 1 when the downstream transmission is smaller than the upstream transmission. In addition, from Fig. 6.4, at a transmission distance of $z = 10$ km, in the case of upstream and downstream, the DoP at the receiver initially decreases, then increases, and approaches 1 with respect to the increase off-axis. This result is consistent with Fig. 3 in [7]. In Fig. 6.5, regardless of the change in the source polarization amplitude ratio A_x/A_y, the change in the DoP of the light field is more concentrated in the downward transmission of the EGSM beams compared to the upstream transmission. By comparing (a), (b), and (c) in Fig. 6.5, it is shown that the DoP variation is greater when $A_x/A_y > 1$, but smaller when $A_x/A_y < 1$, and the DoP variation is close to 1 at almost every point on the receiving plane with increasing polarization axis distance.

Based on Eq. (6.20), the changes in the DoP along the transmission distance at the axis point of an EGSM beam for upstream and downstream transmission to the receiving end are compared and analyzed, as shown in Figs. 6.5 and 6.6.

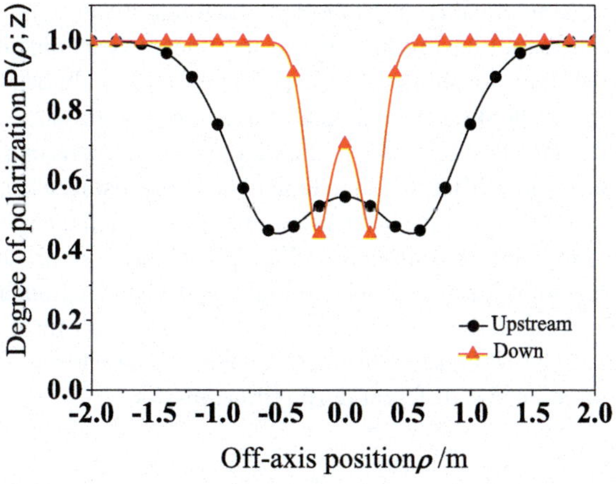

Fig. 6.4 . Distribution curve of DoP in up and down transport

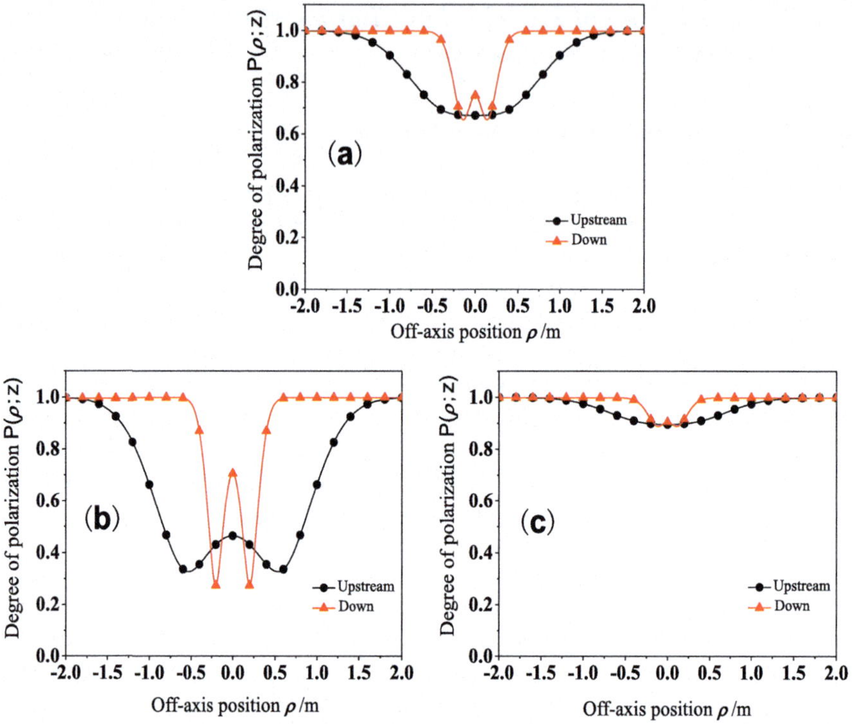

Fig. 6.5 Distribution curve of DoP for upstream and downstream transmission for $\sigma_x = 0.01$ m; $\sigma_y = 0.02$ m; **a**$A_x/A_y = 1$; **b**$A_x/A_y = 2$; **c** $A_x/A_y = 1/2$

Figures 6.6 and 6.7 show the change in the DoP with the transmission distance when EGSM beams are transmitted upstream and downstream to a point on the receiving axis through atmospheric turbulence with the same and different σ_x, σ_y, respectively. In these two Figure, the transmission distance corresponding to the maximum value of the DoP in downlink transmission is greater compared to that of upstream transmission. When the transmission distance is very short or very long, the DoP transmitted to the receiving end is almost the same for the same A_x/A_y case. The difference is that in Fig. 6.6, for the same σ_x, σ_y, when the transmission distance is short or long, the random fluctuations caused by the refractive index are negligible. Therefore, the DoP over a long distance becomes close to the initial value, and the DoP of the beam is better maintained in the upstream transmission than in the downward transmission. In Fig. 6.7, when σ_x, σ_y are different, the DoP approaches a certain value but is not the initial value when the transmission distance is longer. This is because the waist radius of the EGSM beam for the x, y component is different; i.e., the DoP at the distance does not change to the initial value when the effect of turbulence is different during transmission.

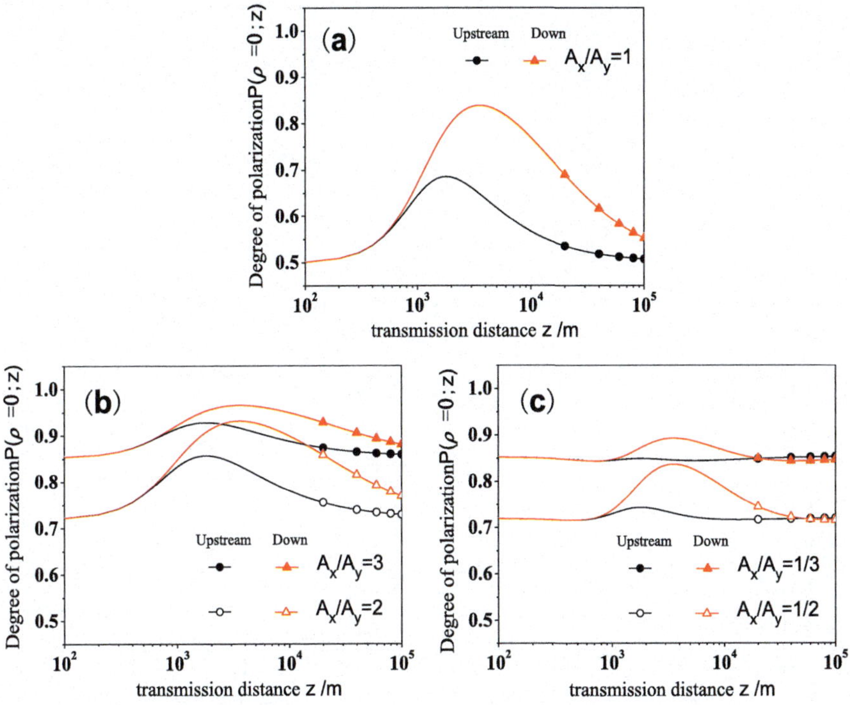

Fig. 6.6 $\sigma_x = \sigma_y = 0.01$ m. Change of DoP at the axis point for the transmission distance for up and down transmission **a** $A_x/A_y = 1$; **b** $A_x/A_y > 1$; **c** $A_x/A_y < 1$

6.3.1.2 Comparison of Polarization Angles

Based on Eq. (6.21), the changes in the polarization angle with the transmission distance of the EGSM beams for upstream and downstream transmission to the receiving end were compared and analyzed, as shown in Figs. 6.7 and 6.8.

Figures 6.8 and 6.9 show the curves of the polarization angle of the EGSM beams that propagate upstream and downstream through atmospheric turbulence as a function of the transmission distance under the same and different conditions, respectively. These two figures show that with an increase in the transmission distance, the polarization angle changes faster for downward transmission compared to upstream transmission. Within the transmission distance range, the polarization angle hardly changes, whether the initial spot is circular or elliptical. When the transmission distance is very long, the polarization angle is almost the same for both upstream and downstream transmissions. The change in the polarization angle is similar to that of the DoP. Figure 6.8 is similar to that of Fig. 6.6 in that the polarization angle is the same as that of the initial beam when the transmission distance is very long, and Fig. 6.9 is similar to Fig. 6.7. Although the polarization angle tends to a certain value for a very long transmission distance, it is not the initial value.

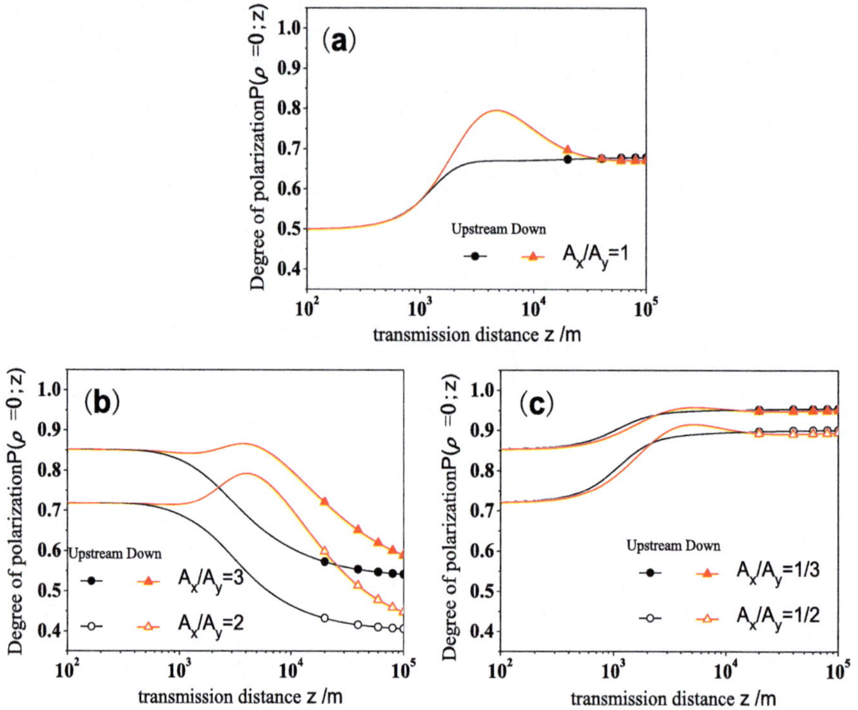

Fig. 6.7 $\sigma_x = 0.01$ m; $\sigma_y = 0.02$ m. The change of the DoP at the axis point with transmission distance for up and down transmission **a** $A_x/A_y = 1$; **b** $A_x/A_y > 1$; **c** $A_x/A_y < 1$

6.3.2 Comparison of Polarization Characteristics Between Horizontal and Slope Transmission

6.3.2.1 Effect of Beam Waist Radius

When the waist radius of the EGSM beam in the x and y direction is different at the source field, the DoP distribution at the receiving end is shown in Fig. 6.10, and the DoP distribution at the receiving end axis is shown in Fig. 6.11.

Figure 6.10 shows the distribution of the EGSM beams through atmospheric turbulence for slope and horizontal transmittance to the receiver DoP at a transmission distance of z = 10 km, when σ_x, and σ_y change. Figure 6.10 shows that no matter how σ_x and σ_y change, the change of DoP in the field at the receiving end is more concentrated for oblique transmission compared to horizontal transmission. This is because the difference between slope and horizontal transmission is that the turbulence term M_3 is smaller compared to horizontal transmission M_3. Therefore, under the same conditions, the DoP of the receiving end approaches 1 when the distance from the axis is small for the slant transmission of EGSM beams. Secondly, in the case of slope and horizontal transmission, the DoP of the receiver initially decreases,

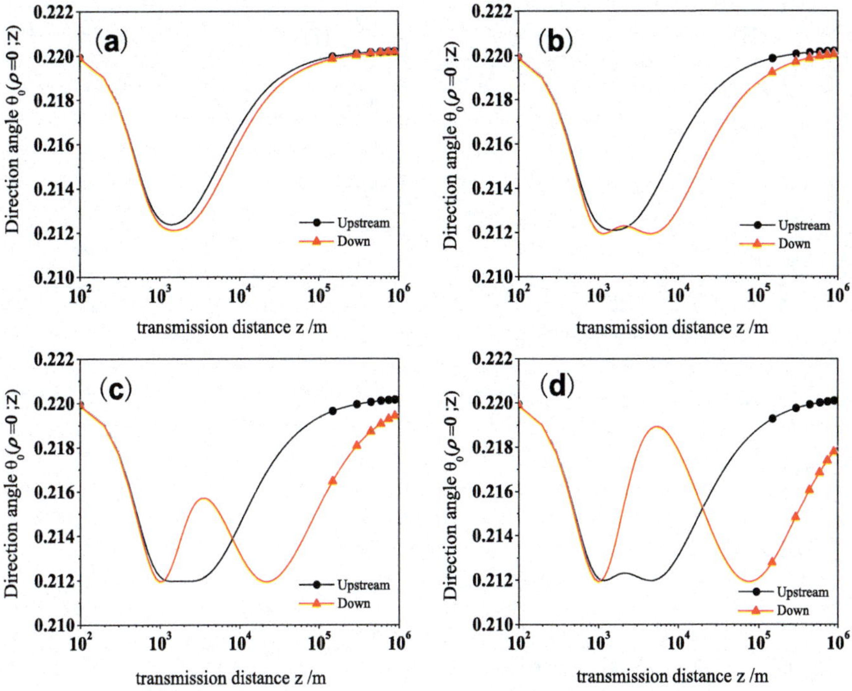

Fig. 6.8 $\sigma_x = \sigma_y = 0.01$ m. Angle changes with transmission distance for up and down transmission **a**$H = 50$ m; **b** $H = 200$ m; **c** $H = 500$ m; **d** $H = 1000$ m

then increases, and tends to a fixed value of 1 as the distance from the axis increases. From the comparison of (a), (b), and (c) in Fig. 6.10, and an increase of the waist radius when $\sigma_x = \sigma_y$, the DoP of the light field is first concentrated, then it diffuses. That is, it can be inferred that when 3 cm $<\sigma_x < 6$ cm, there is a specific value that causes the DoP of the entire light field to be highly concentrated, i.e., when the distance from the axis is very small, the DoP approaches 1. When $\sigma_x < \sigma_y$ is used, the DoP of the light field becomes more concentrated, and when $\sigma_x > \sigma_y$, the DoP of the field diffuses.

Figure 6.11 shows how the DoP of the EGSM beams propagate horizontally through the atmospheric turbulence slope to the receiving axis changes with the transmission distance when σ_x and σ_y change. From this figure, irrespective of the change in σ_x and σ_y, the transmission distance corresponding to the maximum DoP at the receiving end axis is longer than that of the horizontal transmission when the slope transmission is completed. For both slope and horizontal transmission, the transmission distance increases with the increase of the waist radius when the DoP on the receiving end reaches its maximum value. Comparing (a), (b), and (c) in Fig. 6.11, it is evident that when $\sigma_x = \sigma_y$, the DoP at the axis tends to the initial value when the transmission distance is long. When $\sigma_x \neq \sigma_y$, the DoP at the axis point tends to be fixed, but initially, not at the transmission distance. This is because

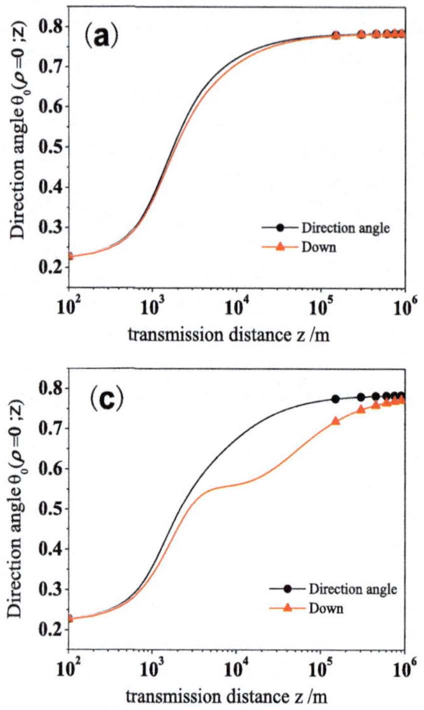

Fig. 6.9 $\sigma_x = 0.01$ m;$\sigma_y = 0.02$ m. Angle changes with transmission distance for upstream and downstream transmission **a** $H = 50$ m; **b** $H = 200$ m; **c** $H = 500$ m; **d** $H = 1000$ m

the waist radii of EGSM beams are different for the *x*, *y*, and components, i.e., they are affected by turbulence during transmission, and the DoP at the axis point at a given distance does not approach the initial value.

6.3.2.2 Effect of Amplitude Ratio

When the amplitude ratio of the EGSM beam in the x and y directions is different at the source field, the DoP distribution at the receiving end is as shown in Fig. 6.12, and the DoP at the point on the receiving end axis changes with the transmission distance, as shown in Fig. 6.13.

Figure 6.12 shows how the DoP of EGSM beams propagating horizontally to a point on the receiving end axis varies with the transmission distance when the amplitude ratio changes. Similar to the conclusion associated with Fig. 6.10, irrespective of the variation of the amplitude, the transmission distance corresponding to the maximum DoP at the receiving end axis is longer than that of the horizontal transmission. Comparing with (a), (b), and (c) in Fig. 6.12, it is determined that when $A_x/A_y = 1$, the DoP at the point on the receiving end axis changes most with the increase of the transmission distance. When $A_x/A_y < 1$, with the decrease of the

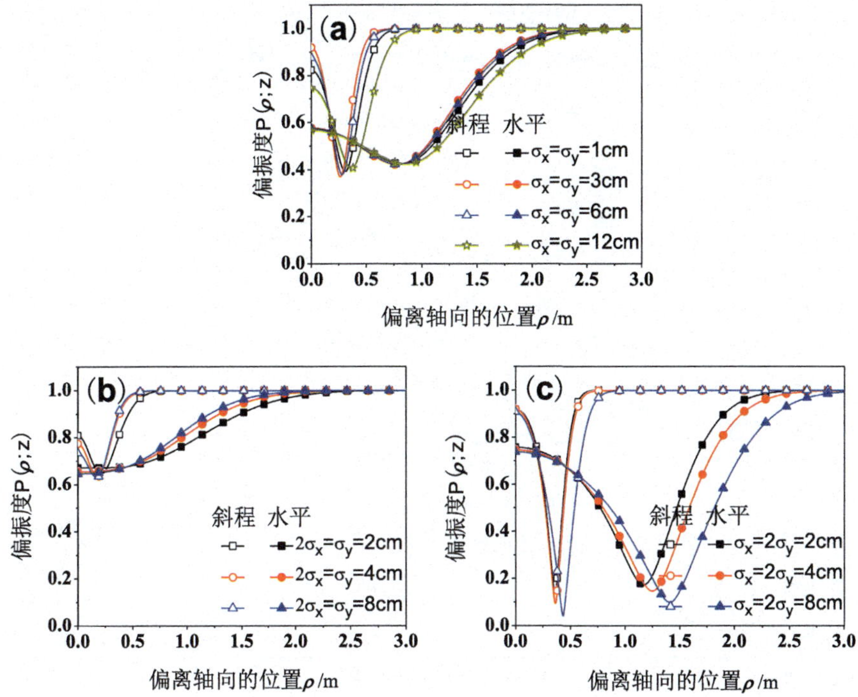

Fig. 6.10 Distribution curves of DoP at different waist radii **a** $\sigma_x = \sigma_y$; **b** $\sigma_x < \sigma_y$; **c**$\sigma_x > \sigma_y$

amplitude ratio, the DoP of the axis point at $z = 0$ increases and the DoP of the receiving end axis point decreases. However, when $A_x/A_y > 1$, with the increase in the amplitude ratio, the DoP of the axis point at $z = 0$ increases, and the DoP change of the receiving endpoint decreases.

6.3.2.3 Effect of the Structure Constant of the Atmospheric Refractive Index

When the structure constants of the atmospheric refractive index are different, the DoP distribution at the receiving end is as shown in Fig. 6.13, and the DoP at the receiving end axis changes with the transmission distance, as shown in Fig. 6.14.

For different values of the atmospheric refractive index structure constant C, Fig. 6.13 shows the distribution of the DoP at the receiving end for an EGSM beam propagating horizontally and with a slope through atmospheric turbulence up to $z = 10$ km. Figure 6.14 shows the variation in the DoP at the receiving end axis as a function of the transmission distance. From Fig. 6.13, for the same C_0 condition, the change in the DoP of the field of the receiving end is more concentrated

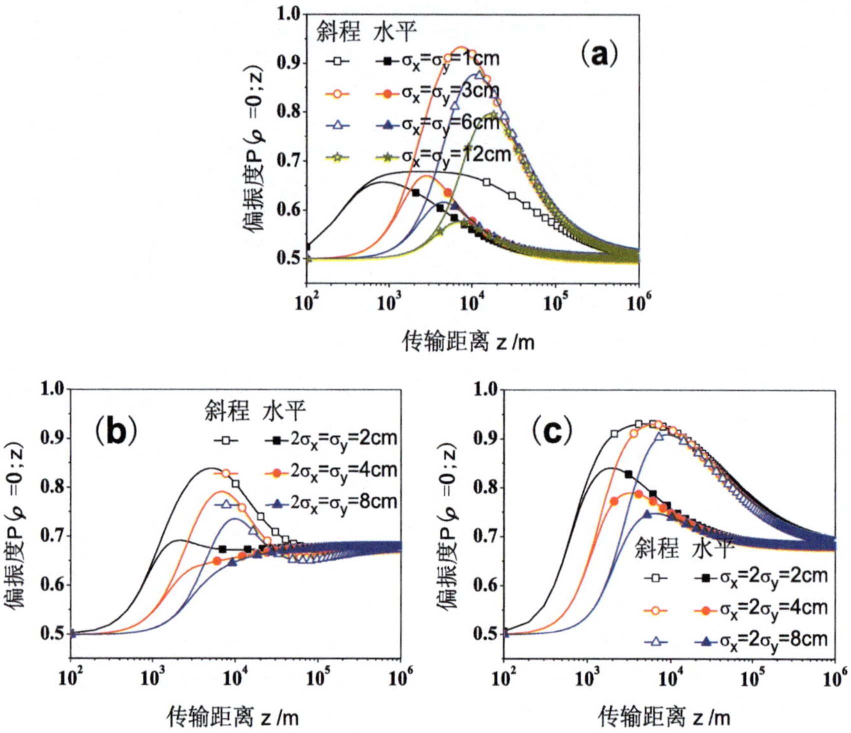

Fig. 6.11 Variation of DoP at the axis point with transmission distance at different waist radii **a** $\sigma_x = \sigma_y$; **b**$\sigma_x < \sigma_y$; **c** $\sigma_x > \sigma_y$

when the EGSM beam is transmitted obliquely compared to horizontal transmission. Moreover, with the decrease in C_0, the DoP distribution of the entire light field becomes more concentrated both horizontally and obliquely, but the change in the speed for horizontal transmission is faster compared to oblique transmission. From Fig. 6.14, for the same C_0 condition, when the DoP of the receiving end axis point is a maximum, the corresponding transmission distance is longer than that of horizontal transmission, and when the DoP of the receiving end axis point is restored to its initial value, the transmission distance is longer than that of the horizontal transmission. Thus, the oblique transmission time detector can receive more distant information compared to horizontal transmission. In addition, with an increase in the transmission distance, the DoP of the point on the axis increases with an increase in C_0.

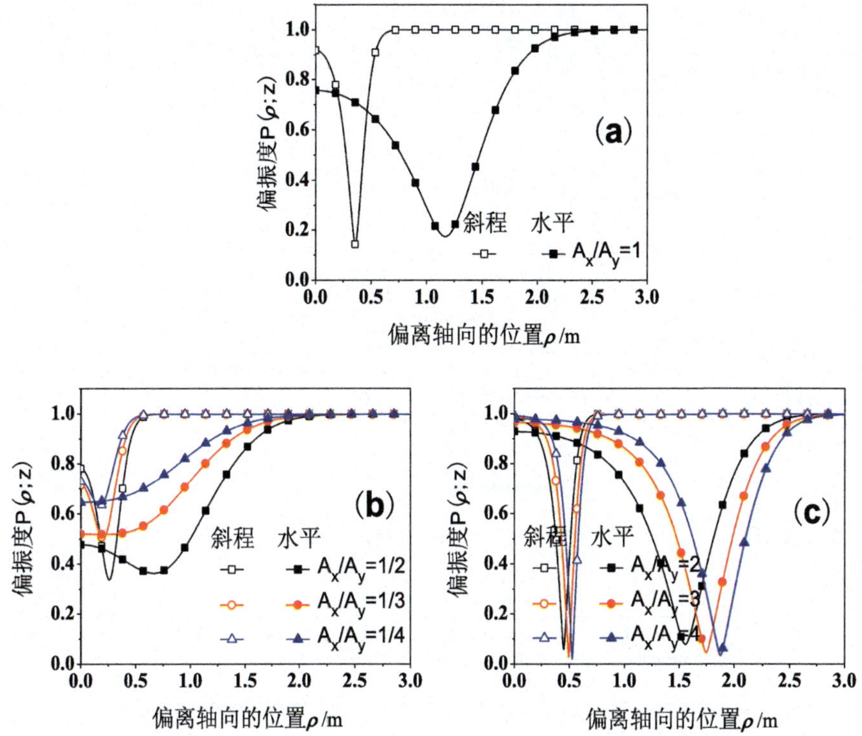

Fig. 6.12 Distribution curves of DoP at different amplitude ratios **a** $A_x/A_y = 1$; **b** $A_x/A_y < 1$; **c** $A_x/A_y > 1$

6.3.2.4 Effect of Beam Wavelength

When the EGSM beam wavelength at the source field is different, the DoP distribution at the receiving end is as shown in Fig. 6.15, and the DoP at the receiving end axis changes with the transmission distance, as shown in Fig. 6.16.

Figure 6.15 shows the distribution of the DoP at the receiving end of an EGSM beam that propagates horizontally and with a slope through atmospheric turbulence up to $z = 10$ km at different beam wavelengths, and Fig. 6.16 shows the change rule of DOP of the point on the axis of the receiving end with the transmission distance. From Fig. 6.15, for the same beam wavelength, the change in the DoP of the field at the receiving end is more concentrated when the EGSM beam is transmitted obliquely compared to horizontal transmission. Moreover, with an increase in the wavelength, the DoP distribution of the light field becomes more concentrated, both horizontally and obliquely. Similar to the conclusion associated with Fig. 6.14, in Fig. 6.16, the detector can receive more distant information when the beam wavelength is the same as that used for horizontal transmission. In addition, with an increase in the

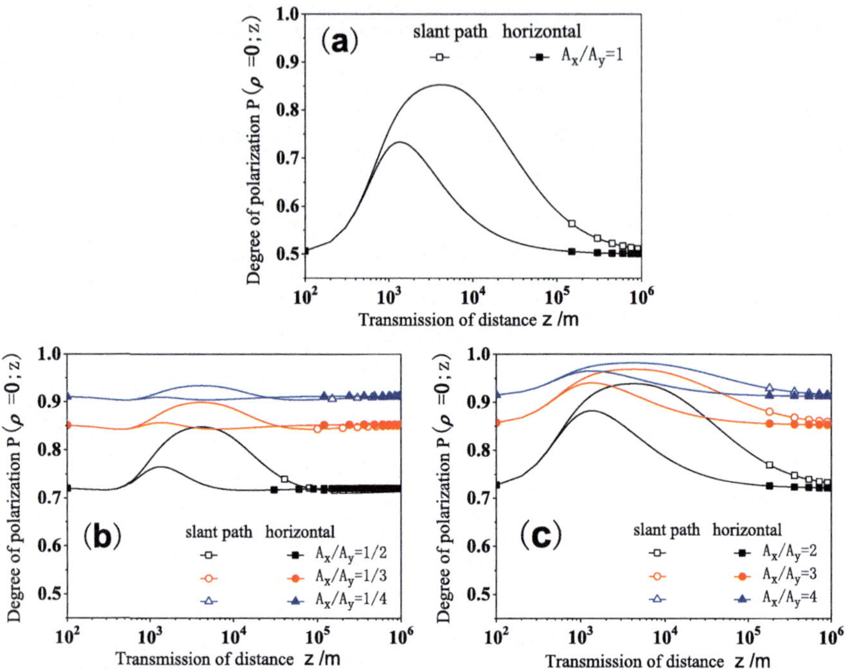

Fig. 6.13 Change of dot Dop on time axis with transmission distance for different amplitude ratios **a** $A_x/A_y = 1$; **b** $A_x/A_y < 1$; **c** $A_x/A_y > 1$

Fig. 6.14 Variation of DoP at axis points with transmission distance for different atmospheric refractivity structure constants

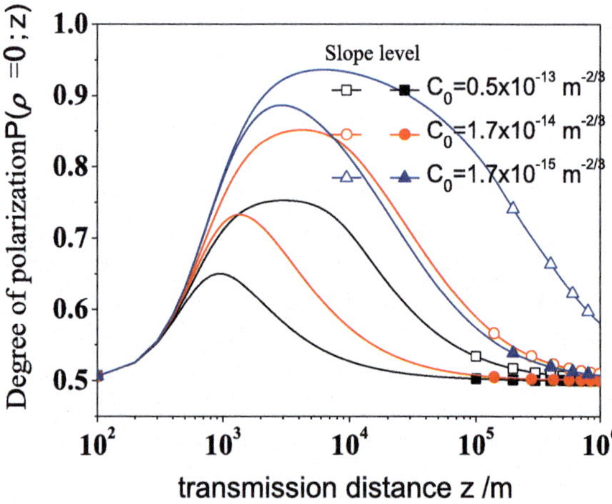

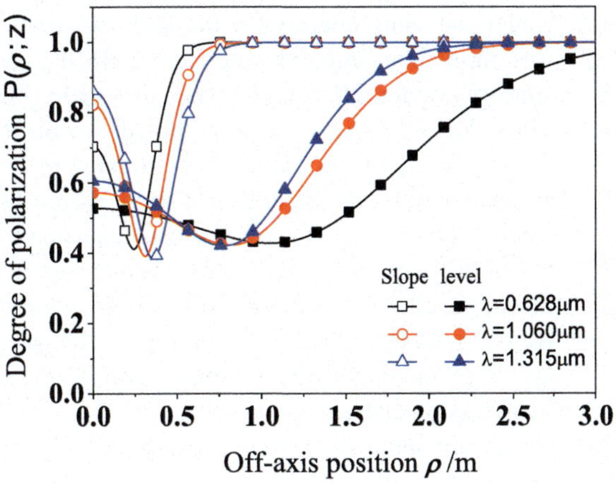

Fig. 6.15 Distribution curves of DoP at different beam wavelengths

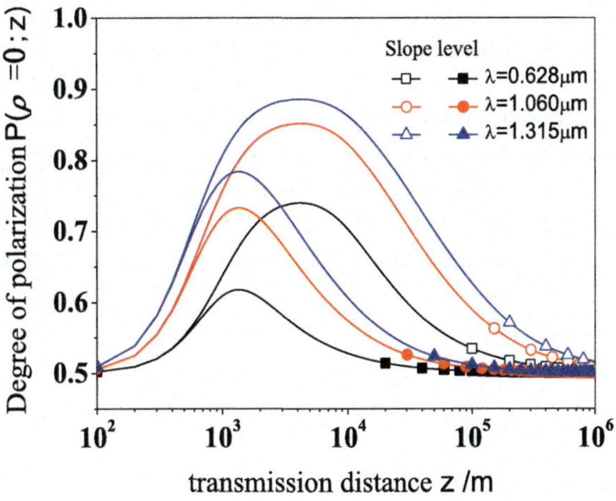

Fig. 6.16 Changes of DoP at axis points with transmission distance at different beam wavelengths

transmission distance, the DoP at the point on the axis changes more with an increase in the beam wavelength.

6.4 Summary

In this section, expressions for the DoP and polarization angle of the EGSM beam for oblique transmission in atmospheric turbulence were derived by theoretically comparing the changes in the DoP and polarization angle for upstream and downstream transmission, in addition to the differences between the oblique and horizontal transmission of the EGSM beam to the receiving end. The main conclusions are as follows:

(1) Under the same conditions, the DoP distribution curve of the receiving end was more concentrated when the EGSM beams propagated downward in atmospheric turbulence than when they propagated downward.
(2) When EGSM beams are transmitted downward in atmospheric turbulence, the distance corresponding to the maximum DoP value at the axis is longer compared to upstream transmittance, which indicates that the detector can receive information at a greater distance when transmitting downward.
(3) In atmospheric turbulence, with an increase in receiver height, the polarization angle changes faster in downward transmission than in upstream transmission. However, regardless of how they change, the polarization angles of the upstream and downstream transmissions to the receiver are the same when the transmission distance is very long.
(4) Under the same light source parameters, the DoP distribution curve at the receiving end of an EGSM beam is more concentrated when it travels obliquely under atmospheric turbulence than when it travels horizontally.
(5) Compared to horizontal transmission, when the EGSM beam travels obliquely in atmospheric turbulence, the distance corresponding to the maximum DoP on the axis is longer, indicating that the detector can receive information at a greater distance for oblique transmission.
(6) The longer the beam wavelength, the smaller the amplitude ratio, and the smaller the structure constant of the atmospheric refractive index, the more concentrated the DoP distribution at the receiving end. When the waist radius of the EGSM beam component is equal, there is an optimal waist radius, which causes the distribution of the DoP at the receiving end to be most concentrated.

References

1. Alvarez-Perez AJ (2011) Coherence, Polarization, and Statistical Independence in Cloude-Pottier's Radar Polarimetry. IEEE Trans Geosci Remote Sens 49(1):426–441
2. Liang S, Wang J, Zong S, Wu R, Magi K, Wang X, Wang L (2013) Laser detection of vessel wake bubbles based on multiple scattering intensity and polarization characteristics. J Phys 62(6):0607041–06070411
3. Li J, Zhu J, Qi C, Zheng C, Gao B, Zhang Y, Hou X (2013) Static Fourier transform hyperspectral full polarization imaging. J Phys 62(4):0442061–0442066
4. Gori F, Santarsiero M, Vicalvi S, Borghi R, Guattari G (1998) Beam coherence-polarization matrix. Pure Appl Opt 7(5):941–951
5. Wolf E (2003) Unified theory of coherence and polarization of random electromagnetic beams. Phys Lett A 312(5):263–267
6. Salem M, Korotkova O, Dogariu A, Wolf E (2004) Polarization changes in partially coherent electromagnetic beams propagating through turbulent atmosphere. Waves Random Media 14(4):513–523
7. Korotkova O, Wolf E (2005) Changes in the state of polarization of a random electromagnetic beam on propagation. Opt Commun 246(1–3):35–43
8. Shirai T, Wolf E (2004) Coherence and polarization of electromagnetic beams modulated by random phase screens and their changes on propagation in free space. J Opt Soc Am A 21(10):1907–1916

9. Shirai T (2005) Polarization properties of a class of electromagnetic Gaussian Schell-model beams which have the same far-zone intensity distribution as a fully coherent laser beam. Opt Commun 256(4):197–209
10. Tao W, Jixiong P, Ziyang C et al (2008) Polarization variation of partially coherent beams when propagating in a turbulent atmosphere. J Overseas Chin Univ 29(2):198–202
11. Zhao X, Yao Y, Sun Y, Liu C (2009) Condition for Gaussian Schell-model beam to maintain the state of polarization on the propagation in free space. Opt Express 17(20):17888–17894
12. Chen K, Cao J, Lu Q (2011) Research progress on correlation between light coherence and polarization. Opt Technol 37(5):607–612
13. Zhang L, Yin X, Zhu Y (2014) Polarization fluctuations of partially coherent Hermite-Gaussian beams in a slant turbulent channel. Optik 125(13):3272–3276
14. Yang A, Lin Q (2014) Polarization characteristics of partially coherent Airy beams propagating in a turbulent atmosphere. J Phys 63(20):2041011–2041017
15. Ming Gao, Lei Gong, Peng-Li Wu. Depolarization characteristics of incompletely polarized and partially coherent laser beams in slant atmospheric turbulence[J]. Optik,2014,125(17): 4860-4863
16. Gao M, Li Y, Lv H, Gong L (2014) Polarization properties of polarized and partially coherent Electromagnetic Gaussian-Schell model pulse beams on slant poth in turbulent atmosphere[J]. Infrared Phys Technol 67:98–106
17. Ou J, Jiang Y, He Y (2015) Intensity and polarization properties of elliptically polarized vortex beams in turbulent atmosphere. Opt Laser Technol 67:1–7
18. Liu J, Chen Z, Zhang L et al (2015) Propagation characteristics and polarization state of angularly polarized non-diffraction beam. J Phys 64(6):0642011–0642018
19. Chen S, Ding P-F, Pu J (2015) Study on coherence in partially coherent radially polarized beam transmission. J Phys 64(13):1342011–13420110
20. Ren H, Liu D (2006) Signal processing for optical communication. Electronic Industry Press, Beijing
21. Walker N (2012) Study on polarization control in coherent optical communication system. Xi'an University of Technology, Xi'an
22. Walker NG, Walker GR (1990) Polarization control for coherent communications. IEEE Lightwave Technology 8:438–458
23. Heismann F (1994) Analysis of a reset-free polarization stabilization in fiber-optic transmission systems. IEEE Lightw Technol 12(4):690–699
24. Hirabayashi K (2003) Feed-forward continuous and complete polarization control with a PLZT rotatable-variable waveplate and inline polarimeter[J]. IEEE Lightwave echnology 21(9):1920–1932
25. Yu J, Lee J, Han M, et al (2003) Ultra-fast acoustically tunable all-fiber polarization controller based on a hollow optical fiber. In: Berthold JF (ed) OFC. Optical Society of America, Georgia, USA, p 573
26. Lyons ER, Lee HP (2002) An electrically tunable all-fiber polarization controller based on deposited thin-film microheaters. IEEE Photon Technol Lett 14(9):1318–1320
27. Terrel M, Digonnet MJ, Fan S (2007) Polarization controller for hollow-core fiber. Opt Lett 32(11):1524–1526
28. Oh Y-II, Kwon M-S, Shin S-Y (2004) In-line polarization controller that uses a hollow optical fiber filled with a liquid crystal. Opt Lett 29(22):2605–2607
29. Li W, Zhang X, Chen K, Zou Y (2006) Application of genetic algorithms in polarization control. J Zhejiang Univ 40(3):443–447
30. Li W (2005) Design and research of transparent electro-optical ceramic polarization controller and its algorithm. Zhejiang University, Zhejiang
31. Wu Z (2010) A high speed polarization control system based on DSP. Zhejiang University, Zhejiang
32. Yao J, Yin Y (2010) Study on Algorithm of delayer cascade polarization controller. J Xi'an Univ Posts Telecommun 15(1):1–4

33. Yanna L, Juan Y, Yafang Y (2010) Error analysis and research of a phase delay polarization controller. Opt Commun Res 3:40–43
34. Zhao N (2008) Research on polarization controller algorithms. Xi'an University of Posts and Telecommunications, Xi'an
35. Zhang X, Duan G, Xia L (2009) The polarization controller completes the minimum degree of freedom for any polarization change. J Opt 29(5):1173–1176
36. Zhang L, Duan G, Zhang X et al (2006) Analysis and research of electronically controlled crystal polarization controller. Photon Technol 3:144–149
37. Wennan W (2008) Research on controlling polarization controller by using field program (GA). Beijing University of Posts and Telecommunications, Beijing
38. Guo Y, Xu H, Tang Q (2009) Research on full-winding algorithm of polarization controller. Opt Instrum 31(5):48–51
39. Yuan X, Sun J, Hu Q, Zhang L (2005) Comfort analysis and implementation of dynamic polarization controller. Opt Photoelectron 3(5):49–52
40. Chen S, Wen PJ, Zhang H et al (2004) Improvement of adaptive extrusion fiber polarization controller. High Tech Commun 2004:1–4
41. Qi L (2006) Study on polarization control in single mode fiber. Beijing Jiaotong University, Beijing
42. Yao Y, Shi K, Lu W, Pang C et al (1995) On-line single mode optical fiber polarization controller. J Opt 5(15):636–640
43. Jin Y, Dou Q, Liu Y et al (2006) Effect of stress on polarization characteristics of single mode optical fibers. Laser Infrared 6(36):498–500
44. Li C, Wang T, Zhang H et al (2014) Effects of light source parameters and atmospheric turbulence on polarization characteristics of electromagnetic beam transmission. J Phys 63(10):1042011–1042018
45. Gao M, Nan L, Lv H et al (2013) Analysis of the effect of turbulence on the polarization characteristics of elliptically polarized laser transmission. J Photon 42(9):1102–1107
46. Wang SCH, Plonus MA (1979) Optical beam propagation for a partially coherent source in the turbulent atmosphere. J Opt Soc Am 69(9):1297–1304
47. Andrews LC, Phillips RL (2005) Laser beam propagation through random media. SPIE Optical Engineering Press, Bellingham, p 195
48. Korotkova O, Wolf E (2004) Generalized Stokes parameters of random electromagnetic beams. Opt Lett 30(2):198–200

Chapter 7
Influence of Atmospheric Turbulence Scale on Partially Coherent Optical Transport Properties

Using the modified von Karman spectral model and considering the inner and outer scales of turbulence, analytical expressions for the average intensity distribution, the beam spread root mean square width, and the drift variance of partially coherent light in atmospheric turbulence can be deduced. The effects of turbulent inner and outer scales on the transmission characteristics of partially coherent light on horizontal and inclined paths in atmospheric turbulence are compared and analyzed for different atmospheric turbulence intensities.

7.1 Introduction

When a beam travels in atmospheric turbulence, it experiences turbulence effects such as a reduction in the peak light intensity, beam expansion, and drift, to varying degrees. Research on the transmission characteristics of partially coherent beams in atmospheric turbulence can be traced back to the late 1970s and the early 1980s. Leader [1, 2] studied the scattering of laser irradiation on rough surfaces and discovered that the spatial coherence of the incident light changed. Subsequently, an increasing number of scholars began investigating the propagation of partially coherent Gaussian beams through atmospheric turbulence [3–6].

In 1983, Baykal et al. [7] of Northwestern University studied the scintillation of partially coherent light, considering the inner and outer scales under weak turbulence, and discovered that when the inner scale was 1 m, the scintillation index also increased as the normalized source parameter increased. In 1985, Baykal et al. [8] of Northwestern University studied the intensity distribution of partially coherent light when considering the inner and outer scales, and discovered that when the turbulent inner scale was considered, the worse the coherence, the more concentrated the light intensity distribution. In 2004, Korotkova et al. [9] from the University of Florida investigated the transmission of partially coherent Gaussian beams in atmospheric

X. Ke, *Partially Coherent Optical Transmission Theory in Optical Wireless Communication*, Optical Wireless Communication Theory and Technology,
https://doi.org/10.1007/978-981-97-7158-5_7

turbulence and discovered that when the internal scale was 1 cm, the scintillation index first increased and then decreased with increasing transmission distance. In October 2004, Zhang et al. [10] derived a simplified dual-scale model for refractive index fluctuations, that included turbulent inner and outer scale factors, and a scintillation aperture smoothing factor was introduced for plane waves propagating in a turbulent atmosphere. The results revealed that the inner and outer scales of finite turbulence are the aperture area, which increases the smoothing effect of scintillation. In March 2005, Zhang et al. [11] studied the atmospheric communication problem using a narrow-band Gaussian shell beam as the light source in a weak atmospheric turbulence environment. They the cross spectral density function of the narrow bandwidth light field instead of the mutual interference function of the light field and the simplified refractive index spectral density function, including the inner and outer scales of atmospheric turbulence to obtain the analytical approximation relationship between the wave structure function (WSF) and the variance of the angle of arrival fluctuation of a Gaussian shell beam propagating in a turbulent atmosphere. In August 2008, Wei et al. [12] deduced the drift variance considering the outer scale according to the Huygens-Fresnel principle and achieved oblique transmission to the same height; the larger the outer scale of the turbulence, the larger the drift variance of the beam. In March 2014, Wei et al. [13] investigated the influence of non-atmospheric turbulence on the transmission factor of partially coherent Hermite-Gaussian beams and discovered that the smaller the outer turbulent scale, the larger the inner turbulent scale and beam order, and the greater the effect of non-atmospheric turbulence on beam transmission. As the turbulence is reduced, the beam quality improves. In 2014, Huang [14] determined the turbulence of partially coherent flat-top beams (PCFT) in non-atmospheric turbulence, considering the inner and outer scales based on the generalized Huygens-Fresnel principle, and the turbulent distance was defined using a relative M^2 factor. It is found that the turbulence distance of PCFT beams in non-k atmospheric turbulence increases with the increase of the inner scale of turbulence and decreases with the increase of the outer scale. In 2014, Li et al. [15] deduced the scintillation variance under the Hill spectrum considering the internal and external scales based on the modified Rotov theory and determined that if the wavelengths are equal to 1.06 and 1.315 μm, in the moderately turbulent region, the effect of the internal scale on the scintillation index is greater than that of the external scale. When the wavelengths are 3.8 and 10.6 μm, the influence of the inner- and outer-scale effects on the scintillation index is greater than the inner-scale effect in the turbulent region.

In this chapter, the modified von Karman spectral model considering the inner and outer scales is mainly used to deduce analytical expressions for the average light intensity distribution, the beam spread root mean square width, and the drift variance of partially coherent light in atmospheric turbulence. Moreover, the effects of turbulent inner and outer scales on the transmission characteristics of partially coherent light in atmospheric turbulence on the horizontal and oblique paths are investigated for different turbulence intensities.

7.2 Theoretical Calculation

When a partially coherent GSM beam travels along a slant distance z under atmospheric turbulence, the intensity distribution is expressed as follows [16, 17]:

$$I(\rho, z) = W(\rho, \rho, z) = \frac{A^2}{\Delta^2(z)} \exp\left[-\frac{|\rho|^2}{2\sigma_s^2 \Delta^2(z)}\right] \tag{7.1}$$

in the equation, $\Delta^2(z) = 1 + \left[\frac{z}{k\sigma_s\delta}\right]^2 + \frac{2Mz^2}{k^2\sigma_s^2}$, $\frac{1}{\delta^2} = \frac{1}{4\sigma_s^2} + \frac{1}{\sigma_g^2}$.

The parameters A, σ_s, and σ_g represent the light intensity, beam waist width, and coherence length of the light source, respectively. The wavenumber $k = 2\pi/\lambda$。 $M = \frac{1}{2}B_0B_3$, $B_0 = 2\pi^2k^2 \sec\theta \int\limits_0^\infty \phi_n(\kappa)\mathrm{d}\kappa$, $B_3 = \int\limits_{h_0}^{H} C_n^2(h)\xi^2\mathrm{d}h$.

In addition, $\xi = 1 - \frac{h-h_0}{H-h_0}$, H represents the vertical height of the receiver from the ground, and h_0 is the vertical height of the transmitter from the ground. $C_n^2(h)$ is the atmospheric refractive index structure model based on ITU-R in 2001 [18], and can be expressed as,

$$C_n^2(h) = 8.148 \times 10^{-56} v_{RMS}^2 h^{10} e^{-h/1000} + 2.7 \times 10^{-16} e^{-h/1500} + C_0 e^{-h/100} \tag{7.2}$$

where $v_{RMS}^2 = \sqrt{v_g^2 + 30.69v_g + 348.91}$ is the vertical path wind speed, v_g is the near-surface wind speed, and C_0 is the near-surface atmospheric refractive index structure constant.

When the atmospheric refractive index structure constant $C_n^2(h)$ is fixed, the transmission is horizontal, and $M = \frac{1}{3}\pi^2k^2z\int_0^\infty \kappa^3\phi_n(\kappa)\mathrm{d}\kappa$. For free space propagation, $M = 0$. In the turbulent power spectrum model for the turbulence term M, the modified von Karman spectrum [19] that considers both the inner and outer scales is used in this study and is expressed as follows:

$$\phi_n(\kappa, h) = 0.033C_n^2(h)\exp(-\kappa^2/\kappa_m^2) \times (\kappa^2 + \kappa_0^2)^{-11/6} \tag{7.3}$$

where $\kappa_m = 5.92/l_0$, $\kappa_0 \approx 2\pi/L_0$, and l_0 and L_0 represent the inner and outer scales of turbulence, respectively.

To analyze the intensity distribution of partially coherent beams when considering the inner and outer scales of atmospheric turbulence, the light intensity is normalized, and the normalized light intensity of the beam is defined as [20],

$$I^N(\rho, z) = \frac{I_{ts}(\rho, z)}{I_{fs}(\rho, z)\big|p = 0, \sigma_{g\to\infty, M=0}} \tag{7.4}$$

where $I_{ts}(\boldsymbol{\rho}, z)$ is the intensity distribution of the beam when it propagates in atmospheric turbulence and $I_{fs}(\boldsymbol{\rho}, z)|_{\rho=0,\,\sigma_g\to\infty, M=0}$ is the on-axis light intensity value of the perfectly coherent beam when it travels obliquely in free space.

The normalized light intensity of the beam can be obtained as:

$$I^N(\rho, z) = \frac{1 + \left(z/2k\sigma_s^2\right)^2}{\Delta^2(z)} \exp\left[-\frac{\rho^2}{2\sigma_s^2\Delta^2(z)}\right] \tag{7.5}$$

According to the definition of the beam expansion root mean square width, we have [21]:

$$\omega(z) = \sqrt{\frac{\int I(\rho, z)|\rho|^2 d^2\rho}{\int I(\rho, z) d^2\rho}} \tag{7.6}$$

Substitute Eq. (7.1) into Eq. (7.6), we obtain:

$$\omega(z) = \left[2\sigma_s^2\Delta^2(z)\right]^{\frac{1}{2}} = \left[2\sigma_s^2 + \frac{2}{k^2\delta^2}z^2 + \frac{4M}{k^2}z^2\right]^{\frac{1}{2}} \tag{7.7}$$

The first two terms of the preceding equation represent the etendue caused by free-space diffraction and the third term represents the etendue caused by turbulence [2].

To intuitively analyze the influence of turbulence on beam expansion, the relative beam width is defined as the ratio of the beam width in turbulence to the free-space beam width. Thus, we obtain [22],

$$\omega_r(z) = \left[1 + \frac{\frac{4M}{k^2}z^2}{2\sigma_s^2 + \frac{2}{k^2\delta^2}z^2}\right]^{\frac{1}{2}} \tag{7.8}$$

Andrews and Philips used geometric optics and Rytov approximations to introduce a Gaussian filter function to obtain the drift model of a light spot under weak turbulence [16]:

$$\left\langle\rho^2 c\right\rangle = 4\pi^2k^2w^2(L)\int_0^L\int_0^\infty \kappa\phi n(\kappa, z)H_{LS}(\kappa, z) \times \left[1 - e^{-\Lambda PL\kappa^2\xi^2/k}\right]d\kappa dz \tag{7.9}$$

where $w(\mathrm{L})$ is the beam radius at the receiving end in the free space. $H_{LS}(\kappa, z)$ is a low-pass filter that acts on inhomogeneous turbulent vortices equal to or greater than the beam size:

$$H_{LS}(\kappa, z) = \exp\left[-\kappa^2 w^2(z)\right] \tag{7.9a}$$

where $w(z)$ is the beam radius at the receiving end in free space. Λ_P is a low-pass filter that acts on inhomogeneous turbulent vortices equal to or greater than the beam size:

$$\Lambda_P = \frac{2L}{kw^2(L)} \tag{7.9b}$$

By performing a geometric optics approximation of the last term of Eq. (7.9), we get:

$$1 - e^{-\Lambda_P L\kappa^2\xi^2/k} \cong \frac{\Lambda_P L\kappa^2\xi^2}{k}, \quad \frac{L\kappa^2}{k} \ll 1 \tag{7.9c}$$

Here, $\phi n(\kappa, z)$ is based on the von Karman spectrum that considers the internal and external scales, and substituting (7.9a)–(7.9c) and (7.3) into (7.9), we have:

$$\langle \rho_c^2 \rangle = 8\pi^2 \times 0.033L^2C_n^2\kappa_0^{1/3}\exp[-k^2w^2(z)]\int_0^L (1-\frac{z}{L})^2U(2;\frac{7}{6};\kappa_0^2/\kappa_m^2)dz \tag{7.10}$$

In this chapter, Eqs. (7.5), (7.8), and (7.10) are combined to analyze the normalized light intensity distribution, the relative beam width, and the drift variance of a GSM beam in atmospheric turbulence for both oblique and horizontal transmission paths, and detailed analysis and discussion are presented.

7.3 Light Intensity Distribution for GSM Beam Transmission in Turbulent Atmosphere

Figure 7.1 shows the influence of the intensity distribution of partially coherent light on the transmission distance for horizontal and oblique paths when the atmospheric turbulence has different inner and outer scales. The parameters are; zenith angle $\theta = \frac{\pi}{4}$, transmitter height $h_0 = 2$ m, near-surface wind speed $v_g = 2.8$ m/s during oblique transmission, atmospheric refractive index structure constant $C_n^2(2) = 1.7 \times 10^{-14}\,\text{m}^{-2/3}$ during horizontal transmission, coherence length $\sigma_g = 0.02$ m, beam waist width $\sigma_s = 0.05$ m, and wavelength $\lambda = 650$ nm. In (a), the values of the inner scales l_0 in the figure are 0.001, 0.01, and 0.1 m, the outer scale L_0 is 10 m, and the near-surface atmospheric refractive index structure constant C_0 is $1.7\times10^{-14}\,\text{m}^{-2/3}$. In (b), the values of the inner scale l_0 in the figure are 0.001, 0.01, and 0.1 m, respectively, the outer scale L_0 is 100 m, and the near-surface atmospheric refractive index structure constant C_0 is $1.7 \times 10^{-14}\,\text{m}^{-2/3}$. For (c), the inner scale l_0 in the figure has values of 0.001, 0.01, and 0.1 m, the outer scale L_0 is 1000 m, and the near-surface atmospheric refractive index structure constant C_0 is $1.7 \times 10^{-14}\,\text{m}^{-2/3}$. In (d), the inner scale

l_0 has a value of 0.001 m, the outer scale L_0 has values of 10, 100, and 1000 m, respectively, and the near-surface atmospheric refractive index structure constant C_0 is $1.7 \times 10^{-13}\ \mathrm{m}^{-2/3}$, $1.7 \times 10^{-14}\ \mathrm{m}^{-2/3}$, and $1.7 \times 10^{-15}\ \mathrm{m}^{-2/3}$, respectively.

As shown in Fig. 7.1, for both horizontal and oblique transmission, as the transmission distance increases, the light intensity distribution gradually disperses. When the transmission distance is less than 1500 m, with an increase in the inner-scale l_0, the light intensity distribution is more concentrated; when the transmission distance is > 1500 m, with the increase of the inner scale l_0, the curve change is small. Compared to horizontal transmission, the light intensity distribution is more concentrated, and the light intensity peak is greater for oblique transmission, which indicates that partially coherent optical transmission is better for suppressing turbulence than horizontal transmission. Comparing the three graphs (a), (b), and (c) in Fig. 7.1, when the outer-scale L_0 increases, the change of the curve is minimal. The main reason for this result is that the smaller the inner scale of the turbulence, the smaller the turbulent

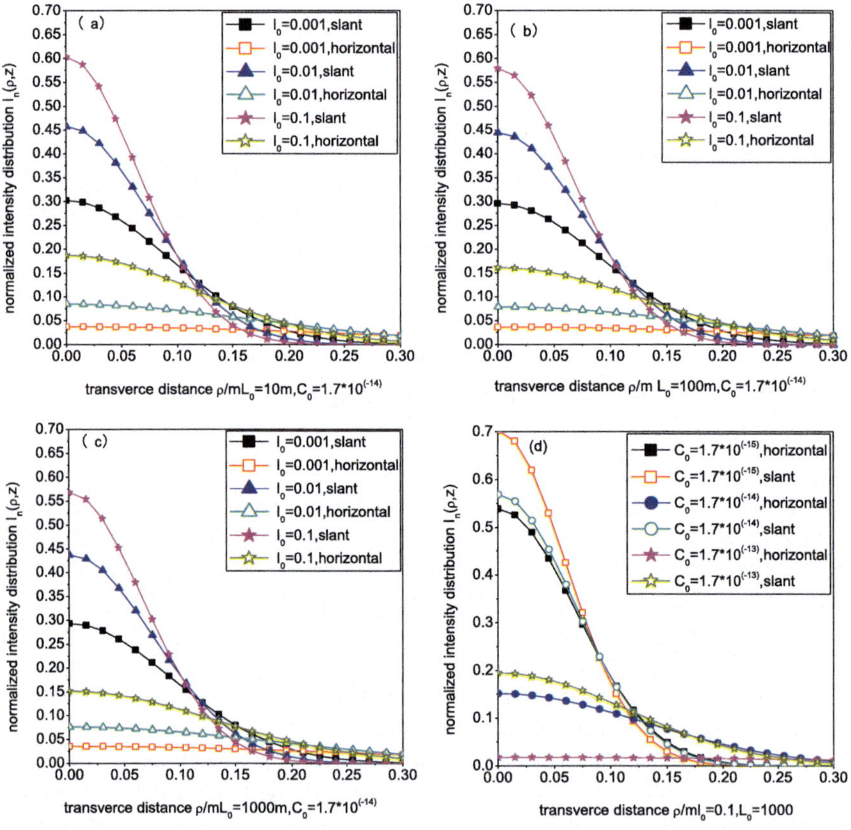

Fig. 7.1 Internal and external scales of atmospheric turbulence **a** $L0 = 10$ m, **b** $L0 = 100$ m, **c** $L0 = 1000$ m and strength **d** $cn^2 = 1.7 \times 10^{-13}$, $cn^2 = 1.7 \times 10^{-14}$, and $cn^2 = 1.7 \times 10^{-15}$ influence on the intensity distribution of partially coherent light

eddies contained in the beam cross-section, and the more significant the diffraction of the beam that irradiates these eddies, which causes the random distribution of the light beam intensity in space and time more serious, The random distribution is more significant, which causes the light intensity to become more dispersed, similar to the results in

Comparing (c) and (d) in Fig. 7.1, it is evident that when the inner scale is $l_0 = 0.1$ m and the outer scale is $L_0 = 1000$ m, under strong turbulence, the normalization of transmission along the horizontal path increases with an increase in the transmission distance. The normalized light intensity does not change significantly, but the light intensity distribution is very dispersed; when transmitting along the inclined path, the normalized light intensity decreases with an increase in the transmission distance. Under moderate turbulence, the normalized light intensity transmitted along the horizontal path decreases with the increase of the transmission distance. When the transmission distance is > 1500 m, it coincides with the curve for oblique transmission under strong turbulence. When the transmission distance is < 1000 m, the normalized light intensity is higher compared to the case for transmission along a horizontal path in weak turbulence, and when the transmission distance is > 1000 m, the two curves coincide.

Under weak turbulence, the normalized light intensity is significantly larger than that in moderately strong turbulent flow, and the normalized light intensity during transmission along an inclined path decreases rapidly when the transmission distance is > 1500 m. This is because when the beam travels in the atmosphere, the random fluctuation of the refractive index causes phase distortion, which causes the beam to expand beyond a certain distance, resulting in a more dispersed light intensity distribution and a weaker peak light intensity. The larger the random fluctuation, the more dispersed the light intensity.

7.4 Beam Spreading of GSM Beams Propagating in a Turbulent Atmosphere

Figure 7.2 shows the effect of beam spread of partially coherent light as a function of the transmission distance for horizontal and oblique paths when the inner and outer scales of atmospheric turbulence are different. The main parameters are: zenith angle $\theta = \frac{\pi}{4}$, transmitter height $h_0 = 2$ m, near-surface wind speed $v_g = 2.8$ m/s during oblique transmission, atmospheric refractive index structure constant $C_n^2(2) = 1.7 \times 10^{-14}\ \mathrm{m}^{-2/3}$ during horizontal transmission, coherence length $\sigma_g = 0.02$ m; beam waist width $\sigma_s = 0.05$ m, and wavelength $\lambda = 650$ nm. (a) In the figure, the inner scale values of l_0 are 0.001, 0.01 and 0.1 m respectively, the outer scale L_0 is 10 m, and the near-surface atmospheric refractive index structure constant C_0 is $1.7 \times 10^{-14}\ \mathrm{m}^{-2/3}$, (b) In the figure, the inner scale values of l_0 are 0.001, 0.01 and 0.1 m, respectively, the outer scale L_0 is 100 m, and the near-surface atmospheric refractive index structure constant C_0 is $1.7 \times 10^{-14}\ \mathrm{m}^{-2/3}$. (c) In the figure, the values of the inner scale of l_0 are 0.001, 0.01 and 0.1 m respectively, the outer scale L_0 is 1000 m, and the near-surface atmospheric refractive index structure constant C_0 is 1.7×10^{-14} $m^{-2/3}$. (d) In the figure, the inner scale l_0 is 0.001 m, the values of the outer scales of L_0 are 10, 100, and 1000 m, respectively, and the near-surface atmospheric refractive index structure constants C_0 are 1.7×10^{-13} $m^{-2/3}$, 1.7×10^{-14} $m^{-2/3}$, and 1.7×10^{-15} $m^{-2/3}$, respectively.

From Fig. 7.2, it is evident that the beam expansion becomes increasingly significant as the transmission distance increases, regardless of horizontal or oblique transmission. When the transmission distance is < 1000 m, with the change of the internal scale l_0, the curve changes very slowly, and the beam expansion phenomenon is not obvious. However, when the transmission distance is > 10000 m, the curve change trend is larger, and as the internal scale l_0 decreases, the beam expands. When considering the inner and outer scales of turbulence, the beam expansion of the oblique transmission is smaller than that of horizontal transmission. This is because the atmospheric structure constant during oblique transmission is smaller than that during horizontal transmission close to the ground. Thus, turbulence is less

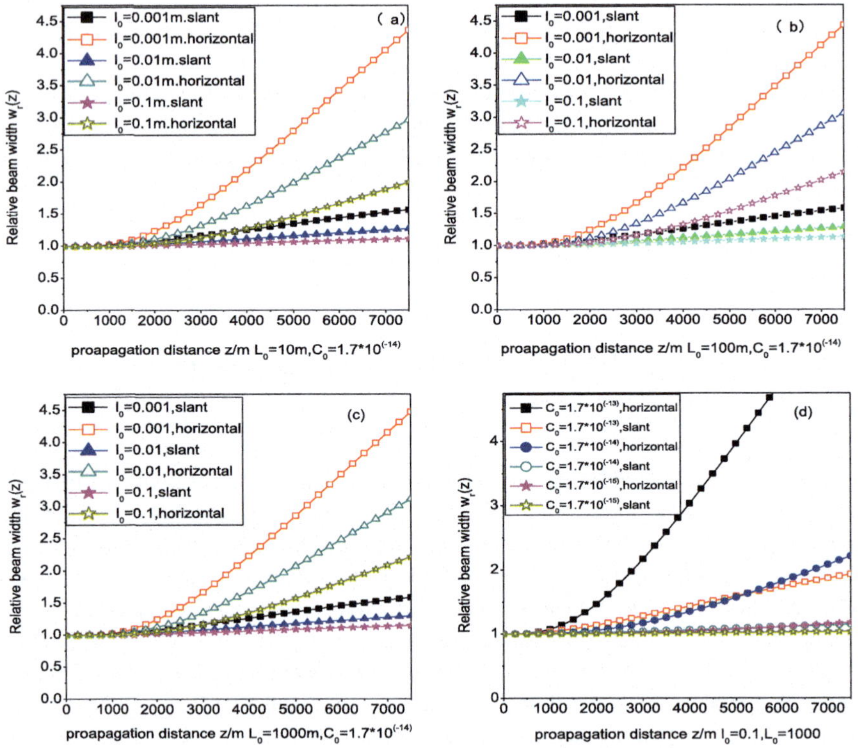

Fig. 7.2 Influence of the inner and outer scales of atmospheric turbulence on the beam expansion of partially coherent light **a** $L_0 = 10$ m, **b** $L_0 = 100$ m, **c** $L_0 = 1000$ m and strength **d** $cn^2 = 1.7\times10^{-13}$, $cn^2 = 1.7 \times 10^{-14}$, $cn^2 = 1.7 \times 10^{-15}$

affected. Comparing graphs (a), (b), and (c) in Fig. 7.2, when the turbulent outer scale L_0 increases, the curves of the three graphs are very close, which indicates that the change in the outer scale has little effect on the beam expansion.

Comparing (c) and (d) in Fig. 7.2, when the inner scale is $l_0 = 0.1$ m and the outer scale is $L_0 = 1000$ m under strong turbulence, with the increase of the transmission distance, the beam spread is significant for transmission along the horizontal path. Under moderate turbulence, the curve for travel along the horizontal path is very close to the curve for propagation along the inclined path under strong turbulence. At a transmission distance of $L = 5000$ m, the two curves cross, and when the transmission distance is > 5000 m, for strong turbulent flow, The beam spread along the horizontal path in strong turbulence is smaller than that in medium turbulence, and the curve for propagation along the inclined path almost coincides with the curve for weak turbulent flow. Figure 7.2 shows that in the case of weak turbulence, irrespective of whether the transmission is along a horizontal or inclined path, the curve does not change significantly, and the beam expansion phenomenon is not obvious. As the turbulence intensity increases, the beam expansion phenomenon becomes more significant. This is because the beam is affected by the refractive index of the atmosphere, and the resulting wavefront phase distortion becomes more severe with an increase in the transmission distance. Thus, the beam is more affected by turbulence. When turbulence is weak, the beam expansion of partially coherent light in atmospheric turbulence is independent of the transmission path.

7.5 Beam Drift of GSM Beams Propagating in a Turbulent Atmosphere

Figure 7.3 shows the effect of beam drift of partially coherent light with the transmission distance for horizontal and oblique paths when the inner and outer scales of the atmospheric turbulence are different. The parameters for oblique transmission are: zenith angle $\theta = \frac{\pi}{4}$, transmitter height $h_0 = 2$ m, and near-surface wind speed $v_g = 2.8$ m/s. During horizontal transmission, the parameters are: atmospheric refractive index structure constant $C_n^2(2) = 1.7 \times 10^{-14}\,\text{m}^{-2/3}$, coherence length $\sigma_g = 0.02$ m, beam waist width $\sigma_s = 0.05$ m, and wavelength $\lambda = 650$ nm. (a) In the figure, the inner scale l_0 is 0.001 m, the outer scale values of L_0 are 10, 100, and 1000 m, and the near-surface atmospheric refractive index structure constant C_0 is $1.7 \times 10^{-14}\,\text{m}^{-2/3}$. (b) In the figure, the inner scale l_0 is 0.01 m, the outer scale values of L_0 are 10, 100, and 1000 m, and the near-surface atmospheric refractive index structure constant C_0 is $1.7 \times 10^{-14}\,\text{m}^{-2/3}$. (c) In the figure, the inner scale l_0 is 0.1 m, the outer scale values of L_0 are 10, 100, and 1000 m, and the near-surface atmospheric refractive index structure constant C_0 is $1.7 \times 10^{-14}\,\text{m}^{-2/3}$. (d) In the figure, the inner scale l_0 is 0.1 m, the outer scale L_0 is 1000 m, and the values of the near-surface atmospheric refractive index structure constant C_0 are $1.7 \times 10^{-13}\,\text{m}^{-2/3}$, $1.7 \times 10^{-14}\,\text{m}^{-2/3}$, and $1.7 \times 10^{-15}\,\text{m}^{-2/3}$, respectively.

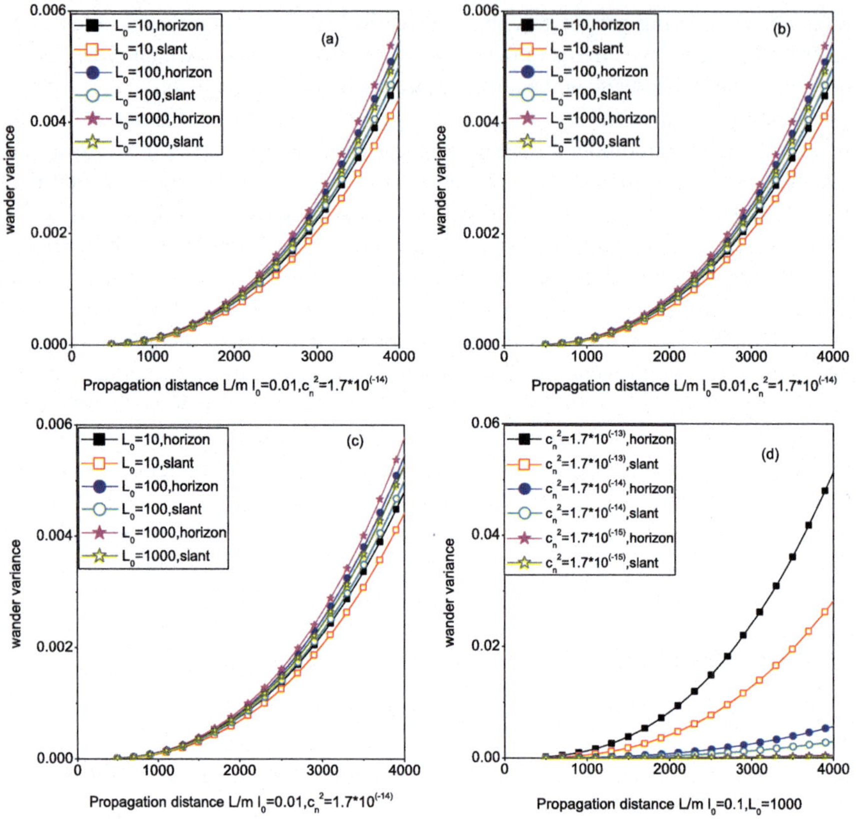

Fig. 7.3 Influence of the inner and outer scales of atmospheric turbulence on the beam drift of partially coherent light **a** $l_0 = 0.001\,\text{m}$, **b** $l_0 = 0.01\,\text{m}$, **c** $l_0 = 0.1\,\text{m}$ and strength **d** $cn^2 = 1.7 \times 10^{-13}$, $cn^2 = 1.7 \times 10^{-14}$, $cn^2 = 1.7 \times 10^{-15}$

From Fig. 7.3, irrespective of whether the transmission is horizontal or oblique, the beam drift becomes increasingly significant with an increase in the transmission distance. When the transmission distance is within 500 m, the beam drifts. When the transmission distance is greater than 500 m, the curve change trend is large, with the outer-scale $L0$ increases, the drift becomes increasingly significant. When the inner and outer scales are constant, the drift caused by the horizontal transmission is more severe than that caused by oblique transmission. This is because the atmospheric structure constant of oblique transmission is smaller than that of horizontal transmission close to the ground, and turbulence has a reduced effect. Comparing the three graphs in Fig. 7.3a–c, when the internal scale l_0 changes, the curves of the three graphs are very similar, which shows that the change in the internal scale does not affect the beam drift. This is similar to previously reported results in the literature [12].

Comparing graphs (c) and (d) of Fig. 7.3, when the inner scale is $l_0 = 0.1\,\text{m}$ and the outer scale is $L_0 = 1000\,\text{m}$, the drift phenomenon is significant for strong turbulence. However, when the transmission distance is < 1000 m, the curve hardly changes. When the transmission distance is > 1000 m, the beam drift increases rapidly, irrespective of whether the transmission is along the horizontal or inclined path. Under moderate turbulence, when the transmission distance is < 3000 m, the change in the curve is minor, and when the transmission distance is > 3000 m, the beam drift increases slowly, irrespective of the transmission path. Under weak turbulence, as the transmission distance increases, the change of the curve is minor, and the transmission curve along the horizontal path almost coincides with that of the inclined path. Moreover, as the turbulence intensity increases, the beam drift phenomenon increases rapidly. This is because the beam is affected by the refractive index of the atmosphere, and the resulting wavefront phase distortion becomes significant with an increase in the transmission distance, such that the beam is affected by turbulent flow. Under strong and moderate turbulence, the drift during oblique transmission is significantly smaller than that associated with horizontal transmission. When the turbulence is weak, the beam drift of partially coherent light in atmospheric turbulence is almost independent of the transmission path. Comparing Figs. 7.1 and 7.3, it is evident that the beam expansion is two orders of magnitude larger than the drift, which indicates that the beam expansion effect in a turbulent atmosphere is more significant than the drift effect.

7.6 Summary

This chapter is based on the modified von Karman spectral model and the atmospheric structural constant model promulgated by ITU-R considering the altitude. The normalized intensity distribution, relative width, and beam drift are numerically calculated and analyzed. The results show that:

(1) Compared to horizontal transmission, oblique transmission is better at suppressing turbulence;
(2) The light intensity distribution becomes more dispersed with the increase in the turbulent intensity, and the inner scale decreases, which leads to an increase in the beam width.
(3) The relative beam width increases with an increase in the turbulence intensity and increases with a decrease in the inner scale, which causes the beam to be more affected by turbulence.
(4) The beam drift increases with an increase in the turbulence intensity and outer scale.
(5) The outer scale has little effect on the light intensity distribution and beam expansion, and the inner scale has little effect on the beam drift.

References

1. Leader JC (1978) Atmospheric propagation of partially irradiation. Opt Soc Am 68:175–185
2. Ouyang CF (1979) Irradiance scintillation of a partially coherent source in extremely strongly turbulence. Appl Opt 18:1133–1135
3. Wei HY, Wu ZS (2008) Study on the effect of laser beam propagation on the slant path through atmospheric turbulence. J Electromagn Waves Appl 22(5–6):787–802
4. Wei L, Liu LR, Sun JF et al (2007) Change in degree of coherence of partially coherent electromagnetic beams propagating through atmospheric turbulence. Opt Commun 271:1–8
5. Dan YQ, Zhang B (2008) Beam propagation factor of partially coherent flat-topped beams in a turbulent atmosphere. Opt Express 16(20):15563
6. Ji X, Huang T, Lv B (2006) Spreading of partially coherent cosh-Gaussian beams propagating through turbulent atmosphere. Chin Phys B 55:978–982
7. Baykal Y, Plonus MA, Wang SJ (2012) The scintillations for weak atmospheric turbulence using a partially coherent source. Radio Sci 18(4):551–556
8. Baykal Y, Plonus MA, Intensity fluctuations due to a spatially partially coherent source in atmospheric turbulence as predicted by Rytov's method. Optical Society of America
9. Korotkova O, Andrews LC, Phillips RL (2004) Model for a partially coherent Gaussian beam in atmospheric turbulence with application in Lasercom. Opt Eng 43(2):330–341
10. Zhang Y, Zhu T, Tao C (2004) Two-scales model of the refractive-index fluctuations for laser propagation in an atmosphere turbulence. J Optoeletron Laser 15(10)
11. Zhang Y, Tao C (2005) Arrival angle fluctuations of Gaussian Schell beams propagating in turbulent atmosphere. Chin J Photon 34(3)
12. Wei H, Wu Z (2008) Spreading and drift characteristics of laser beam slope transmission in atmospheric turbulence. J Radio Wave Sci
13. Wang W, Li J, Lai Y, Wei J (2014) Influence of non-Kolmogorov atmospheric turbulence on the transmission factor of partially coherent Hermitian Gaussian beams. Intense Laser Particle Beam 26(3):31010
14. Huang Y, Duan Z, He Y, Zhang B (2014) Variation of turbulent distance of partially coherent flat-topped beams in non-Kolmogorov atmospheric turbulence. China J Laser 41:0913002
15. Li Y, Wu Z (2014) Scintillation of partially coherent Gaussian-Schell model beam propagation in slant atmospheric turbulence considering inner- and outer-scale effects. Chin Phys B 23(7):074202
16. Li CQ, Zhang HY, Wang TF, Liu LS, Guo J (2013) Investigation on coherence characteristics of Gauss-Schell model beam propagating in atmospheric turbulence. Acta Phys Sinica 62(22):224203
17. Andrews LC, Phillips RL (2005) Laser beam propagation through random media. SPIE optical Engineering Press, Bellingham, p 195
18. Wu J, Yang C, Liu J (2005) Optical transport theory in atmosphere. Beijing University of Posts and Telecommunications Press, Beijing, pp 142–147
19. Wei H (2006) Research on the propagation characteristics of laser beams in sloping turbulent atmosphere. Xidian University, Xi'an, pp 5–12, 17–22
20. Tomohiro S, Aristide D, Emil W (2003) Mode analysis of spreading of partially coherent beams propagating through atmospheric turbulence. J Opt Soc Am A-Opt Image Sci Vis 20(6):1094–1102
21. Yang AL, Li JH, Lü BD (2009) A comparative study of the beam-width spreading and angular spreading in atmosphere turbulence. Acta Phys Sin 58(4):2451–2460
22. Ji X, Chen X, Lu B (2008) Spreading and directionality of partially coherent Hermite-Gaussian beams propagating through atmospheric turbulence. J Opt Soc Am A 25(1):21–28

Chapter 8
The Broadening of Pulsed Beams Traveling Horizontally in Atmospheric Turbulence

Based on the Huygens-Fresnel principle and the modified von Karman spectral model, a detailed numerical analysis of the pulse broadening of partially coherent Gaussian-Schell Model pulse (GSMP) beams propagating horizontally in atmospheric turbulence is conducted. An expression for the average light intensity of the GSMP beam in the frequency and time domains in atmospheric turbulence is derived, and an expression describing the pulse broadening associated with horizontal transmission in a turbulent atmosphere is derived. Moreover, an expression for the structure constant (D_{12}) in isotropic turbulent flow is obtained using the Markov approximation, and the pulse-broadening characteristics of partially coherent GSMP beams are discussed for different turbulent intensities.

8.1 Introduction

In recent years, the problem of pulse-signal transmission through turbulent atmospheric conditions has gradually attracted attention [1]. Given that a high bandwidth facilitates high data rates in laser communication, interest has gradually focused on the transmission of ultrashort pulsed beams in random media [2–5]. Hong et al. discussed the transmission of Gaussian pulses in a turbulent atmosphere using the dual-frequency mutual coherence function (MCF), which can be applied to any turbulent region [6, 7]. Mintzer [8] investigated the statistical properties of acoustic pulse signal transmission in ocean turbulence, and Su and Plous [9] subsequently utilized the Rytov approximation to study the transmission of optical pulses through a turbulent atmosphere and The mean square amplitude and phase fluctuation of the short pulse signal are derived when the refractive index changes obviously. In 1974, Liu [10] used a parabolic approximation to obtain the mutual intensity functions of symmetric and antisymmetric matrices of light waves propagating through random

X. Ke, *Partially Coherent Optical Transmission Theory in Optical Wireless Communication*, Optical Wireless Communication Theory and Technology,
https://doi.org/10.1007/978-981-97-7158-5_8

media and discussed the time-domain broadening properties using Markov approximation assuming forward scattering (parabolic approximation). In 1977, Hong [11] discussed the problem of plane pulse waves propagating through a random medium based on the assumption of forward-scattering and the relationship between dual-frequency mutual coherence function (MCF) and different parameters (operating frequency, transmission distance, turbulence intensity and scatterer density, etc.) is given. In 1979, Sreenivasiah [12] obtained a general analytical solution to the parabolic equation for an incident wave using the Rytov approximation method, and used the result to study the transmission characteristics of the cross-correlation function for single-frequency and dual-frequency beams; Yeh [13] using the time-moment and the perturbation methods of the two-frequency mutual interference function, the mean square pulse width of electromagnetic waves passing through random and dispersive media has been investigated. In 1980, Liu [14] calculated the average intensity of a pulse signal propagating through a random medium based on the concept of time-moment. This simplified the solution process of the dual-frequency mutual coherence function, which is not only applicable to a pulse plane beam, but can also be extended to other complex beams. In 1981, Fante [15] used the Huygens-Fresnel principle to obtain the two-point dual-frequency cross-correlation function (mutual coherence function, MCF) of light beams in turbulent media and investigated this function for different source coherence and turbulent conditions. In 1996, Young [16] used the ABCD matrix to deduce the dual-frequency mutual coherence function (mutual coherence function, MCF) of a weakly turbulent Gaussian pulsed beam, derived an arbitrary power spectral density function for a weakly turbulent medium, and also defined the limit case of dual-frequency mutual coherence function (MCF) of plane and spherical waves. In 1998, Young [17] obtained an analytical expression for the long-term time expansion of a collimated space-time Gaussian pulse propagating along a horizontal path in weak turbulence and presented general results for the standard parameter that characterizes atmospheric laser communication. In 2005, Lajunen et al. [18] derived the coherent mode representation of partially spatial and partially spectral coherent pulses, and analyzed the corresponding coherent Gaussian-Schell model pulsed (GSMP) beams. In 2008, Chaoliang et al. [19] investigated the temporal characteristics of Gauss-Schell model pulsed beams propagating in free space, and the results proved that partial spatial and partial spectral coherence Gauss-Schell model pulsed beams occurred in free-space propagation. The spectral shift was found to be related to the position of the field point, the spatial correlation length, and the temporal coherence length. In 2012, Chen et al. [20] derived an analytical expression for the time half-width of spherical wave Gaussian (SWG) pulses propagating in non-Kolmogorov turbulence. In 2014, Li Yaqing et al. [17] studied the transmission of partially coherent GSMP beams in a sloping turbulent atmosphere and obtained expressions for the average intensity, beam radius, and dual-frequency complex coherence.

In this chapter, using the Gauss-Schell model for pulsed beams as an example and based on the Huygens-Fresnel principle and the modified von Karman turbulence spectrum model, a detailed numerical analysis of the propagation of partially coherent GSMP beams in atmospheric turbulence is presented. The problem of pulse broadening after transmission and the comparative analysis of pulse broadening in the frequency and time domains provide a theoretical basis for the second-order statistical properties of GSMP beams in a turbulent atmosphere. The discussion in this chapter is based on the frequency limit $(\omega_1 - \omega_2)/(\omega_1 + \omega_2) \leq 1$ (ω_1, ω_2) are the angular wave frequencies), and the results apply to turbulent regions of arbitrary intensity.

8.2 Frequency Domain—Beam Radius

The GSM beam is the most convenient mathematical model for describing spatial partial coherence. It can also be applied to pulsed partially coherent spectral plane waves. The combination of the two can yield spatial and spectral partially coherent unstable fields of GSMP beams. The propagation properties of this unsteady field will be investigated in the space-frequency domain and their spectral variation and temporal characteristics will be discussed. In the mutually independent space–time domain, the mutual coherence function (MCF) of a partially space coherence time GSMP beam at the source plane ($z = 0$) can be expressed as [17]:

$$\Gamma_0^2\left(\vec{r_1}, \vec{r_2}, t_1, t_2\right) = \Gamma_0 \exp\left[-H\left(\vec{r_1}, \vec{r_2}\right) - G(t_1, t_2)\right] \tag{8.1}$$

where $H\left(\vec{r_1}, r\vec{_2}\right) = \frac{r_1^2+r_2^2}{w_0^2} + \frac{\left|\vec{r_1}-\vec{r_2}\right|^2}{2\sigma_g^2}$, $G(t_1, t_2) = \frac{t_1^2+t_2^2}{2T_0^2} + \frac{(t_1-t_2)^2}{2T_c^2} + i\omega_0(t_1 - t_2)$

where

- $\vec{r_1}, \vec{r_2}$ are the position vectors of the source plane,
- Γ_0 is a constant,
- w_0 is the beam waist radius,
- σ_g is the spatial lateral correlation length of the pulse,
- ω is the angular frequency,
- ω_0 is the central angular frequency of the pulse,
- T_0 is the initial pulse width, and
- T_c is the temporal coherence length.

Based on the Fourier transform in the mutually independent space-frequency domain, the cross-spectral density function (CSDF) of a partially coherent GSMP beam at the source plane can be expressed as [17]:

$$\begin{aligned} W^0\left(\vec{r_1}, \vec{r_2}, \omega_1, \omega_2\right) &= \Gamma^0\left(\vec{r_1}, \vec{r_2}, \omega_1, \omega_2\right) \\ &= \frac{\Gamma_0 T_0}{2\pi\Omega_0} \exp\left[-H\left(\vec{r_1}, \vec{r_2}\right) - G(\omega_1 - \omega_2)\right] \end{aligned} \tag{8.2}$$

where $H(\vec{r_1}, \vec{r_2}) = \frac{r_1^2+r_2^2}{w_0^2} + \frac{|\vec{r_1}-\vec{r_2}|^2}{2\sigma_g^2}$, $G(\omega_1, \omega_2) = \frac{(\omega_1-\omega_0)^2+(\omega_2-\omega_0)^2}{2\Omega_0^2} + \frac{(\omega_1-\omega_2)^2}{2\Omega_c^2}$, represents the pulse spectral width, and $\Omega_c = T_c\Omega_0/T_0$ represents the pulse spectrum coherence width.

Assuming that the light source fluctuation is independent of the turbulent atmospheric fluctuation, according to the Huygens-Fresnel principle, in the $L > 0$ half-space, the cross-spectral density function of the GSMP beam can be expressed as:

$$\begin{aligned}\Gamma_2(\vec{\rho_1}, \vec{\rho_2}, L, \omega_1, \omega_2) = {} & \frac{\omega_1\omega_2}{4\pi^2L^2c^2}\exp[-iL(\omega_1-\omega_2)/c] \\ & \times \int d\vec{r_1} \int d\vec{r_2}\,\Gamma^0(\vec{r_1}, \vec{r_2}, \omega_1, \omega_2) \\ & \times \exp\left[\frac{i\omega_2(\vec{\rho_1}-\vec{r_2})^2}{2cL} - \frac{i\omega_1(\vec{\rho_2}-\vec{r_1})^2}{2cL}\right] \\ & \times \langle\exp[\psi(\vec{r_1}, \vec{\rho_1}, L, \omega_1) + \psi^*(\vec{r_2}, \vec{\rho_2}, L, \omega_2)]\rangle \end{aligned} \tag{8.3}$$

where $\rho = |\vec{\rho}| = (x^2+y^2)^{1/2}$ is the lateral off-axis distance perpendicular to the transport direction, $*$ is the complex conjugate, $\prec\succ$ is the ensemble mean of the turbulent flow, and c is the speed of light in a vacuum.

Assuming that the turbulence is statistically uniform and static and the complex phase fluctuation ψ caused by the turbulence is a Gaussian random variable, the cross-correlation function ψ of the complex phase of the spherical wave satisfies [21]

$$\langle\exp[\psi(\vec{r_1}, \vec{\rho_1}, L, \omega_1) + \psi^*(\vec{r_2}, \vec{\rho_2}, L, \omega_2)]\rangle = \exp[-1/2D_{12}(\vec{\rho_1}-\vec{\rho_2}, \vec{r_1}-\vec{r_2})] \tag{8.4}$$

Ishimaru [22] obtained an expression for D_{12} in isotropic turbulent flow using the smooth perturbation method which is expressed as follows:

$$\begin{aligned}D_{12}(\vec{\rho_1}-\vec{\rho_2}, \vec{r_1}-\vec{r_2}) = (2\pi)^2\int_0^\infty kdk\int_0^z dz\prime\Phi_n(k, z\prime)\left\{\frac{\omega_1^2+\omega_2^2}{c^2}\right. \\ \left. -2J_0(k\beta)\frac{\omega_1\omega_2}{c^2}\cos\left[\frac{z\prime c(z-z\prime)}{2z}k^2\left(\frac{1}{\omega_1}-\frac{1}{\omega_2}\right)\right]\right\}\end{aligned} \tag{8.5}$$

where $\beta = \left|\frac{z\prime}{z}(z-z\prime) + \left(1-\frac{z\prime}{z}\right)(\vec{r_1}-\vec{r_2})\right|$, $\Phi_n(k, z\prime)$ represents the turbulent refractive index fluctuation spectrum, $J_0(x)$ represents the zero-order Bessel function, and k represents the wavenumber ($\omega = ck$). The preceding equations assume that the wavelength of the pulsed signal is smaller than the inner scale of the turbulent vortex, and the utilized smooth perturbation method is only limited to the weakly turbulent region. In the limit case $\omega_1 = \omega_2$, Fante [23] discovered that the results obtained based on Eqs. (8.8)–(8.10) are the same as those obtained using the Markov approximation method, and are suitable for any turbulent flow region. Therefore, the frequency limit $(\omega_1 - \omega_2)/(\omega_1 + \omega_2) \le 1$ is set, which is suitable for any turbulent intensity and narrow-band signal source.

Here, we use the modified von Karman turbulence spectrum, which is given as follows:

$$\Phi_n(\kappa, z\prime) = 0.033 C_n^2 \frac{\exp\left(-\kappa^2/\kappa_m^2\right)}{\left(\kappa^2 + \kappa_0^2\right)^{11/6}} \tag{8.6}$$

where C_n^2 represents the refractive index structure constant and $\kappa_m = 5.92/l_0$, $\kappa_0 = 1/L_0$, and l_0, and L_0 represent the inner and outer scales of the turbulent flow, respectively. Using the approximation process [20], Eq. (8.6) can be substituted into Eq. (8.5) to obtain:

$$\begin{aligned} D_{12}\left(\vec{\rho_1} - \vec{\rho_2}, \vec{r_1} - \vec{r_2}\right) &= 0.782\Delta^2 L_0^{5/3} z \int_0^1 C_n^2(t)dt \\ &+ 2.92\omega_1\omega_2 z \Big/ c \int_0^1 C_n^2(t)\left|t\left(\vec{\rho_1} - \vec{\rho_2}\right) + (1-t)\left(\vec{r_1} - \vec{r}\right)\right|^{5/3} dt \end{aligned} \tag{8.7}$$

Considering horizontal transmission and substituting Eq. (8.7) into (8.4), we obtain:

$$\begin{aligned} &\left\langle \exp\left[\psi\left(\vec{r_1}, \vec{\rho_1}, L, \omega_1\right) + \psi^*\left(\vec{r_2}, \vec{\rho_2}, L, \omega_2\right)\right]\right\rangle \\ &\approx \Gamma \exp\left[-\frac{\left(\vec{\rho_1} - \vec{\rho_2}\right)^2 + \left(\vec{\rho_1} - \vec{\rho_2}\right)\cdot\left(\vec{r_1} - \vec{r_2}\right) + \left(\vec{r_1} - \vec{r_2}\right)^2}{\rho_0^2}\right] \end{aligned} \tag{8.8}$$

where $\Gamma = \exp\left(-0.391 C_n^2 L k_d^2 L_0^{5/3} / c^2\right)$, $\rho_0 = \left(0.55 C_n^2 k_c^2 l_0^{-1/3} L\right)^{-1/2}$, $k_d = (\omega_1 - \omega_2)/c$, $k_c = (\omega_1 + \omega_2)/2c$, and ρ_0 represents the spatial coherence length of spherical waves in atmospheric turbulence.

When $\vec{\rho_1} = \vec{\rho_2}$, in the limit case $\omega_1 = \omega_2$, substituting Eqs. (8.2) and (8.8) into Eq. (8.3) and integrating the result, the spectral intensity of the GSMP at any point on the observation surface can be obtained:

$$S\left(\vec{\rho}, L, \omega\right) = \frac{\Gamma_0 T_0}{2\pi\Omega_0} \frac{1}{\delta(L, \omega)} \exp\left[-\frac{2\rho^2}{w_0^2 \delta(L, \omega)}\right] \exp\left[-\frac{(\omega - \omega_0)^2}{\Omega_0^2}\right] \tag{8.9}$$

where the expansion factor is given by:

$$\delta(L,\omega) = 1 + \left(\frac{2cL}{w_0\omega}\right)^2\left(\frac{1}{w_0^2} + \frac{1}{\sigma_g^2} + \frac{2}{\rho_0^2}\right) \tag{8.10}$$

As shown in Eq. (8.9), the average spectral intensity of the GSMP beam is the product of its average spectral density and the Gaussian pulsed spectrum. Thus, the beam radius at the receiving end can be expressed as:

$$W_c = w_0 \times \delta(L,\omega)^{1/2} = w_0\left[1 + \left(\frac{2cL}{w_0\omega}\right)^2\left(\frac{1}{w_0^2} + \frac{1}{\sigma_g^2} + \frac{2}{\rho_0^2}\right)\right]^{1/2} \tag{8.11}$$

The atmospheric turbulence model for horizontal transmission sets the atmospheric refractive index structure constant and turbulent outer scale as fixed values, which are not available for oblique transmission. The coherence length of spherical waves in atmospheric turbulence for oblique propagation of a GSMP beam can be expressed as:

$$\rho_0 = \left(1.46k^2 \int_0^L C_n^2(z)(1 - z/L)^{5/3} dz\right)^{-3/5} \tag{8.12}$$

where the angular frequency $\omega = ck$ and the wavenumber $k = 2\pi/\lambda$. In this case, the atmospheric structure constant model H-V, published by the International Telecommunication Union-Radiocommunication Department ITU-R is adopted. Assuming $\xi = h/H = z/L$, the atmospheric structure constant can be expressed as:

$$\begin{aligned} C_n^2(z) = {} & 8.148 \times 10^{-56} v_{RMS}^2 (\xi H/\cos\theta)^{10} \\ & \times \exp(\xi H/\cos\theta/1000) + 2.7 \times 10^{-16} \\ & \times \exp(-\xi H/\cos\theta/1500) \\ & + C_0^2 \exp(-\xi H/\cos\theta/100) \end{aligned} \tag{8.13}$$

where $v_{RMS}^2 = \sqrt{v_g^2 + 30.69v_g + 348.91}$ is the vertical-path wind speed, v_g is the near-surface wind speed (typically $v_g = 21$ m/s), and C_0 is the atmospheric structure constant near the surface (typically $C_0^2 = 1.7 \times 10^{-14}\,\text{m}^{-2/3}$).

As shown in Figs. 8.1 and 8.2, the radius of the partially coherent GSMP beam at the receiving end is proportional to the transmission distance; i.e., as the transmission distance increases, the beam radius at the receiving end gradually increases. The beam radius in atmospheric turbulence is larger than that in free space; i.e., the beam expands under the influence of atmospheric turbulence. Partially coherent GSMP beams have a larger beam radius as the transverse correlation length of the pulse space decreases. However, compared to free space, the beam radius expansion caused by turbulence becomes increasingly smaller, i.e., The influence of turbulence on the beam radius spread with poor coherence is smaller. Under the same conditions, as the transmission distance increased, the larger the angular frequency, the smaller the beam radius spread at the receiving end.

In Figs. 8.3 and 8.4, the transmission distance $L = 2$ km, the beam waist radius $w_0 = 0.05$ m, and the pulse space lateral correlation length $\sigma_g = 0.01$ m. As the transmission distance increases, the beam radius expansion becomes increasingly significant. The smaller the turbulent inner scale, the larger the beam radius expansion. Under the same conditions, the beam radius at the receiving end gradually expands with an increase in the turbulence intensity. Under weak turbulence, the beam radius expansion effect is not obvious, whereas in the case of strong turbulence, the beam radius increases, and the expansion effect is notable.

As shown in Figs. 8.5 and 8.6, as the turbulence intensity increases, the beam radius at the receiving end increases, and the partially coherent GSMP beam is less affected by turbulence compared to the fully coherent GSMP beam. For a weak turbulence intensity, the beam radius remains almost constant, and as the turbulence intensity increases, the beam radius increases rapidly under moderate and strong turbulence. Figure 8.5 shows that the smaller the initial radius of the beam, the greater the effect of turbulence on the beam radius. As shown in Fig. 8.6, under weak turbulence, the influence of the pulse spatial lateral correlation length on the beam radius was not significant. However, as the turbulence intensity increases, the spatial lateral correlation length of the pulse increases under moderate and strong

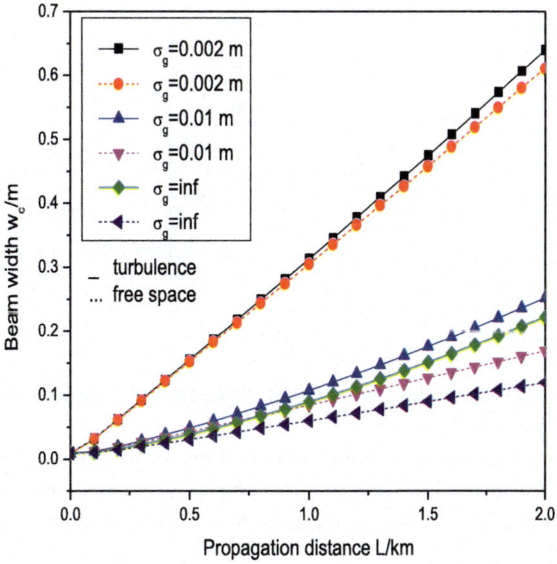

Fig. 8.1 Variation of beam radius with different coherence lengths

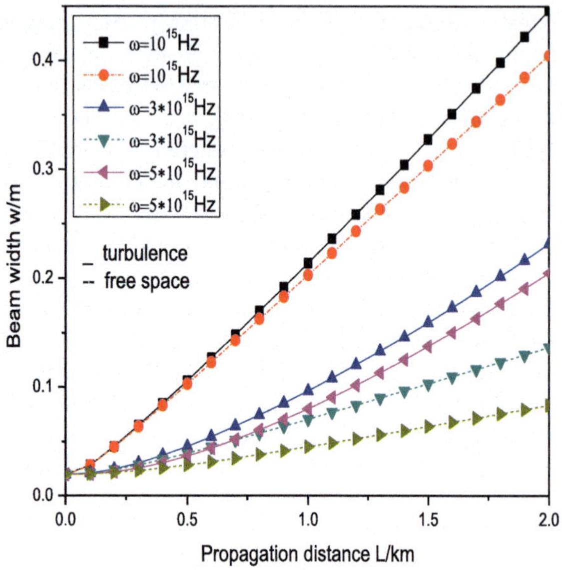

Fig. 8.2 Variation of beam radius at different angular frequencies

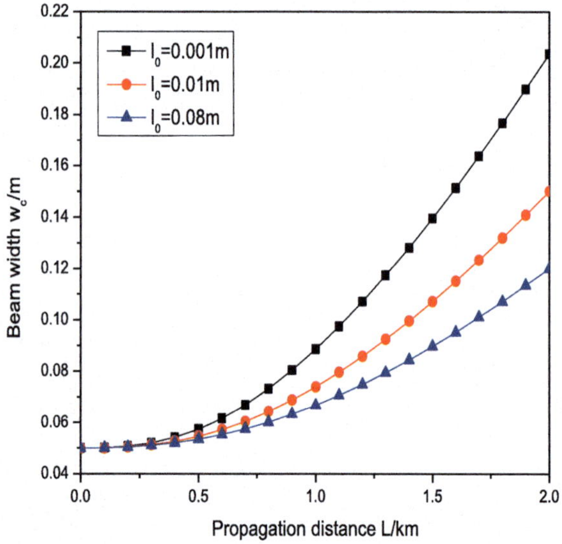

Fig. 8.3 Variation of beam radius at different inner scales

turbulence, the influence on the beam radius is more pronounced, and the expansion is notable.

Figures 8.7 and 8.8 show that when the transmission distance $L = 2$ km for horizontal transmission in atmospheric turbulence, the larger the beam waist radius w_0, the larger the beam radius at the receiving end. In atmospheric turbulence, the beam

Fig. 8.4 Variation of beam radius under different turbulence intensities

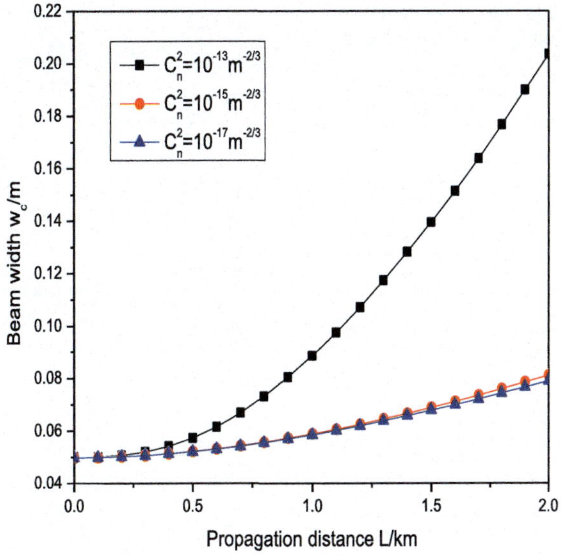

Fig. 8.5 Beam radius for different initial radii

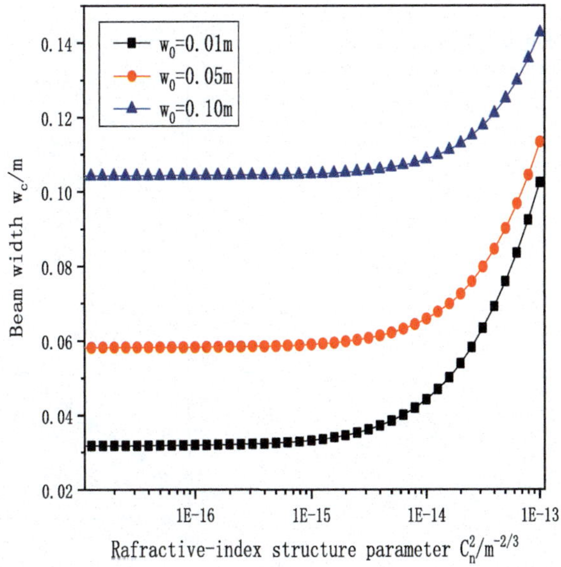

radius is smaller compared to that in free space. The expansion, in this case, is significant. After the beam is transmitted through the turbulent atmosphere, the change in the beam radius with frequency is not obvious; i.e., in a turbulent atmosphere, the influence of the frequency on the beam radius is small.

In Figs. 8.9 and 8.10, the initial radius $w_0 = 0.05$ m, the spatial coherence length $\sigma_g = 0.05$ m, the transmission distance $L = 2$ km, and the transmission height

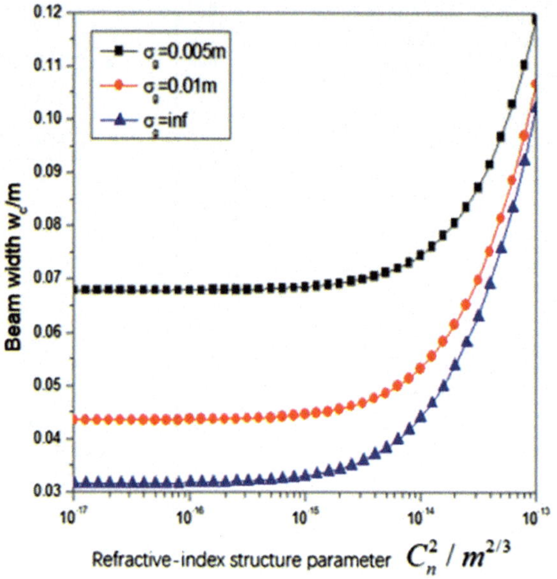

Fig. 8.6 Beam radius for different coherence lengths

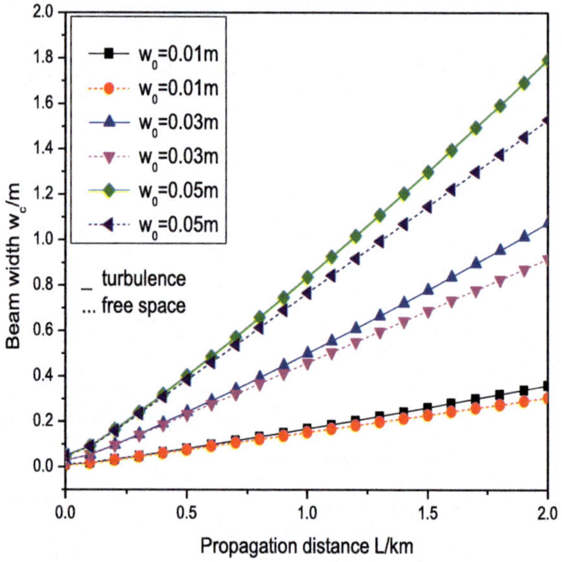

Fig. 8.7 Beam radius for different initial radii

$H = 500$ m. As the transmission distance increases, the beam radius at the receiving end gradually increases, and when the transmission distance is constant, the longer the wavelength, the larger the beam radius. When a certain wavelength $\lambda = 0.63$ μm and beam waist radius $w_0 = 0.02$ m, the beam radius of the receiving end increases gradually due to the influence of the spatial coherence length. the beam radius at the

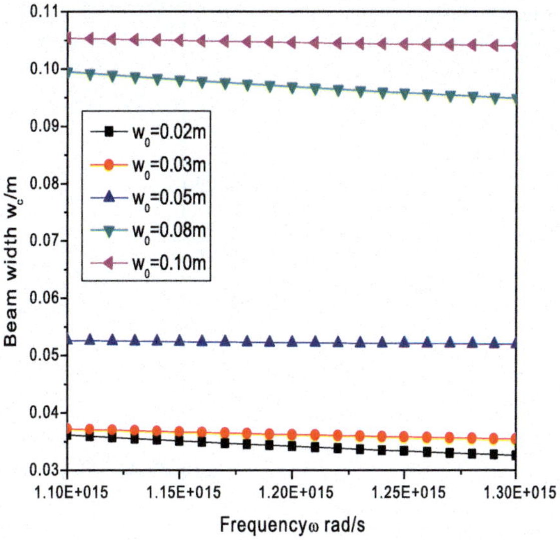

Fig. 8.8 Beam radius for different beam waist radii

receiving end gradually increases, and for the same transmission height, the partially coherent GSMP beam expands more compared to a fully coherent GSMP beam. The larger the radius, i.e., the smaller the spatial coherence length σ_g, the more pronounced the influence of turbulence on the GSMP beam, and the more distinct the expansion. Moreover, when $H = 0$ for horizontal transmission, the beam radius of the GSMP beam at the receiving end is larger compared to the case of oblique transmission under the same conditions. For the same spatial coherence length, when the transmission distance is constant, the higher the receiver height, the greater the radius expansion.

For a transmission distance $L = 5$ km, wavelength $\lambda = 0.63\,\mu$m, and transmission heights $H = 0$ and $H = 500$ m, respectively, Figs. 8.11 and 8.12 show the variation of the beam radius at the receiving end during horizontal and oblique transmissions. The plots show that the beam radius at the receiving end is significantly larger during horizontal transmission compared to the case of oblique transmission. Moreover, the effect of oblique transmission is better than that of horizontal transmission. For a transmission distance $L = 1$ km, transmission height $H = 500$ m, and spatial coherence length $\sigma_g = 0.05$ m, after the GSMP beam is transmitted through the turbulent atmosphere, the change in the beam radius with the frequency is not obvious. Thus, in an inclined turbulent atmosphere, the influence of the frequency on the beam radius is small, and the larger the initial radius, the larger the beam radius.

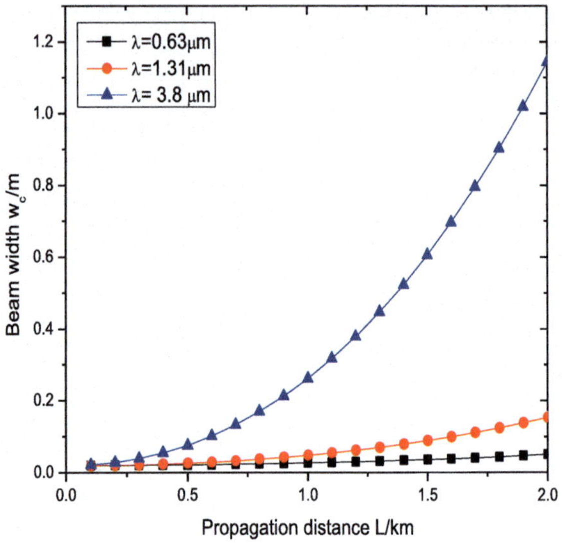

Fig. 8.9 Beam radius at different wavelengths

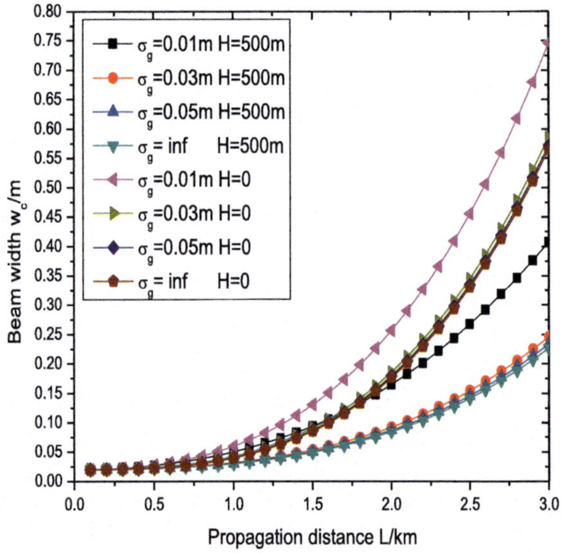

Fig. 8.10 Beam radius at different coherence lengths

8.3 Time Domain—Pulse Stretching

The second-order moment of the field, called the mutual coherence function, predicts the additional beam broadening and spatial coherence loss of the light waves caused by turbulence. The bi-frequency mutual interference function can be expressed as [24]

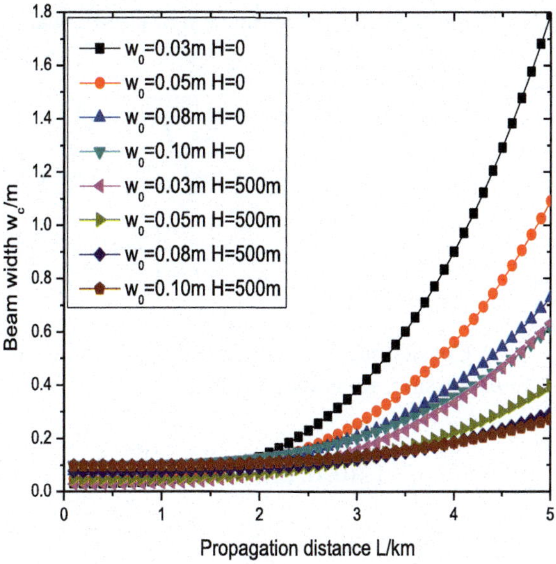

Fig. 8.11 Beam radius as a function of transmission distance

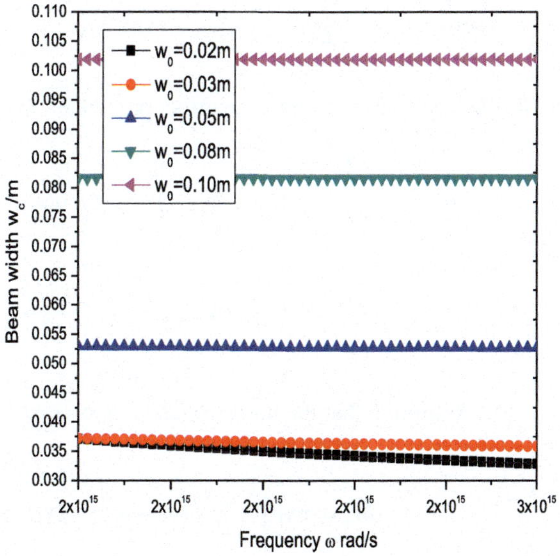

Fig. 8.12 Beam radius as a function of frequency

$$
\begin{aligned}
\Gamma_2(\vec{\rho_1}, \vec{\rho_2}, L; \omega_0 + \omega_1, \omega_0 + \omega_2) &= \langle u(\vec{\rho_1}, L; \omega_0 + \omega_1) u^*(\vec{\rho_2}, L; \omega_0 + \omega_2) \rangle \\
&= \Gamma_2^0(\vec{\rho_1}, \vec{\rho_2}, L; k_1, k_2) M_2(\vec{\rho_1}, \vec{\rho_2}, L; k_1, k_2)
\end{aligned}
\tag{8.14}
$$

where Γ_2^0 represents the dual-frequency mutual coherence function in free space and M_2 is the second order moment associated with turbulence.

(1) The mutual coherence function of GSMP beams in free space can be expressed as

$$\Gamma_2^0(\vec{\rho_1}, \vec{\rho_2}, L, \omega_1, \omega_2) = \frac{\omega_1\omega_2}{4\pi^2L^2c^2}\exp[-iL(\omega_1 - \omega_2)/c] \times \int d\vec{r_1} \int d\vec{r_2}\Gamma^0(\vec{r_1}, \vec{r_2}, \omega_1, \omega_2) \times \exp\left[\frac{i\omega_2(\vec{\rho_1} - \vec{r_2})^2}{2cL} - \frac{i\omega_1(\vec{\rho_2} - \vec{r_1})^2}{2cL}\right] \quad (8.15)$$

Considering the on-axis case ($\vec{\rho_1} = \vec{\rho_2} = 0$) and integrating the above expression, we have:

$$\Gamma_2^0(0, 0, L, \omega_1, \omega_2) = \frac{W_0}{cL}\sqrt{\frac{\omega_c^2}{4\alpha\beta - \sigma_g^{-4}}}\exp[-iL\omega_d^2/c] \times \exp\left[-\frac{(\omega_c - \omega_0)^2}{\Omega_0^2} - \frac{\omega_d^2}{2\Omega_c^2}\right] \quad (8.16)$$

where $W_0 = \frac{\Gamma_0 T_0}{2\pi\Omega_0}$, $\omega_d = \omega_1 - \omega_2$, $2\omega_c = \omega_1 + \omega_2$, $\alpha = \frac{1}{w_0^2} + \frac{1}{2\sigma_g^2} + \frac{i\omega_1}{2cL}$, $\beta = \frac{1}{w_0^2} + \frac{1}{2\sigma_g^2} - \frac{i\omega_2}{2cL}$.

(2) We consider the turbulence effect parameter. M_2, which is given by:

$$M_2 = \exp\left\{-2\pi\left(k_1^2 + k_2^2\right)z\int_0^\infty \kappa\Phi_n(\kappa)d\kappa + 4\pi^2 k_1 k_2 \int_0^1\int_0^\infty \kappa\Phi_n(\kappa)\exp\left[-\frac{i\kappa^2 L\xi}{2}\left(\frac{1}{k_1} - \frac{1}{k_2}\right)\right]d\kappa d\xi\right\} \quad (8.17)$$

Assuming that the narrowband source satisfies $\frac{1}{k_1} - \frac{1}{k_2} \approx 0$, then we have:

$$M_2 = \exp\left[-2\pi^2\left(k_1^2 + k_2^2 - 2k_1k_2\right)L\int_0^\infty \kappa\Phi_n(\kappa)d\kappa\right] \quad (8.18)$$

and the modified von Karman spectrum can be expressed as:

$$\Phi_n(\kappa) = 0.033C_n^2\frac{\exp(-\kappa^2/\kappa_m^2)}{\left(\kappa^2 + \kappa_0^2\right)^{11/6}} \quad (8.19)$$

where $\kappa_m = 5.92/l_0$, $\kappa_0 = 1/L_0$, and L_0, and l_0 represent the outer and inner turbulence scales, respectively. Using Eq. (8.19) and the parameter $\omega_d = \frac{k_d}{c} = \frac{k_1 - k_2}{c}$ and substituting into Eq. (8.18), we obtain:

$$M_2 = \exp\left(-0.3908 C_n^2 L L_0^{5/3} \Big/ c^2 \omega_d^2\right) = \exp\left(-\gamma \omega_d^2\right) \tag{8.20}$$

Substituting Eqs. 8.16 and 8.20 into Eq. (8.14) yields,

$$\Gamma_2(0, 0, L, \omega_1, \omega_2) = \frac{W_0}{cL}\sqrt{\frac{\omega_c^2}{4\alpha\beta - \sigma_g^{-4}}} \exp\left[-\frac{iL\omega_d}{c}\right] \times \exp\left[-\frac{(\omega_c - \omega_0)^2}{\Omega_0^2} - \frac{\omega_d^2}{2\Omega_c^2}\right] \exp\left(-\gamma \omega_d^2\right) \tag{8.21}$$

There is a Fourier transform relationship between the time-averaged intensity and the cross-spectral density function, which is given as:

$$\begin{aligned}\left\langle I\left(\vec{\rho}, \vec{\rho}, L; t, t\right)\right\rangle &= \Gamma\left(\vec{\rho}, \vec{\rho}, L; t, t\right) \\ &= \iint W\left(\vec{\rho}, \vec{\rho}, L; \omega_1, \omega_2\right) \\ &\quad \exp(-i\omega_1 t + i\omega_2 t) d\omega_1 d\omega_2\end{aligned} \tag{8.22}$$

The time-averaged intensity on the optical axis ($\vec{\rho} = 0$) is expressed as:

$$\begin{aligned}\langle I(0, 0, L; t, t)\rangle &= \iint \Gamma_2\left(\vec{\rho_1}, \vec{\rho_2}, L; \omega_0 + \omega_1, \omega_0 + \omega_2\right) \exp(-i\omega_1 t + i\omega_2 t) d\omega_1 d\omega_2 \\ &= \frac{W_0}{cL}\sqrt{\frac{\pi}{a}} \exp\left[-\frac{2\left(t - \frac{L}{c}\right)^2}{8a}\right] \\ &\quad \times \int_{-\infty}^{\infty} \sqrt{\frac{\omega_c^2}{4\alpha\beta - \sigma_g^{-4}}} \exp\left[-\frac{(\omega_c - \omega_0)^2}{\Omega_0^2}\right] d\omega_c\end{aligned} \tag{8.23}$$

where $a = \frac{1}{2\Omega_c^2} + \gamma$, $\gamma = 0.3908 C_n^2 L L_0^{5/3} \Big/ c^2$, $\alpha = \frac{1}{w_0^2} + \frac{1}{2\sigma_g^2} + \frac{i\omega_1}{2cL}$, $\beta = \frac{1}{w_0^2} + \frac{1}{2\sigma_g^2} - \frac{i\omega_2}{2cL}$.

(3) Yeh and Liu used time-moments to study the temporal characteristics of random medium-wave propagation [10–13], and extended their results to the study of pulsed-wave propagation in atmospheric turbulence.

The *nth* time moment of a pulse is defined as:

$$\left\langle M^{(n)}\right\rangle = \int_{-\infty}^{\infty} t^n \left\langle I\left(z, \vec{\rho}, t\right)\right\rangle dt \tag{8.24}$$

Each moment equation has a certain physical meaning; the zero-order moment is related to the total energy, the first-order time moment corresponds to the average arrival time t_a of a pulse, and the second-order time-moment corresponds to the mean square pulse width τ^2:

$$\langle t_a \rangle = \frac{\left\langle M^{(1)}\right\rangle}{\left\langle M^{(0)}\right\rangle} \tag{8.25}$$

$$\left\langle t_a^2 \right\rangle = \frac{\left\langle M^{(2)}\right\rangle}{\left\langle M^{(0)}\right\rangle} \tag{8.26}$$

$$\tau^2 = \left\langle t_a^2 \right\rangle - \langle t_a \rangle^2 \tag{8.27}$$

Substituting Eqs. 8.24, 8.25, and 8.26 into Eq. 8.27 yields the following:

$$\tau^2 = \frac{\int_{-\infty}^{\infty} \langle I \rangle (t - t_a)^2 dt}{\left\langle M^{(0)}\right\rangle} - \langle t_a \rangle^2 \tag{8.28}$$

According to the following relationship:

$$\int_{-\infty}^{\infty} t^n \exp(i\omega_d t) dt = 2\pi (-i)^n \delta^{(n)}(\omega_d) \tag{8.29}$$

$$\int_{-\infty}^{\infty} \delta^{(n)}(\omega_d) f(\omega_d) d\omega_d = (-1)^n \frac{\partial^n}{\partial \omega_d^n} f(\omega_d)|\omega_d = 0 \tag{8.30}$$

$$\left\langle M^{(n)}\right\rangle = \frac{i^n}{2\pi} \int_{-\infty}^{\infty} \frac{\partial^n}{\partial \omega_d} \Gamma_2(0, 0, \omega_c + \omega_0)\big|_{\omega_d = 0} d\omega_c \tag{8.31}$$

$\delta^{(n)}$ is a Dirac δ function. Substituting Eq. 8.21 into Eqs. 8.31, 8.25, and (8.28) to obtain the average arrival time t_a and the mean square pulse width τ^2, we have:

$$t_a = \frac{\left\langle M^{(1)}\right\rangle}{\left\langle M^{(0)}\right\rangle} = L/c \tag{8.32}$$

$$\tau^2 = \frac{1}{\Omega_c^2} + 2\gamma \tag{8.33}$$

where $\Omega_c = \frac{T_c\Omega_0}{T_0}$, $\Omega_0 = \left(\frac{1}{T_0^2} + \frac{2}{T_c^2}\right)^{1/2}$.

Considering a narrow-band beam that maintains a Gaussian shape after transmission, the pulse at the receiving end is delayed by time L/c.

The half-width of the pulse at the receiver is denoted by T_2:

$$T_2 = \sqrt{8\left(\frac{1}{2\Omega_c^2} + \gamma\right)} \tag{8.34}$$

Let $n = (T_2 - T_0)/T_0$ be the relative pulse width, i.e., the percentage of pulse broadening between the receiver and the transmitter. When the turbulence parameter is zero, Eq. 8.34 can degenerate to the pulse width in a vacuum. In this equation, the pulse width at the receiving end is related to the initial pulse width, the temporal coherence length, the turbulence intensity, and the outer scale, but is unrelated to the initial radius and the wavelength.

In the case of sloping transmission in atmospheric turbulence, the turbulence parameter term can be expressed as follows:

$$\gamma = \frac{0.391\left(1 + 0.171\delta^2 - 0.287\delta^{5/3}\right)\mu_1 \sec\theta}{c^2} \tag{8.35}$$

$$\mu_1 = \int_{h_0}^{H} C_n^2(h)\left[1/L_0(h)\right]^{-5/3} dh \tag{8.36}$$

$$L_0(h) = \frac{5}{1 + \left[(h - 7500)/2500\right]^2} \tag{8.37}$$

$$l_0(h) = \delta L_0(h), 0 < \delta < 1 \tag{8.38}$$

where L_0 represents the turbulent outer scale and l_0 represents the turbulent inner scale.

Considering a transmission distance $L = 3$ km, as shown in Figs. 8.13 and 8.14, with an increase in the pulse transmission distance, the pulse width at the receiving end becomes larger; the larger the turbulence intensity and outer scale, the larger the pulse width at the receiving end, and the more distinct the broadening.

In Figs. 8.15 and 8.16, the transmission distances is $L = 1$ km, $L_0 = 5$ m, and $T_0 = 20 \times 10^{-15}$ s. As the turbulence intensity increases, the pulse width at the receiving end gradually increases. For weak turbulence, the pulse width does not change significantly, but for moderate and strong turbulence, the pulse width increases rapidly, and the pulse broadens significantly. The smaller the initial pulse width, the greater the influence of turbulence, and the more obvious the pulse broadening. Moreover, under the same conditions, the larger the coherence length of the pulse, the greater the influence of turbulence, the more pronounced the pulse broadening. It can also be seen from Fig. 8.16 that the partially coherent GSMP beam is

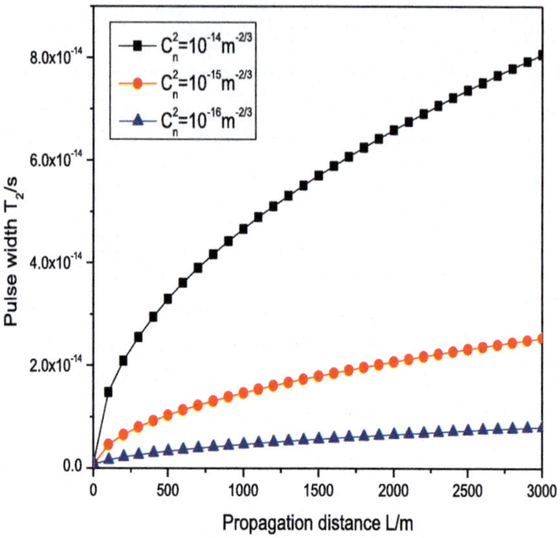

Fig. 8.13 Pulse width at different turbulence intensities

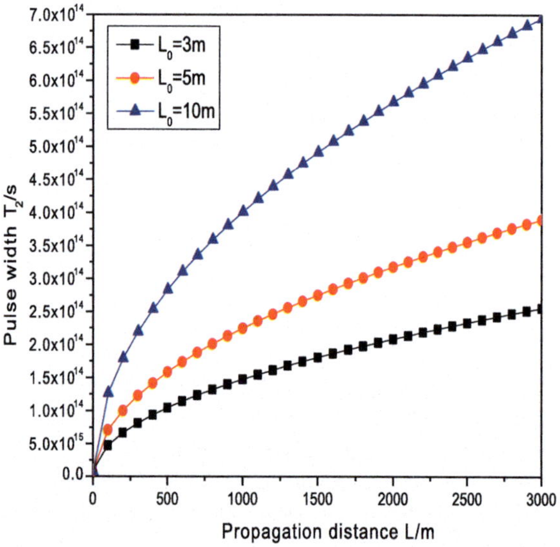

Fig. 8.14 Pulse width at different outer scales

affected by atmospheric turbulence. This effect is less than that of a fully coherent GSMP beam.

In Figs. 8.17 and 8.18, the relative pulse width changes significantly within the first 40 fs. When the initial pulse width is greater than 40 fs, the relative pulse width approaches zero. After transmission through atmospheric turbulence, the pulse width at the receiving end is not significantly different from the initial pulse width, indicating that the pulse broadening is quite distinct in the first 40 fs. When the

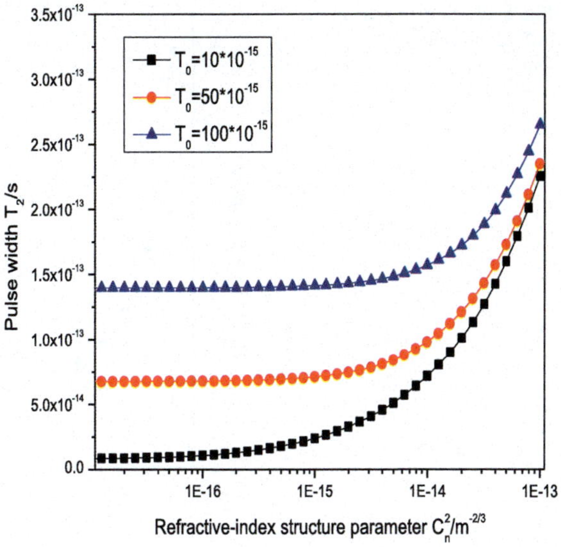

Fig. 8.15 Pulse width for different initial pulse widths

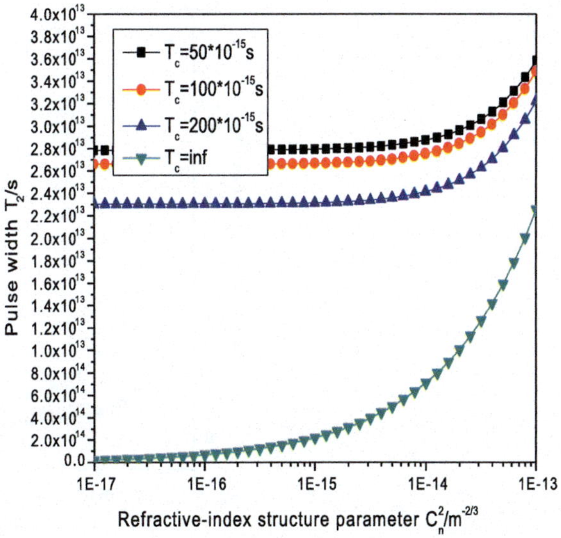

Fig. 8.16 Pulse width at different coherence times

initial pulse width is greater than 40 fs, the pulse width does not change significantly. Moreover, the stronger the intensity and the larger the outer scale, the more distinct the pulse broadening, and the larger the relative pulse width. The larger the relative pulse width, the faster the relative pulse width decreases with the increase of turbulence intensity and initial pulse width, indicating that the pulse width at the receiving end becomes larger.

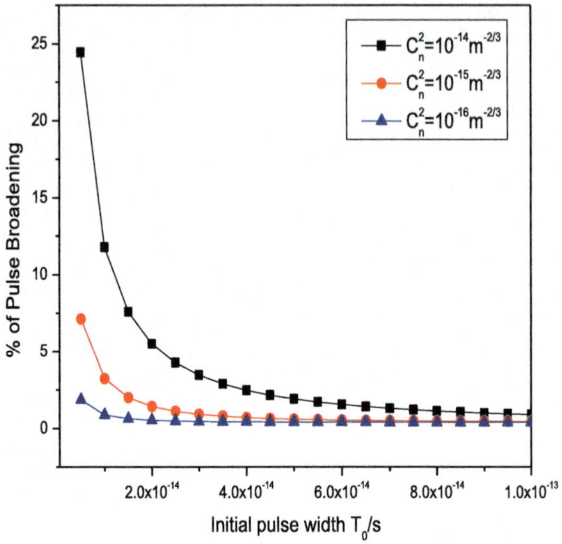

Fig. 8.17 Spread ratio at different turbulence intensities

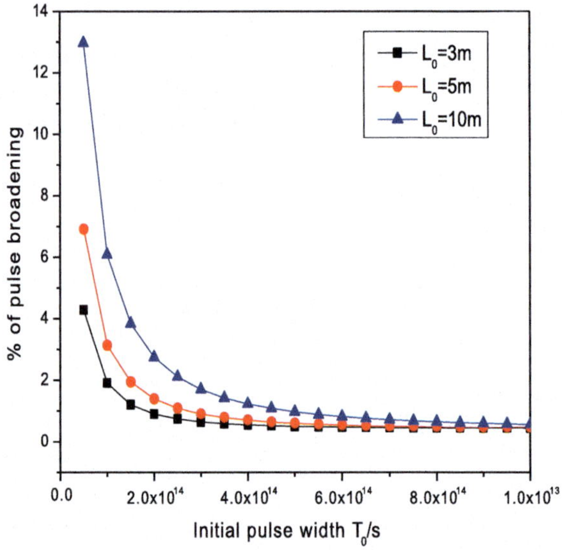

Fig. 8.18 Aspect ratios at different outer scales

In Figs. 8.19 and 8.20, the coherence time $T_c = 100$ fs, the zenith angle $\theta = \pi/4$, and the transmission distance $L = 5$ km. The higher the transmitter height, the larger the pulse width at the receiving end. When the other conditions remain unchanged, and the transmission distance $L = 10$ km, the larger the zenith angle, the larger the pulse width at the receiving end. For the top angle $\theta = \pi/2$ of the sky, which represents horizontal transmission, the pulse width is larger compared to the case of the inclined transmission, indicating that the former is more susceptible to turbulence than the latter.

In Figs. 8.21 and 8.22, the height of the transmitter from the ground is $h_0 = 2$ km, and the zenith angle is $\theta = \pi/4$. The longer the transmission distance, the greater the pulse broadening. When the transmission distance is $L = 5$ km, the pulse width increases with an increase in the initial pulse width. When this parameter is constant, the larger the temporal coherence length, the smaller the pulse broadening,

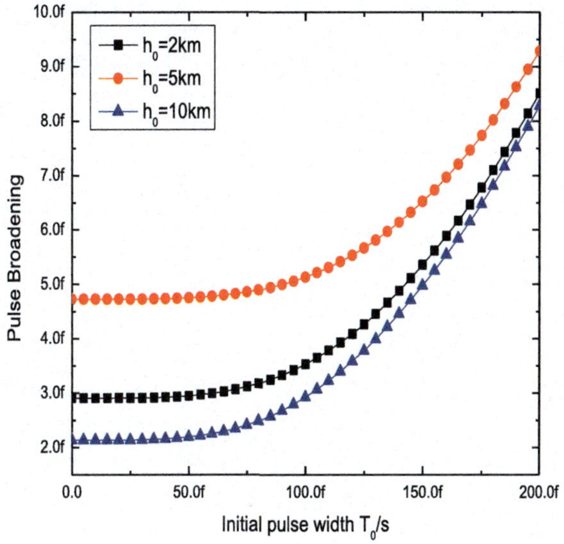

Fig. 8.19 Pulse width at different transmission heights

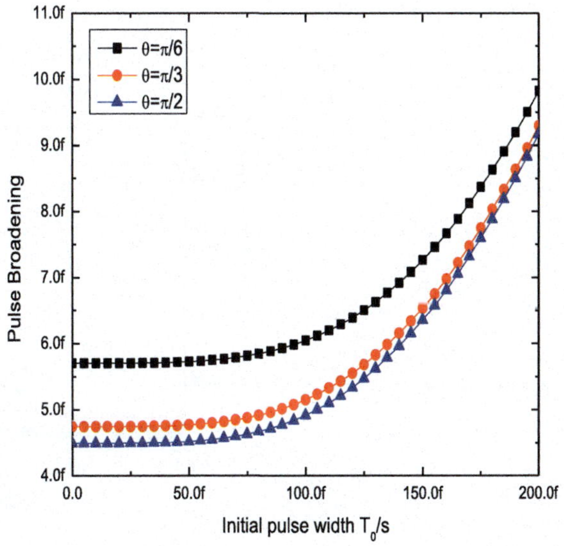

Fig. 8.20 Pulse width at different zenith angles

indicating that the larger the coherence length, the less susceptible is the beam to turbulent flow.

In Figs. 8.23 and 8.24, the height of the transmitter from the ground is $h_0 = 2$ km, the coherence time is $T_c = 500$ fs, the zenith angle is $\theta = \pi/4$, the transmission distance is $L = 5$ km, and the vertical path root mean square wind speed is $v_{rms} = 21$ m/s. When the initial pulse width is constant, the higher the height of the

Fig. 8.21 Pulse width at different transmission distances

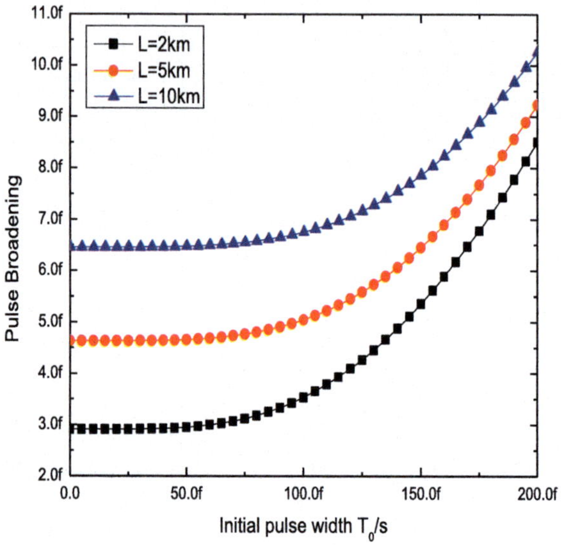

Fig. 8.22 Pulse width at different coherence times

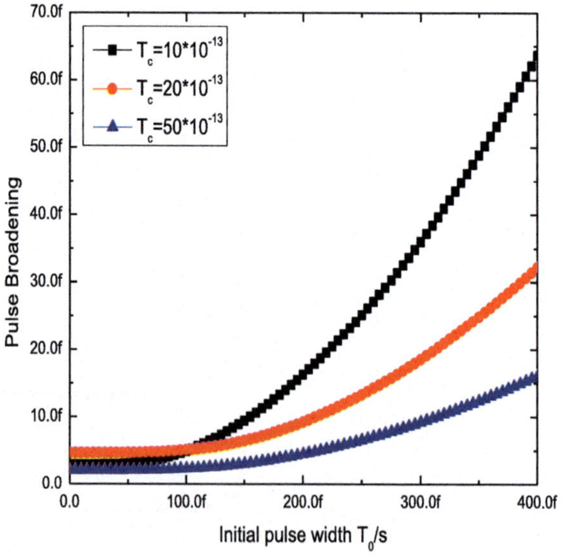

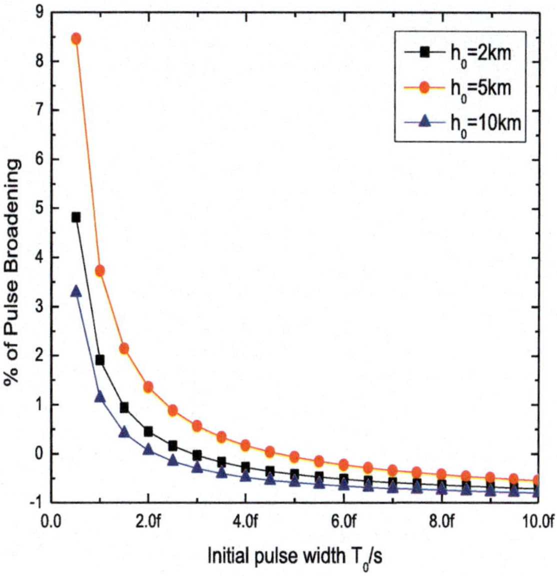

Fig. 8.23 Aspect ratio at different emitter heights

transmitter, the smaller the pulse width ratio. Moreover, the turbulence intensity gradually decreases with an increase in height. When the zenith angle is $\theta = \pi/3$, the laser pulse broadening ratio associated with atmospheric turbulence during oblique transmission increases with an increase in the initial pulse width, and the slope of the curve of the pulse broadening ratio decreases. When the initial pulse width is constant, the greater the transmission distance, the greater the pulse stretching ratio.

In Fig. 8.25, the coherence time is $T_c = 500$ fs, the transmitter height is $h_0 = 2$ km, and the transmission distance is $L = 10$ km. When the initial pulse width is constant, the larger the zenith angle, the smaller the pulse broadening ratio, indicating that the turbulence intensity gradually decreases with an increase in height. Moreover, when the zenith angle is $\theta = \pi/2$, i.e., horizontal transmission, the pulse broadening ratio is larger compared to the case of oblique transmission.

8.4 Summary

Based on the coherence theory of an unstable light field, the Huygens-Fresnel principle and the modified von Karman model were used to obtain expressions for the average light intensity and pulse broadening for the horizontal transmission of a partially coherent GSMP beam in atmospheric turbulence. The variation of the pulse broadening when a partially coherent GSMP beam propagates horizontally through atmospheric turbulence was investigated. Moreover, the effects of spatial coherence parameters on pulse broadening such as the spatial lateral coherence length of the pulse, the coherence time, the refractive index structure constant of atmospheric

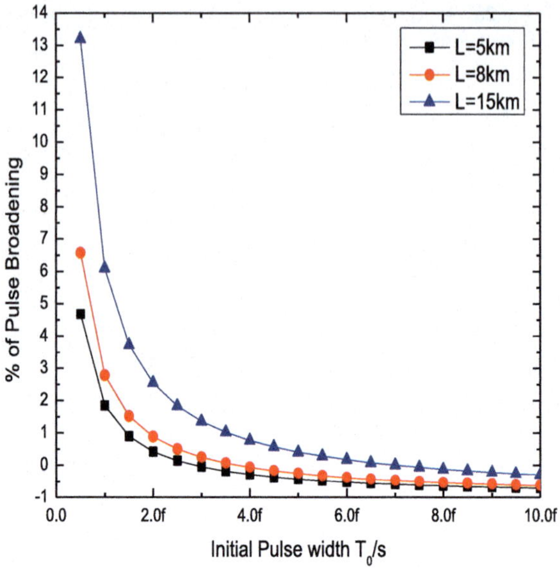

Fig. 8.24 Aspect ratio at different transmission distances

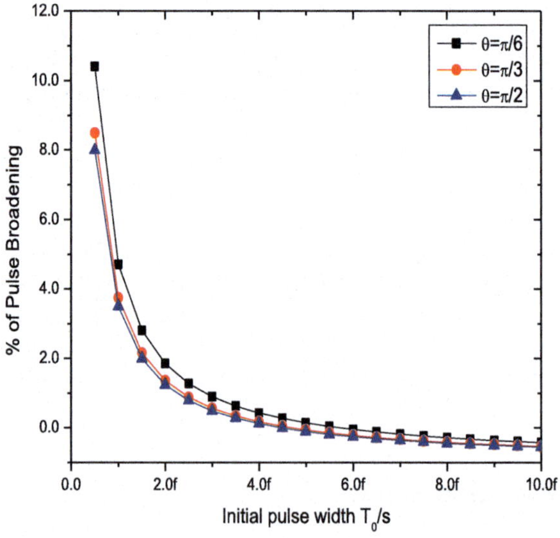

Fig. 8.25 Aspect ratios at different zenith angles

turbulence, and the inner and outer scales, were analyzed in detail. The main results can be summarized as:

(1) A partially coherent GSMP beam undergoes broadening of the pulse width during horizontal transmission in atmospheric turbulence, whereas a partial GSMP beam is less affected by turbulence.

(2) Under weak turbulence, the pulse width of a partially coherent GSMP beam increases with a decrease in the spatial lateral correlation length of the pulse. Under moderate and strong turbulence, the larger the spatial lateral correlation length, the greater the influence of the pulse width. Therefore, the same parameters had different effects on the pulse width for different turbulence intensities.

(3) The time waveform of the pulse had a delay with the increase in the transmission distance, and the delay time was $t = L/c$. Considering the narrow-band beam, the waveform after transmission maintained a Gaussian shape.

(4) In the time domain, when the initial pulse width was constant, the pulse width ratio decreased with an increase in the pulse transmission height, indicating that the turbulence intensity also gradually decreases with an increase in height.

(5) Comparing the effects of oblique transmission and horizontal transmission, the beam radius in the former is less susceptible to atmospheric turbulence; thus, oblique transmission is generally used in practice.

References

1. Sereda L, Bertolotti M, Ferrari A (1998) Coherence properties of nonstationary light wave fields. J Opt Soc Am A 15(3):695–705
2. Gozani J (1979) Pulsed beam propagation through random media. Opt Lett 21:1712–1714
3. Zou Q, Lv B (2005) Transmission properties of ultrashort-pulsed Bessel-Gaussian beams in free space. Intense Laser Particle Beam 17(12):1769–1772
4. Sheppard CJR (2002) Generalized Bessel pulse beams. J Opt Soc Am A 19:2218–2222
5. Ji X, Tang M, Zhang T (2007) Spectral shifts and spectral jumps of ultrashort-pulsed Hermitian-Gaussian beams in a turbulent atmosphere. Acta Physica Sinica 56(07)
6. Hong ST, Ishimaru A (1976) Two-frequency mutual coherence function, coherence bandwidth, and coherence time of millimeter and optical waves in rain, fog, and turbulence. Radio Science 11(6):551–559
7. Sreenivasiah I, Ishimaru A, Hong ST (1976) Two-frequency mutual coherence function and pulse propagation in a random medium: an analytic solution to the plane wave case. Radio Sci 11(10):775–778
8. Suand HH, Plonus A (1971) Optical-pulse propagation in a turbulent medium. J Opt Soc Am 61:256–260
9. Liu C, Wernik A, Yeh K (1974) Propagation of pulse trains through a random medium. IEEE Trans Antennas Propag 22:624–627
10. Shin H, Sreenivashiah I, Ishimaru A (1977) Plane wave pulse propagation through random media. IEEE Trans Antennas Propag 25:822–828
11. Sreenivasiah I, Ishimaru A (1979) Beam wave two-frequency mutual-coherence function and pulse propagation in random media: an analytic solution. Appl Opt 18(10):1613–1618
12. Liu CH, Yeh CK (1979) Pulse spreading and wandering in random media. Radio Sci 14(5):925–931
13. Liu CH, Yeh CK (1980) Statistics of pulse arrival time in turbulent media. J Opt Soc Am A 70:168–172
14. Fante RL (1981) Two-position, two-frequency mutual-coherence function in turbulence. J Opt Soc Am 71:1446–1451

15. Young CY, Ishimaru A, Andrews LC (1996) Two-frequency mutual coherence function of a Gaussian beam pulse in weak optical turbulence: an analytic solution. Appl Opt 35(33):6522–6526
16. Young CY, Andrews LC, Ishimaru A (1988) Time-of-arrival fluctuations of a space-time Gaussian pulse in weak optical turbulence: an analytic solution. Appl Opt 37:7655–7660
17. Lajunen H, Vahimaa P, Tervo J (2005) Theory of spatially and spectrally partially coherent pulses. J Opt Soc Am A 22(8):1536–1546
18. Ding C, Lu B (2008) Spectral and temporal properties of partially spatial and partially spectral coherent Gaussian-Shell model pulsed beams in free space. Intense Laser Particle Beam 20:1855–1860
19. Chen C, Yang H, Lou Y (2012) Temporal broadening of optical pulses propagating through non-Kolmogorov turbulence. Opt Soc Am 20(7)
20. Li Y-Q, Wu Z, Wang M-J (2014) Partially coherent Gaussian-Schell model pulse beam propagation in slant atmospheric turbulence. Chin Phys B 23(6)
21. Fante RL (1981) Two-position, two-frequency mutual-coherence function in trubulence. J Opt Soc Am 71:1446–1451
22. Ishimaru A (1972) Temporal frequency spectra of multifrequency waves in turbulent atmosphere. IEEE Trans Antennas Propag 20:10–19
23. Fante R, Poirier J (1973) Mutual coherence function of a finite optical beam in a turbulent medium. Appl Opt 12:2247–2248
24. Young CY, Andrews LC, Ishimaru A (1998) Time-of-arrival fluctuations of a space-time Gaussian pulse in weak optical turbulence: an analytic solution. J Opt Soc Am 37(33):7655–7660

Chapter 9
Propagation of Partially Coherent Vortex Beams in Atmospheric Turbulence

Based on the Huygens-Fresnel principle, the cross-spectral **density** function is used to derive an analytic expression for the transmission time of a partially coherent Gaussian-Cher vortex beam in atmospheric turbulence. Then the effects of propagation distance, turbulence intensity, topological charge number, beam coherence length, beam waist width and wavelength on the beam intensity distribution are analyzed and discussed.

9.1 Introduction

The transmission of a laser beam through atmospheric turbulence is of great significance in remote sensing, tracking, and wireless optical communication [1, 2]. In the early 1990s, Wu et al. theoretically proved that "partially coherent light is less affected by atmospheric turbulence" [3] Thus, the transmission of partially coherent light in atmospheric turbulence has developed into an area of active research. A vortex beam has a continuous spiral wavefront, zero central intensity, and a phase singularity. A partially coherent vortex beam combines the characteristics of a vortex beam and that of a partially coherent beam. As such, the investigation of the propagation of partially coherent vortex beams in turbulent medium has high scientific value.

In 2001, Ponomarenko et al. [4] studied the transmission characteristics of a new class of partially coherent beams in free space. This beam had a separable phase and could be classified as an optical vortex, which can be expressed as the superposition of Laguerre-Gaussian (LG) beams. In 2009, Dijk et al. [5] used a spatial light modulator to generate a partially coherent vortex beam and investigated its intensity distribution and spectral coherence. In 2010, Liu et al. [6] utilized a Gauss-Schell model vortex beam to analyze the evolution of coherent vortices in the focusing field, and discovered that truncation parameters, normalized coherence parameters, and the

X. Ke, *Partially Coherent Optical Transmission Theory in Optical Wireless Communication*, Optical Wireless Communication Theory and Technology,
https://doi.org/10.1007/978-981-97-7158-5_9

topological charge number all affect the position and number of coherent vortices. In 2011, Chen et al. [7] studied the deep focusing characteristics of partially coherent circularly polarized vortex beams using Debye theory. It was determined that both the coherence of the incident beam and the numerical aperture of the focusing lens affect the intensity and coherence of the deep focusing plane. In 2012, Ding et al. [8] derived an analytical expression for the cross-spectral density matrix element of a partially coherent vortex beam during transmission in free space, and analyzed the change in the intensity at the center of the spot during transmission. In 2012, Fang et al. [9] performed a detailed analysis of the intensity distribution of partially coherent double-vortex beams during propagation in atmospheric turbulence. In 2013, Wang et al. [10] derived an expression for the spectral density (intensity) of a partially coherent flat-top vortex beam using the cross-spectral density function. It was established that as the beam propagates through atmospheric turbulence, the effect of the transmission distance, turbulence intensity, topological charge number, beam waist width, and coherence length on the spectral characteristics of the beam, (including spectral density, spectral polarization and spectral coherence) of a beam propagating in atmospheric turbulence are analyzed. In 2013, Alieva [11] et al. conducted an experimental study on the intensity distribution of a fully coherent vortex beam and a partially coherent vortex beam, and concluded that the latter can reduce speckle and increase the capacity for information coding. In 2014, Qin et al. [12] investigated the intensity characteristics of a partially phase-dry Besser-Gaussian vortex beam by using the Huygens-Fresnel principle, and determined the transmission distance, topological charge number, coherence length, and turbulence intensity of the beam affect the intensity distribution.

This chapter considers the Gaussian-Schell (GSM) vortex beam and utilizes the Huygens-Fresnel principle and the cross-spectral density function to derive an expression for the time-dependent intensity distribution of a partially coherent Gaussian-Schell vortex beam propagating in atmospheric turbulence. In addition, the variation law associated with its propagation is examined. The effects of the transmission distance, turbulence intensity, topological charge number, coherence length, beam waist width, and wavelength on the intensity distribution are also considered. Finally, the results are compared and analyzed.

9.2 Intensity Distribution of Partially Coherent Vortex Beams

Assuming that a light source is located in the plane $z = 0$, and the beam emitted by the source propagates through a turbulent medium along the positive direction of the z-axis, then the field component of the light beam in the source plane is given by:

$$E(\boldsymbol{r}, 0) = I_0 \exp\left(-\frac{\boldsymbol{r}^2}{\omega^2}\right) \exp(-in\theta) \exp(i\beta) \tag{9.1}$$

Without loss of generality, it is assumed that the statistical distribution of the phase β satisfies the Schell model [13], i.e.

$$C(\boldsymbol{r}_1, \boldsymbol{r}_2) = C(|\boldsymbol{r}_1 - \boldsymbol{r}_2|) = \exp\left(-|\boldsymbol{r}_1 - \boldsymbol{r}_2|^2 / \sigma^2\right) \tag{9.2}$$

The cross-spectral density function of the source plane is then given by: [14]

$$\begin{aligned} W(\boldsymbol{r}_1, \boldsymbol{r}_2, 0) &= \langle E(\boldsymbol{r}_1, 0) E^*(\boldsymbol{r}_2, 0) \rangle \\ &= I_0^2 \exp\left(-\frac{r_1^2 + r_2^2}{\omega^2}\right) \exp[-in(\theta_1 - \theta_2)] \exp\left[-\frac{(\boldsymbol{r}_1 - \boldsymbol{r}_2)^2}{\sigma^2}\right] \end{aligned} \tag{9.3}$$

where, β is the phase, $\boldsymbol{r}_1$, and $\boldsymbol{r}_2$ are any two-dimensional position vectors of the source plane, θ_1, and θ_2 are the phase angles associated with the source plane position vectors $\boldsymbol{r}_1$, and $\boldsymbol{r}_2$, respectively, $\langle\cdot\rangle$ is the ensemble average, $*$ is the complex conjugate, I_0 is a constant, ω is the beam waist width, σ is the coherence length, and n is the topological charge number.

According to the Huygens-Fresnel principle, when the beam is transmitted over a certain distance in a turbulent medium, the cross-spectral density can be expressed as follows [14, 15]:

$$\begin{aligned} W(\boldsymbol{\rho}_1, \boldsymbol{\rho}_2, z) = \left(\frac{k}{2\pi z}\right)^2 \int_{-\infty}^{\infty}\int_{-\infty}^{\infty}\int_{-\infty}^{\infty}\int_{-\infty}^{\infty} W(\boldsymbol{r}_1, \boldsymbol{r}_2, 0) \\ \times \exp\left[-\frac{ik}{2z}(\boldsymbol{r}_1 - \boldsymbol{\rho}_1) + \frac{ik}{2z}(\boldsymbol{r}_2 - \boldsymbol{\rho}_2)\right] \\ \times \langle \exp[\Psi(\boldsymbol{r}_1, \boldsymbol{\rho}_1, z) + \Psi^*(\boldsymbol{r}_2, \boldsymbol{\rho}_2, z)] \rangle d^2\boldsymbol{r}_1 d^2\boldsymbol{r}_2 \end{aligned} \tag{9.4}$$

where, z is the transmission distance, and $\boldsymbol{\rho}_1$, and $\boldsymbol{\rho}_2$ are any two-dimensional position vectors at the receiving plane. Using the quadratic approximation of the Rytov phase structure function, the spherical wave complex phase cross-correlation function can be expressed as, [9]

$$\begin{aligned} \langle \exp[\Psi(\boldsymbol{r}_1, \boldsymbol{\rho}_1, z) + \Psi^*(\boldsymbol{r}_2, \boldsymbol{\rho}_2, z)] \rangle &= \exp[-0.5 D_\Psi(\boldsymbol{r}_1 - \boldsymbol{r}_2)] \\ &= \exp\left[-\frac{1}{\rho_0^2}(\boldsymbol{r}_1 - \boldsymbol{r}_2)^2\right] \end{aligned} \tag{9.5}$$

where, $\Psi(\boldsymbol{r}_1, \boldsymbol{\rho}_1, z)$ is a random phase factor that describes the influence of a turbulent atmosphere on a spherical wave, $D_\Psi(\boldsymbol{r}_1 - \boldsymbol{r}_2)$ is the phase structure function, $\rho_0 = \left(0.545 C_n^2 k^2 z\right)^{-3/5}$ is the coherence length of a spherical wave transmitted through a turbulent medium, and C_n^2 is the structure constant of the atmospheric refractive index. Substituting Eqs. 9.3 and 9.5 into Eq. 9.4, the following is obtained after simplification

$$W(\rho_1, \rho_2, \varphi_1, \varphi_{2,}z) = I_0^2\left(\frac{k}{2\pi z}\right)^2 \exp\left[-\frac{ik}{2z}\left(\rho_1^2 - \rho_1^2\right)\right]$$
$$\times \int_{-\infty}^{\infty}\int_{-\infty}^{\infty}\int_{-\infty}^{\infty}\int_{-\infty}^{\infty} \exp\left[-\frac{ik}{2z}\left(r_1^2 - r_2^2\right)\right]$$
$$\times \exp\left[-\left(\frac{1}{\omega^2 + \sigma^2 + \rho_0^2}\right)\left(r_1^2 + r_2^2\right)\right]$$
$$\times \exp\left[\frac{ikr_1\rho_1}{z}\cos(\varphi_1 - \theta_1)\right] \times \exp\left[-\frac{ikr_2\rho_2}{z}\cos(\varphi_2 - \theta_2)\right]$$
$$\times \exp\left(2r_1r_2\left(\frac{1}{\sigma^2} + \frac{1}{\rho_0^2}\right)\right)\cos\left(\theta_1 - \theta_2\right))$$
$$\times \exp(-in(\theta_1 - \theta_2))r_1r_2dr_1dr_2d\theta_1d\theta_2 \tag{9.6}$$

In the Eq. 9.6, any two-dimensional position vectors of the source plane can be expressed as $\boldsymbol{r} = (r, \theta)$, the vector differential $d\boldsymbol{r}_1 d\boldsymbol{r}_2 = r_1r_2dr_1dr_2d\theta_1d\theta_2$ [16], the position vector $\boldsymbol{\rho}$ of the receiving plane can be expressed as $\boldsymbol{\rho} = (\rho, \varphi)$ [16], and φ_1, and φ_2 are the phase angles of the position vectors $\boldsymbol{\rho}_1$, and $\boldsymbol{\rho}_2$ of the receiving plane, respectively.

Equation 9.6 is the cross-spectral density function of a partially coherent vortex beam propagating through atmospheric turbulence, which can be simplified using the mutual transformation of special functions. A Bessel function can be expressed as, [17]

$$\exp(it\cos\beta) = \sum_{n=-\infty}^{+\infty} J_n(t)i^n\exp(in\beta) \tag{9.7}$$

and the Imaginary scalar Bezier equation can be expressed as [17]

$$I_n(z) = \frac{1}{2\pi i}\int_0^{2\pi i} \exp\left(z\frac{e^\omega + e^{-\omega}}{2} - n\omega\right)d\omega \tag{9.8}$$

From Eq. 9.7, we have:

$$\exp\left[\frac{ikr\rho}{z}\cos(\varphi - \theta)\right] = \sum_{l=-\infty}^{+\infty} J_l\left(\frac{kr\rho}{z}\right)i^l\exp[il(\varphi - \theta)] \tag{9.9}$$

and from Eq. 9.8, we have:

$$I_n\left(\frac{2r_1r_2}{\sigma^2}\right) = \frac{1}{2\pi}\exp(in\theta_2)\int_0^{2\pi}\exp\left[\frac{2r_1r_2}{\sigma^2}\cos(\theta_1 - \theta_2) - in\theta_1\right]d\theta_1$$

for which,

$$\int_0^{2\pi} \exp\left[\frac{2r_1r_2}{\sigma^2}\cos(\theta_1-\theta_2)-in\theta_1\right]d\theta_1$$
$$=2\pi\exp(-in\theta_2)I_n\left(\frac{2r_1r_2}{\sigma^2}\right) \tag{9.10}$$

In addition

$$\int_0^{2\pi}\exp(im\theta)d\theta=\begin{cases}2\pi & (\mathrm{m}=0)\\ 0 & (\mathrm{m}\neq 0)\end{cases} \tag{9.11}$$

Substituting Eqs. 9.9, 9.10, and 9.11 into Eq. 9.6, then assuming $\rho_1=\rho_2,\varphi_1=\varphi_2$, the intensity of partially coherent vortex beams transmitted through atmospheric turbulence can be expressed as follows:

$$I(\rho,z)=I_0^2\left(\frac{k}{z}\right)^2\sum_{l=-\infty}^{+\infty}\int_{-\infty}^{\infty}\int_{-\infty}^{\infty}\exp\left[-\left(\frac{1}{\omega^2+\sigma^2+\rho_0^2}\right)\left(r_1^2+r_2^2\right)\right]$$
$$\times\exp\left[-\frac{ik}{2z}\left(r_1^2+r_2^2\right)\right]J_l\left(\frac{k\rho r_1}{z}\right)J_l\left(\frac{k\rho r_2}{z}\right)$$
$$I_{l+n}\left(\frac{2r_1r_2}{\sigma^2}+\frac{2r_1r_2}{\rho_0^2}\right)r_1r_2dr_1dr_2 \tag{9.12}$$

Simplify

$$I(\rho,z)=I_0^2\left(\frac{k}{z}\right)^2\sum_{l=-\infty}^{+\infty}\int_{-\infty}^{\infty}\int_{-\infty}^{\infty}\exp\left[-(\alpha_1+\alpha_2)r_1^2\right]$$
$$\times\exp\left[-(\alpha_1-\alpha_2)r_2^2\right]$$
$$J_l(\eta r_1)J_l(\eta r_2)I_{l+n}(\beta r_1r_2)r_1r_2dr_1dr_2 \tag{9.13}$$

In the equation, $\alpha_1=\frac{1}{\omega^2+\sigma^2+\rho_0^2}$, $\alpha_2=\frac{ik}{2z}$, $\eta=\frac{k\rho}{z}$, $\beta=\frac{2(\sigma^2+\rho_0^2)}{\sigma^2\rho_0^2}$, $J_l(\eta r_1)$ are Bessel functions of order l, and $I_{l+n}(\beta r_1r_2)$ is a Bessel function of order $l+n$. Thus we obtain [16]:

$$I_v(z)=\sum_{k=0}^{+\infty}\frac{1}{k!\Gamma(v+k+1)}\left(\frac{z}{2}\right)^{v+2k} \tag{9.14}$$

$$\int_{-\infty}^{\infty} x^{\mu} \exp(-\alpha x^2) J_v(\beta x) dx = \frac{\beta^v \Gamma\left(\frac{v+\mu+1}{2}\right)}{2^{v+1} \alpha^{\frac{v+\mu+1}{2}} \Gamma(v+1)} \times {}_1F_1\left(\frac{v+\mu+1}{2}; v+1; -\frac{\beta^2}{4\alpha}\right)$$

$$(\mathrm{Re}\alpha > 0,\ \mathrm{Re}(\mu + \mathrm{v}) > -1) \tag{9.15}$$

where gamma function is given by:

$$\Gamma(n) = (n-1)! \tag{9.16}$$

The confluent hypergeometric function is given by:

$${}_1F_1(a; b; z) = \sum_{k=0}^{+\infty} \frac{(a+k-1)!(b-1)!z^k}{(a-1)!(b+k-1)!k!} \tag{9.17}$$

By substituting Eqs. 9.14 and 9.15 into (9.13), the intensity expression for a partially coherent vortex beam can be obtained as:

$$I(\rho, z) = I_1(\rho, z) + I_2(\rho, z) \tag{9.18}$$

wherein,

$$\begin{aligned}
I_1(\rho, z) = &\sum_{k=0}^{\infty}\sum_{l=0}^{\infty} \frac{1}{k!(|l+n|+k)!}\left(\frac{\beta}{4}\right)^{|l+n|+2k} \\
&\times {}_1F_1\left(\frac{|l+n|+2k+l+2}{2}; l+1; -\frac{\eta^2}{4(\alpha_1+\alpha_2)}\right) \\
&\times {}_1F_1\left(\frac{|l+n|+2k+l+2}{2}; l+1; -\frac{\eta^2}{4(\alpha_1-\alpha_2)}\right) \\
&\times \eta^{2l} \cdot \left(\Gamma\left(\frac{|l+n|+2k+l+2}{2}\right)\right)^2 \\
&\times 1 \Big/ \left[2^{2l+1} \cdot (\alpha_1+\alpha_2)^{\frac{|l+n|+2k+l+2}{2}} \cdot (\alpha_1-\alpha_2)^{\frac{|l+n|+2k+l+2}{2}} \cdot [\Gamma(l+1)]^2\right] \\
I_2(\rho, z) = &\sum_{k=0}^{\infty}\sum_{l=1}^{\infty} \frac{1}{k!(|l+n|+k)!}\left(\frac{\beta}{4}\right)^{|-l+n|+2k} \\
&\times {}_1F_1\left(\frac{|-l+n|+2k+l+2}{2}; l+1; -\frac{\eta^2}{4(\alpha_1+\alpha_2)}\right) \\
&\times {}_1F_1\left(\frac{|-l+n|+2k+l+2}{2}; l+1; -\frac{\eta^2}{4(\alpha_1+\alpha_2)}\right) \\
&\times \eta^{2l} \cdot \left(\Gamma\left(\frac{|-l+n|+2k+l+2}{2}\right)\right)^2
\end{aligned}$$

$$\times 1 \Big/ \left[2^{2l+1} \cdot (\alpha_1 + \alpha_2)^{\frac{|-l+n|+2k+l+2}{2}} \cdot (\alpha_1 + \alpha_2)^{\frac{|-l+n|+2k+l+2}{2}} \cdot [\Gamma(l+1)]^2 \right]$$

When the beam is transmitted through free space, the structure constant of the atmospheric refractive index is $C_n^2 = 0$, thus, $\rho_0 = 0$, and Eq. 9.13 α_1 can be expressed as,

$$\alpha_1 = \frac{1}{\omega^2 + \sigma^2}. \tag{9.19}$$

Equation 9.13β can also be expressed as:

$$\beta = \frac{2}{\sigma^2} \tag{9.20}$$

When the partially coherent beam is a completely coherent vortex beam, $\sigma \to \infty$.

9.3 Numerical Analysis

9.3.1 Comparison of Intensity Distribution of Two Vortex Beams

When a beam is transmitted through atmospheric turbulence, the intensity on the receiving plane is reduced owing to random fluctuations of the atmospheric refractive index. The intensity distribution on the receiving surface varies with the coherence of the beam. According to Eq. 9.18 and the conditions of transmission of completely coherent light in free space, the normalized intensity distribution of the partially coherent vortex beam and completely coherent vortex beam in atmospheric turbulence and free space are analyzed. Figure 9.1 shows the normalized intensity distribution of a fully coherent vortex beam and a partially coherent vortex beam in atmospheric turbulence and free space.

The parameters used are constant $I_0 = 1$, wavelength $\lambda = 632.8$ nm, wave number $k = 2\pi / \lambda$, beam waist width $\omega = 0.01$ m, coherence length $\sigma = 0.03$ m, topological charge $n = 2$, atmospheric refractive index structure constant $C_n^2 = 10^{-15}$ m$^{-2/3}$, and transmission distance z = 2 km. I_{pt} and I_{pf} represent the normalized intensity of the partially coherent vortex beam propagating in atmospheric turbulence and free space, respectively. I_{ft} and I_{ff} represent the normalized intensity of a fully coherent vortex beam propagating in a turbulent atmosphere and free space, respectively. Figure 9.1 shows that when the fully coherent vortex beam is transmitted in free space, the intensity distribution is always hollow and the central intensity is always zero. This is because it is not affected by the beam's coherence and atmospheric turbulence, which is consistent with the conclusion in reference [7]. When a fully coherent vortex

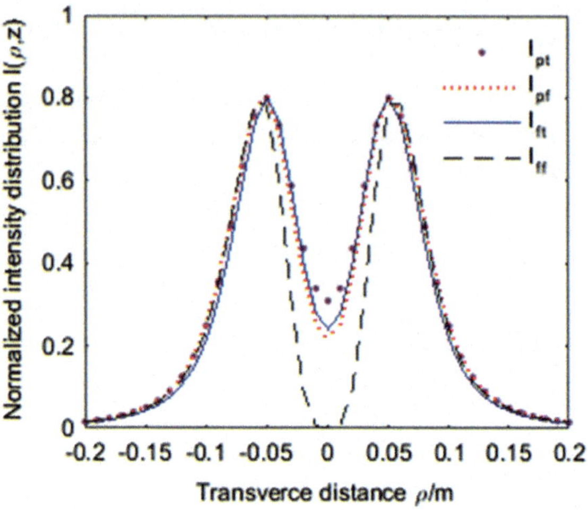

Fig. 9.1 Intensity distribution of partially and completely coherent vortex beams in atmospheric turbulence

beam propagates in a turbulent atmosphere, it is significantly affected compared to free space propagation, and the central intensity increases with the increase of the propagation distance. When the partially coherent vortex beam propagates in free space, it is significantly affected by the coherence of the beam, and the central intensity also increases. For propagation in a turbulent atmosphere, the vortex beam is affected by both the beam's coherence and atmospheric turbulence. The increase in the central intensity is the most prominent, and the greater the influence of the factors, the more salient the change in the intensity on the receiving surface, which is consistent with observation. Moreover, by comparing the intensity distribution of a partially coherent vortex beam and a fully coherent vortex beam propagating through atmospheric turbulence with their respective intensity distributions in free space, it can be determined that for the same transmission distance, the central intensity value of the partially coherent vortex beam has little effect. Therefore, the results show that the partially coherent vortex beam has better turbulence suppression compared to the completely coherent vortex beam. In the literature [3], it is established that "partially coherent light is less affected by atmospheric turbulence than fully coherent light," and a partially coherent vortex beam has similar properties to partially coherent light.

9.3.2 *Effects of Propagation Distance and Turbulence Intensity on Beam Intensity Distribution*

Figures 9.2 and 9.3 show the intensity distribution of partially coherent vortex beams in atmospheric turbulence with transmission distance when the atmospheric refractive index structure constant is $C_n^2 = 10^{-14}\,\mathrm{m}^{-2/3}$ and $C_n^2 = 10^{-15}\,\mathrm{m}^{-2/3}$, respectively. The selected parameters are constant $I_0 = 1$, wavelength $\lambda = 632.8\,\mathrm{nm}$, wave number $k = 2\pi / \lambda$, beam waist width $\omega = 0.01$ m, coherence length $\sigma = 0.03$ m, and topological charge $n = 2$. In Figs. 9.2 and 9.3, owing to the influence of atmospheric turbulence, the central intensity of the partially coherent vortex beam increases gradually with the increase of the transmission distance. When the transmission distance reaches a certain value, the intensity distribution of the beam changes from hollow to Gaussian. For a short transmission distance, the influence of atmospheric turbulence on the beam intensity distribution is not significant, and the central intensity of the beam increases slowly. As the transmission distance increases, the influence of turbulence on the beam intensity gradually increases. Figures 9.2 and 9.3 show that the larger the atmospheric refractive index structure constant, the greater the influence of atmospheric turbulence on the propagation of a light beam in a turbulent medium, and the shorter the transmission distance required for the transformation of intensity distribution from hollow to Gaussian beam. In Fig. 9.2, the structure constant of the atmospheric refractive index is $C_n^2 = 10^{-14}\,\mathrm{m}^{-2/3}$. When the transmission distance is $z = 1$ km, the intensity distribution is no longer hollow, and two points with minimum intensity are observed. When the transmission distance is $z = 4$ km, the intensity distribution is Gaussian. In Fig. 9.3, the structure constant of the atmospheric refractive index is $C_n^2 = 10^{-15}\,\mathrm{m}^{-2/3}$. When the transmission distance is $z = 8$ km, the intensity distribution is no longer hollow, and two points with the lowest intensity are observed. When the transmission distance is $z = 10$ km, the central intensity of the beam is the strongest. When the transmission distance is $z = 24$ km, the intensity distribution is transformed from hollow to Gaussian, and the Gaussian distribution is unchanged.

Owing to the random fluctuation of the atmospheric refractive index, phase distortion of the beam is observed after a certain distance as it propagates in atmospheric turbulence, and the intensity of the center of the beam increases gradually. To study the influence of atmospheric turbulence on beam propagation, the time-intensity

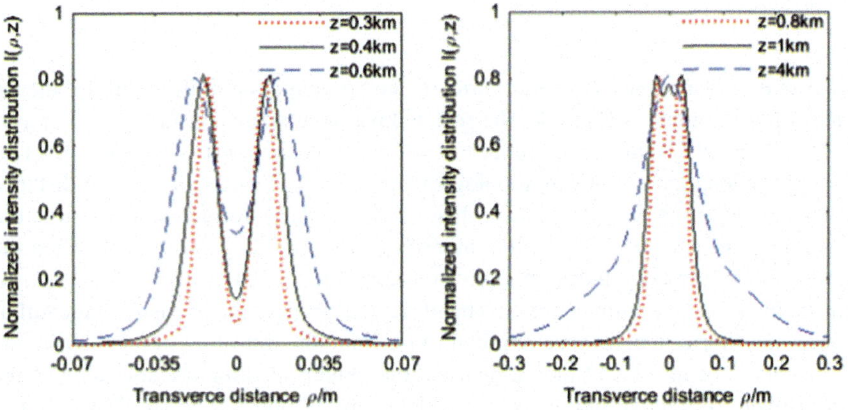

Fig. 9.2 Intensity distribution of partially coherent vortex beams at different transmission distances ($C_n^2 = 10^{-14}\,\mathrm{m}^{-2/3}$)

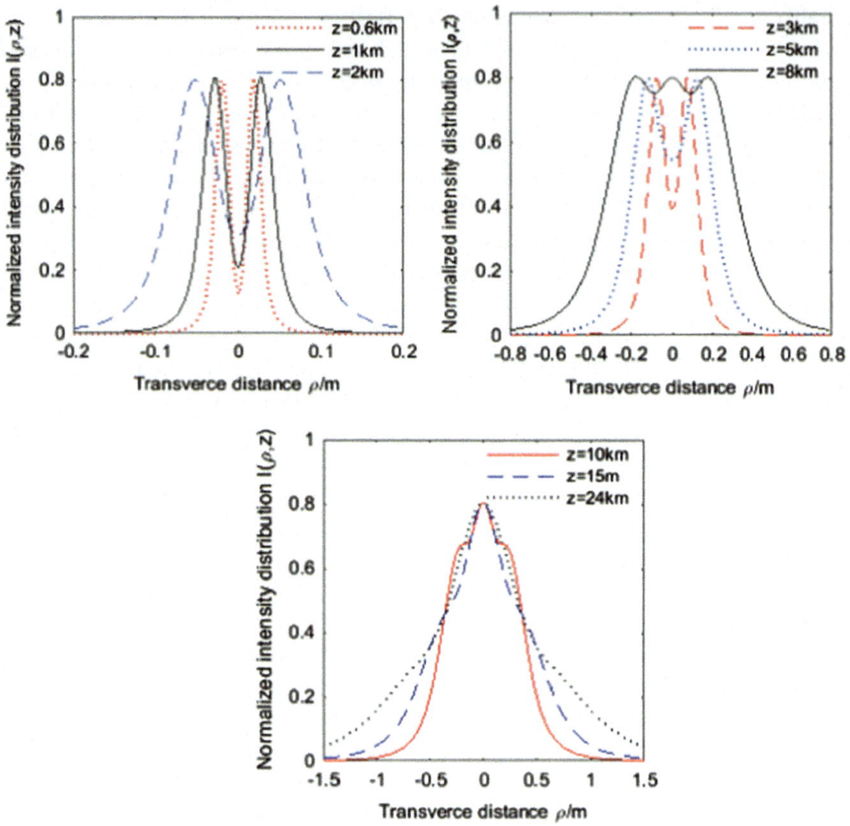

Fig. 9.3 Normalized intensity distribution of partially coherent vortex beams at different propagation distances ($C_n^2 = 10^{-15}\,\mathrm{m}^{-2/3}$), **a** z = 0.6, 1, 2 km; **b** z = 3, 5, 8 km; **c** z = 10, 15, 24 km

distribution of partially coherent vortex beams propagation in different turbulence intensities is shown in Fig. 9.4. The selected parameters are constant $I_0 = 1$, wavelength $\lambda = 632.8\,\mathrm{nm}$, wave number $k = 2\pi/\lambda$, beam waist width $\omega = 0.01\,\mathrm{m}$, coherence length $\sigma = 0.03\,\mathrm{m}$, topological charge $n = 2$, and transmission distance $z = 0.8\,\mathrm{km}$. Figure 9.4 shows that when the atmospheric refractive index structure constant is $C_n^2 = 10^{-13}\,\mathrm{m}^{-2/3}$, the intensity distribution is gaussian, and when the atmospheric refractive index structure constant is $C_n^2 = 10^{-16}\,\mathrm{m}^{-2/3}$, the center of the beam has a minimum intensity. Therefore, the stronger the intensity of the turbulence, the shorter the distance required to facilitate the transition of the distribution of a hollow beam to a Gaussian distribution. When the structure constants of the atmospheric refractive index are $C_n^2 = 10^{-15}\,\mathrm{m}^{-2/3}$ and $C_n^2 = 10^{-16}\,\mathrm{m}^{-2/3}$, The magnitude of the central light intensity of the beam is very close, i.e., the weaker the atmospheric turbulence, the closer the influence on the central intensity of the

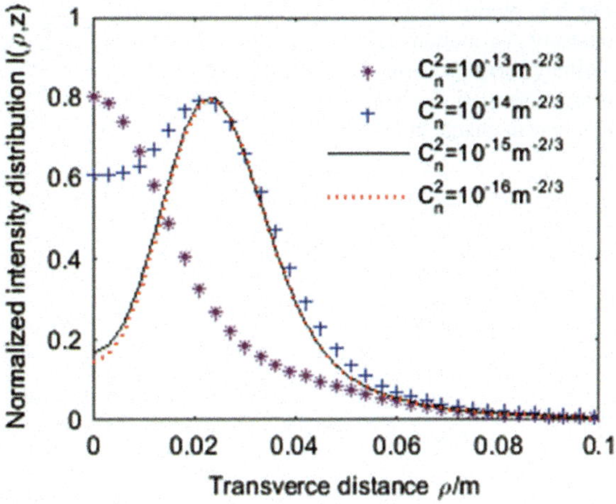

Fig. 9.4 Normalized intensity distribution of partially coherent vortex beams at different turbulence intensities

beam, and the shorter the transmission distance required to transform the intensity distribution from hollow to Gaussian.

9.3.3 The Influence of Beam Parameters on Beam Intensity Distribution

When the topological charge number is $n = 0$, as the transmission distance increases, the transmission-time intensity distribution of partially coherent vortex beams in atmospheric turbulence is shown in Fig. 9.5. The selected parameters are constant $I_0 = 1$, wavelength $\lambda = 632.8$ nm, wave number $k = 2\pi / \lambda$, beam waist width $\omega = 0.01$ m, coherence length $\sigma = 0.03m$, topological charge $n = 0$, and atmospheric refractive index structure constant $C_n^2 = 10^{-15}\,\mathrm{m}^{-2/3}$. Figure 9.5 shows the light intensity distribution with the propagation distance of 0.8, 1.5, 2, 3 km when the light beam is transmitted in the turbulent medium. When the topological charge number is zero, with the increase of the transmission distance, the intensity distribution of a partially coherent vortex beam maintains a Gaussian distribution. This is because when the topological charge number is zero, the partially coherent vortex beam is partially coherent, and no longer has a singularity. The intensity of the light then exhibits a Gaussian distribution.

Figure 9.6 shows the transmission time intensity distribution of some coherent vortex beams in atmospheric turbulence under different topological charges. The selected parameters are constant $I_0 = 1$, wavelength $\lambda = 632.8$ nm, wave number $k = 2\pi / \lambda$, beam waist width $\omega = 0.01$ m, coherence length $\sigma = 0.03$ m, atmospheric refractive index structure constant $C_n^2 = 10^{-15}\,\mathrm{m}^{-2/3}$, and transmission distance $z = 0.8$ km. Figure 9.6 shows that for the same transmission distance, the greater the topological charge number, the smaller the central light intensity of

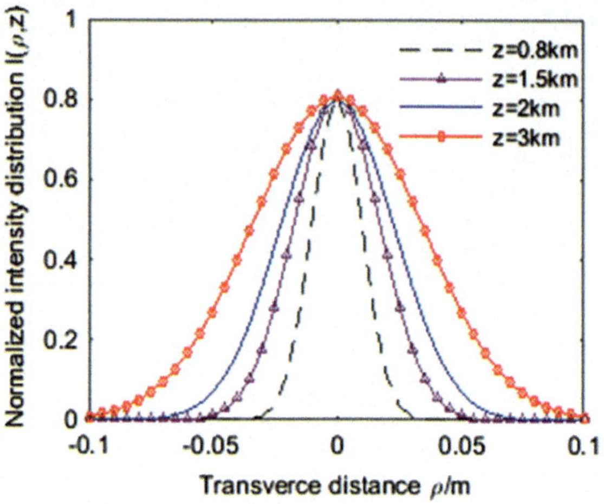

Fig. 9.5 Normalized intensity distribution of partially coherent vortex beam transmission when topological charge number is zero

the beam, the stronger the ability to suppress turbulence, and the longer the transmission distance required for the conversion of the light intensity distribution from the hollow distribution to the Gaussian distribution, this is consistent with the actual situation. The larger the topological charge, the larger the hollow area of the beam. The speed at which the center of the beam becomes solid and achieves a maximum intensity is slower compared to the case of a beam with a smaller topological charge.

Figure 9.7 shows the transmission time intensity distribution of partially coherent vortex beams in atmospheric turbulence and free space for different coherence lengths. The selected parameters are constant $I_0 = 1$, wavelength $\lambda = 632.8\,\mathrm{nm}$, wave number $k = 2\pi / \lambda$, beam waist width $\omega = 0.01$ m, topological charge $n = 2$, atmospheric refractive index structure constant $C_n^2 = 10^{-15}\,\mathrm{m}^{-2/3}$, and transmission distance $z = 1$ km. Figure 9.7 shows that the smaller the coherence length of the

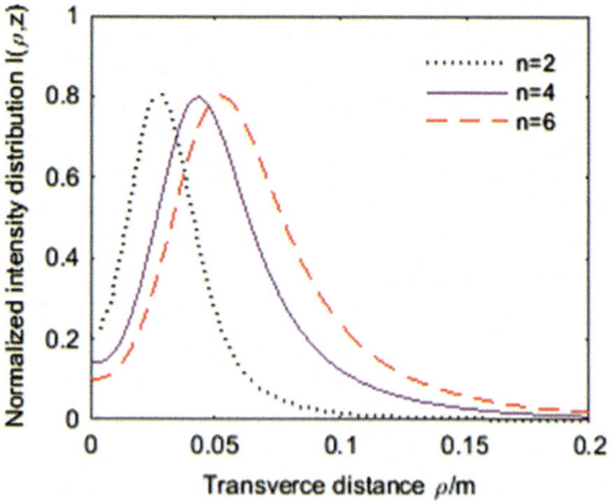

Fig. 9.6 Normalized intensity distribution of partially coherent vortex beams for different topological charge numbers

beam, the closer the transmission characteristics of the central light in turbulence are to those propagating in free space. This indicates that the smaller the coherence length of the beam, the greater its capacity to suppress turbulence. Moreover, the better the coherence of the beam, the weaker the central intensity of the beam, i.e., the better the coherence, the more distinct the singularity of the beam, the longer the distance required for the intensity distribution to transform from hollow to Gaussian. This is because the better the coherence of the light beam, the more coherent the light source of each point on the source plane that is superimposed at the center of the observation plane, the more significant the cancellation, and the longer the transmission distance required, which is consistent with the main conclusion in the literature [7]. Therefore, for practical applications, the coherence length of the beam should be carefully chosen. The beam should have good anti-turbulence characteristics, and the intensity should maintain a hollow distribution.

Figure 9.8 shows the transmission time intensity distribution of some coherent vortex beams propagating in atmospheric turbulence for different beam waist widths. The selected parameters are constant $I_0 = 1$, wavelength $\lambda = 632.8\,\text{nm}$, wave number $k = 2\pi/\lambda$, coherence length $\sigma = 0.03\,\text{m}$, topological charge $n = 2$, atmospheric refractive index structure constant $C_n^2 = 10^{-15}\,\text{m}^{-2/3}$, and transmission distance $z = 1$ km. Figure 9.8 shows that the larger the beam waist width, the higher the central intensity of the beam, and the shorter the transmission distance required for the intensity distribution of the beam in a turbulent medium to transform from hollow to Gaussian.

Figure 9.9 shows the transmission time intensity distribution of partially coherent vortex beams in atmospheric turbulence for different wavelengths. The selected parameters are constant $I_0 = 1$, beam waist width $\omega = 0.01\,\text{m}$, coherence length $\sigma = 0.03\,\text{m}$, topological charge number $n = 2$, atmospheric refractive index structure constant $C_n^2 = 10^{-15}\,\text{m}^{-2/3}$, transmission distance $z = 1$ km, and wavelength $\lambda = 532, 632.8, 1060\,\text{nm}$. Figure 9.9 shows that the wavelength has little influence

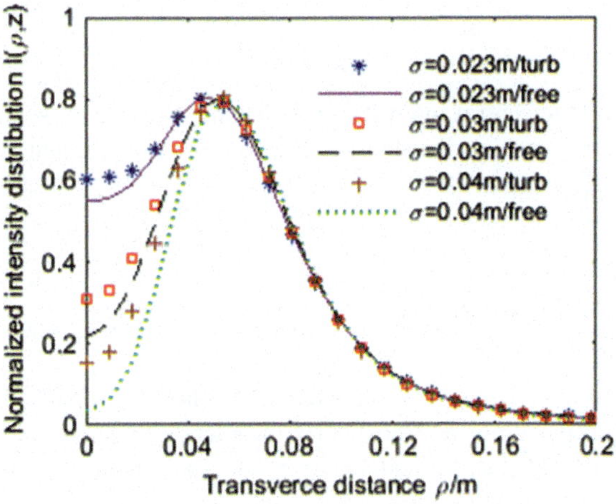

Fig. 9.7 Normalized intensity distribution of partially coherent vortex beams with different coherence lengths

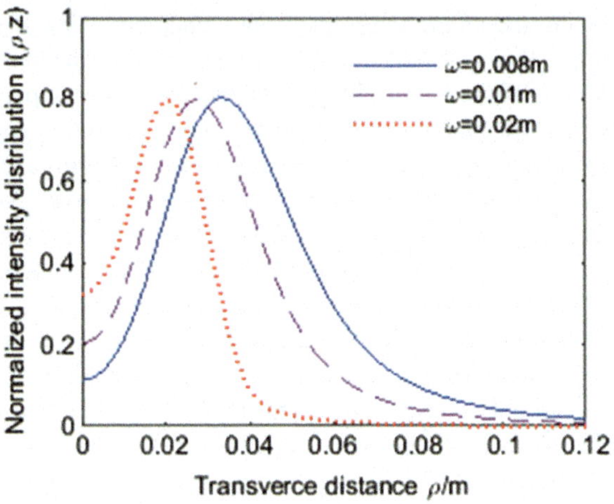

Fig. 9.8 Normalized intensity distribution of partially coherent vortex beams for different waist widths

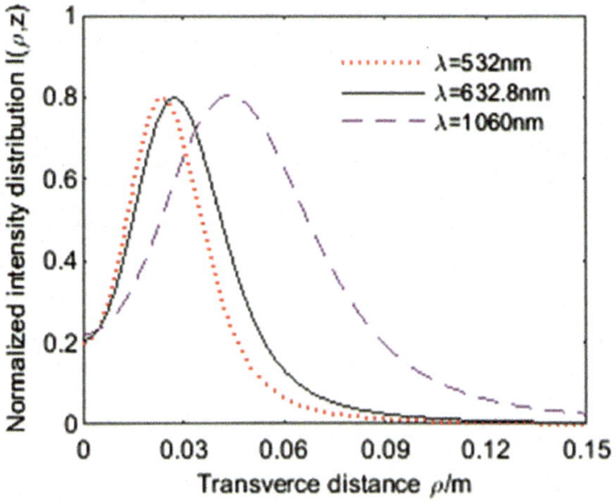

Fig. 9.9 Normalized intensity distribution of partially coherent vortex beams at different wavelengths

on the central intensity of the beam. The transmission distance for the transition of the intensity distribution from hollow to Gaussian is generally independent of the wavelength.

9.4 Summary

In this chapter, the intensity expression for a partially coherent GSM vortex beam propagating in atmospheric turbulence is derived, and the intensity distribution is simulated numerically. The results show that:

1. When a completely coherent GSM vortex beam is transmitted in free space, the central intensity of the beam is always zero because it is not affected by the beam's coherence and the atmospheric turbulence.
2. When a partially coherent GSM vortex beam propagates in atmospheric turbulence, the central intensity increases gradually with the increase of the transmission distance, and the intensity distribution changes from hollow to Gaussian.
3. The central intensity of a partially coherent GSM vortex beam increases with the increase of the turbulence intensity and beam waist width and increases with the decrease of the coherence length and the topological charge. When the topological charge is zero, the intensity distribution of the beam is Gaussian. The wavelength has little effect on the central intensity of the beam.

References

1. Gbur G, Wolf E (2002) Spreading of partially coherent beams in random media. J Opt Soc Am A 19(8):1592–1598
2. Shirai T, Dogariu A, Wolf E (2003) Directionality of Gaussian Schell—model beams propagating in atmospheric turbulence. Opt Lett 28(8):610–612
3. Wu J (1990) Propagation of Gaussian-Schell beam through turbulent media. J Mod Opt 37(4):671–684
4. Ponomarenko SA (2001) A class of partially coherent beams carrying optical vortices. J Opt Soc Am A 18(1):150–156
5. van Dijk T, Visser TD (2009) Evolution of singularities in a partially coherent vortex beam. J Opt Soc Am 26(4):741–744
6. Liu P, Yang H et al (2010) Coherence vortex evolution of partially coherent vortex beams in the focal region. Opt Laser Technol 42:99–104
7. Chen B, Pu J (2009) Tight focusing of partially coherent and circularly polarized vortex beams. J Opt Soc Am A 26(4):862–869
8. Ding PF, Pu JX (2012) Spot analysis of partially coherent vortex beam propagation. J Phys 17(61):174–201
9. Fang G, Pu J (2012) Propagation of partially coherent double-vortex beams in turbulent atmosphere. Opt Laser Technol 44:1780–1785
10. Wang H, Qian X (2013) Spectral properties of a random electromagnetic partially coherent flat-topped vortex beam in turbulent atmosphere. Opt Laser Technol 291:38–47
11. Alieva J, Rodrigo A, Cámara A, Abramochkin E (2013) Partially coherent stable and spiral beams. Opt Soc Am A 30(11):2237–2243
12. Qin Z, Tao R, Zhou P, Xu X (2013) Propagation of partially coherent Bessel–Gaussian beams carrying optical vortices in non-Kolmogorov turbulence. Opt Laser Technol 56:182–188
13. Cheng K, Lü B (2009) Spatial correlation properties and correlation vortices of partially coherent vortex beams diffracted by an aperture. J Mod Opt 56(9):1119–1127
14. Wang T, Pu JX (2007) Propagation characteristics of partially coherent vortex beams in turbulent atmosphere. Opt Technol 11(33):12–15
15. Cai Y, He S (2006) Propagation of a partially coherent twisted anisotropic Gaussian Schell-model beam in a turbulent atmosphere. Appl Phys Lett 89(04):11–17
16. Cang J, Zhang Y (2009) Propagation properties of focused partially coherent vortex beams in turbulent atmosphere. ACTA PHOTONICA SINCA 5(38):1277–1282

17. Wang JJ, Rong J (2012) Influence of atmospheric turbulence on spatial coherence of partially coherent light. University of Electronic Science and Technology of China, 5

Chapter 10
Propagation of Array Beam in Atmospheric Turbulence

A combined beam array can improve the transmitting power of a wireless optical communication antenna and the ability of the beam to suppress atmospheric turbulence. Array beam synthesis can increase beam transmission distance, which is also needed in an optical MIMO (Multiple-Input Multiple-Output) system, which is beneficial to suppressing atmospheric turbulence and increasing channel capacity. This chapter discusses the propagation characteristics of array beam synthesis in atmospheric turbulence.

10.1 Introduction

The rapid advancement of deep space exploration and the development of satellite-earth communication has created a need for lasers of increasingly higher power output. Although the output power of a single fiber laser is in the kilowatt range [1], owing to the restricted relationship between the transmitting power of a laser and its repetition frequency, it is difficult to simultaneously achieve high power and high-speed modulation by only increasing the transmitting power of a single laser. To address this problem, beam synthesis technology has been used to obtain high-power and high-quality laser outputs. The propagation characteristics and beam quality of synthetic beams in free space have been extensively reported [2–5]. Li and Luo [5] investigated the light intensity distribution and Power in the bucket (PIB) of multiple radial arrays of completely coherent non-paraxial Hermite-Gaussian beams synthesized using two approaches in free space. The results showed that the $P \times Q$ radial array beam in the far-field is the average light intensity P^2Q^2 and PQ times (as shown in Fig. 10.1, where P is the number of rings, and Q is the number of ring beams) after coherent and noncoherent synthesis of a single beam of light. The PIB increases with the decrease of the beam pattern number m. When the bucket radius is small, the PIB increases with an increase of $P(P < Q)$. In recent years, the

X. Ke, *Partially Coherent Optical Transmission Theory in Optical Wireless Communication*, Optical Wireless Communication Theory and Technology,
https://doi.org/10.1007/978-981-97-7158-5_10

propagation characteristics of fully coherent array beams in atmospheric turbulence have been investigated. Ji [6] analyzed the beam propagation and directivity of a radially distributed Gaussian array, and the results showed that in the case of coherent synthesis, as the number of sub-beams decreased and the radial distribution radius increased, the less the array beam propagation was affected by turbulence. Ji and Li [7] also compared the average light intensity, second-moment beam width, and the PIB of off-axis Gaussian array beams synthesized using two ways approaches. The results showed that turbulence resulted in an expansion and decreased the peak light intensity of the synthesized beams in both coherent and incoherent synthesis. Moreover, the greater the number of beams in the synthesized beams, the less the effect of turbulence. Lu et al. [8] investigated Gaussian array beam propagation under non-Kolmogorov turbulence and maximum beam spread of Gaussian array when the generalized exponent was $\alpha = 3.108$. Zhou et al. [9] investigated the beam propagation of a linear Gaussian array beam under non-Kolmogorov turbulence, and the results showed that the smaller the inner scale and the larger the outer scale of turbulence, the larger the beam spread of the array. The intensity distribution, beam propagation and PIB of fully coherent array beams have been reported in the literature.

In 1990, Wu et al. [10] discovered that "compared with fully coherent light, partially coherent light is less affected by turbulence." Thus, research on the transmission characteristics of partially coherent array beams in atmospheric turbulence has attracted increasing attention [11–19]. Zhou et al. [20] investigated the light intensity distribution of partially coherent flat-top array beams in atmospheric turbulence but did not consider the variation of the light intensity distribution with the location of the center of the array beam and the internal and external scales of turbulence.

Fig. 10.1 Multiple radial laser beam array

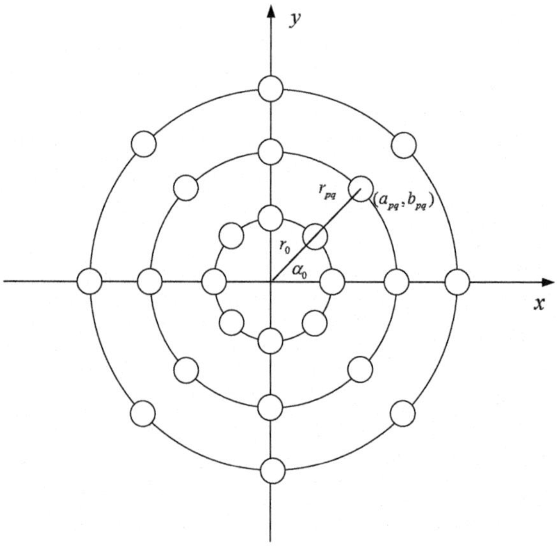

Liu [21] et al. investigated the light intensity distribution of partially coherent Gaussian vortex array beams and discovered that they could be synthesized as Gaussian beams in the far-field owing to the influence of atmospheric turbulence and their coherence length. Liu et al. [22] conducted a quantitative analysis of beam propagation of linearly distributed partially coherent hyperbolic cosine-Gaussian array beams in atmospheric turbulence based on the turbulence distance, and found that the use of appropriate beam parameters and beam stacking methods can reduce the impact of turbulence on beam propagation. However, the variation of beam propagation with coherence length and the internal and external scales of turbulence were not considered. Wang et al. [23] investigated the light intensity distribution and PIB of radial partially coherent flat-top array beams, but did not consider the influence of the array beam center position, the turbulence intensity, and other influencing factors.

In this section, we consider a partially coherent GSM beam array model using an integral transform array to investigate the effect of propagation through atmospheric turbulence on the RMS of the intensity distribution, beam width, and PIB. This involves the use of analytical changes under different impact factors in numerical analysis. The main results are considered to be reasonable physical interpretation.

10.2 Theoretical Calculation

10.2.1 The Light Intensity Distribution

The cross-spectral density function of a partially coherent single GSM beam in the $z = 0$ plane is [24, 25]:

$$W^{(0)}(\boldsymbol{\rho}_{s1}, \boldsymbol{\rho}_{s2}; 0) = A\exp\left(-\frac{|\boldsymbol{\rho}_{s1}|^2 + |\boldsymbol{\rho}_{s2}|^2}{4\sigma_s^2}\right) \cdot \exp\left(-\frac{|\boldsymbol{\rho}_{s1} - \boldsymbol{\rho}_{s2}|^2}{2\sigma_g^2}\right) \tag{10.1}$$

where, $\boldsymbol{\rho}_{s1}, \boldsymbol{\rho}_{s2}$ are the coordinate vectors of two points in the source plane; A, σ_s and σ_g represent the light intensity, beam waist width, and coherence length of the light source, respectively. The larger the value of σ_g, the greater the coherence of the beam, and the worse the coherence. When $\sigma_g \to \infty$, the beam is a completely coherent GSM beam.

Assuming that the beams of a partially coherent GSM array are independent and quasi-direct, their cross-spectral density function in the $z = 0$ plane can be expressed as [26]:

$$W^{(0)}(\boldsymbol{\rho}_{s1}, \boldsymbol{\rho}_{s2}; 0) = A\sum_{i=1}^{N}\exp\left(-\frac{|\boldsymbol{\rho}_{s1} - \mathbf{s}_i|^2 + |\boldsymbol{\rho}_{s2} - \mathbf{s}_i|^2}{4\sigma_s^2}\right)\exp\left(-\frac{|\boldsymbol{\rho}_{s1} - \boldsymbol{\rho}_{s2}|^2}{2\sigma_g^2}\right) \tag{10.2}$$

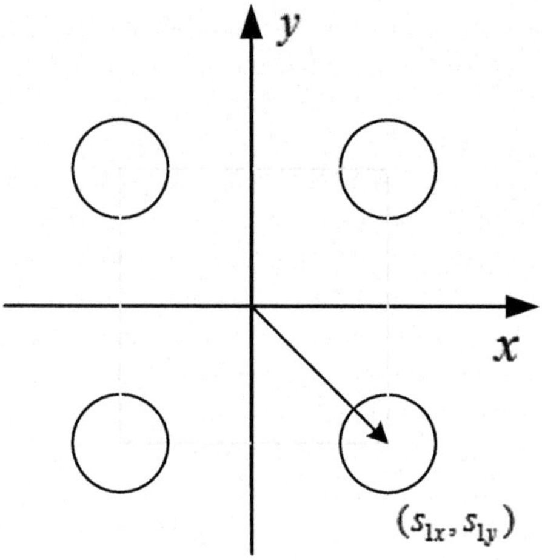

Fig. 10.2 Coordinate diagram of array beam

where, $\mathbf{s}_i = (s_{ix}, s_{iy})(i = 1, 2, ..., N)$ is a vector that represents the center position of the i-th sub-beam. Since we mainly investigate the square array composed of four partially coherent GSM beams, $N = 4$, as shown in Fig. 10.2.

According to the Huygens-Fresnel principle, the cross-spectral densities of points (ρ_1, Z) and (ρ_2, Z) at the cross-section $Z > 0$ can be expressed as [25]:

$$W(\boldsymbol{\rho}_1, \boldsymbol{\rho}_2; z) = \left(\frac{k}{2\pi z}\right)^2 \iint d^2\boldsymbol{\rho}_{s1} \iint d^2\boldsymbol{\rho}_{s2} W^{(0)}(\boldsymbol{\rho}_{s1}, \boldsymbol{\rho}_{s2}; 0) \times \exp\left\{\frac{-ik}{2z}\left[(\boldsymbol{\rho}_1 - \boldsymbol{\rho}_{s1})^2 - (\boldsymbol{\rho}_2 - \boldsymbol{\rho}_{s2})^2\right]\right\} \left\langle \exp\left[\phi^*(\boldsymbol{\rho}_1, \boldsymbol{\rho}_{s1}, z) + \phi(\boldsymbol{\rho}_2, \boldsymbol{\rho}_{s2}, z)\right]\right\rangle \tag{10.3}$$

where, $\langle \exp[\phi^*(\boldsymbol{\rho}_1, \boldsymbol{\rho}_{s1}, z) + \phi(\boldsymbol{\rho}_2, \boldsymbol{\rho}_{s2}, z)]\rangle$ represents the total statistical average value of the atmospheric turbulence, and $\phi(\boldsymbol{\rho}, \boldsymbol{\rho}_s)$ represents the complex random perturbation of spherical waves transmitted from the emission aperture $(\boldsymbol{\rho}_s; 0)$ to the target point $(\boldsymbol{\rho}; \mathbf{Z})$ in the turbulent medium.

Assuming that the fluctuations of a light wave and the atmospheric turbulence are statistically independent, the mean values of these parameters can be calculated separately, and the angle brackets that describe the impact of turbulence in the preceding equation can be expressed as [27]:

$$\left\langle \exp\left[\phi^*(\boldsymbol{\rho}_1, \boldsymbol{\rho}_{s1}, z) + \phi(\boldsymbol{\rho}_2, \boldsymbol{\rho}_{s2}, z)\right]\right\rangle \cong \exp\left[\frac{-\pi^2\kappa^2 z}{3}\left[(\boldsymbol{\rho}_1 - \boldsymbol{\rho}_2)^2 + (\boldsymbol{\rho}_1 - \boldsymbol{\rho}_2)\cdot(\boldsymbol{\rho}_{s1} - \boldsymbol{\rho}_{s2}) + (\boldsymbol{\rho}_{s1} - \boldsymbol{\rho}_{s2})^2\right]\int_0^\infty \kappa^3\phi_n(\kappa)d\kappa\right] \tag{10.4}$$

where $\int_0^\infty \kappa^3 \phi_n(\kappa) d\kappa$ represents the influence of turbulence, $\phi_n(\kappa)$ is the power spectrum of refractive index fluctuations, which can be used to characterize the Tatarskii model or Kolmogorov model, and the wave number is $\kappa = 2\pi/\lambda$ (λ is the wavelength).

Letting $\boldsymbol{\rho}_1 = \rho_2 = \boldsymbol{\rho}$ and substituting Eqs. (10.2) and (10.4) into Eq. (10.3) to obtain an integral solution, the average light intensity distribution of a partially coherent GSM array beam is given as:

$$\langle I(\boldsymbol{\rho}, z)\rangle = W(\boldsymbol{\rho}, \boldsymbol{\rho}, z) = \frac{A}{\Delta^2(z)} \sum_{i=1}^{4} \exp\left[c_1 \boldsymbol{\rho}^2 - 2c_1 \mathbf{s}_i \boldsymbol{\rho} + c_2 \mathbf{s}_i^2\right] \tag{10.5}$$

where we have,

$$c_1 = \frac{1}{2\Delta^2(z)\sigma_s^2}$$

$$c_2 = \frac{z^2}{2k^2\Delta^2(z)\sigma_s^4} + \frac{Mz^2}{k^2\Delta^2(z)\sigma_s^4} - \frac{1}{2\sigma_s^2}$$

$$\Delta^2(z) = 1 + \left[\frac{z}{k\sigma_s\delta}\right]^2 + \frac{2Mz^2}{k^2\sigma_s^2}$$

$$\frac{1}{\delta^2} = \frac{1}{4\sigma_s^2} + \frac{1}{\sigma_g^2}$$

When $s_i = 0$, an analytical expression or the light intensity distribution of a partially coherent single GSM beam located at the center coordinate in atmospheric turbulence can be obtained, which is consistent with the results of reference [25].

For the turbulent power spectrum model and the turbulence entity ψ, we adopt the modified Von Karmon spectrum that considers both internal and external scales. Thus, we have [28]:

$$\phi_n(\kappa, h) = 0.033C_n^2(h)\exp\left(-\kappa^2/\kappa_m^2\right) \times \left(\kappa^2 + \kappa_0^2\right)^{-116} \tag{10.6}$$

where, $\kappa_m = 5.92/l_0, \kappa_0 \approx 2\pi/L_0$, and l_0 and L_0 are the inner and outer scales of turbulence, respectively.

Considering horizontal beam transmission, the atmospheric refractive index structure constant $C_n^2(h)$ is a constant represented by c_0, $M = \frac{1}{3}\pi^2 k^2 z \int_0^\infty \kappa^3 \phi_n(\kappa) \mathrm{d}\kappa$, and $M = 0$ for free space transmission.

Equations (10.5) and (10.6) show that the light intensity distribution is related to the atmospheric refractive index structure constant, beam coherence length, array beam center position, and turbulence intensity.

10.2.2 Root Mean Square Beam Width

The root mean square beam width of a partially coherent single GSM beam can be defined as [26]:

$$\omega(z) = \sqrt{\frac{\int I(\boldsymbol{\rho}, z)|\boldsymbol{\rho}|^2 d^2\boldsymbol{\rho}}{\int I(\boldsymbol{\rho}, z) d^2\boldsymbol{\rho}}} \tag{10.7}$$

Thus, the root mean square beam width of an array beam in the x and y directions can be expressed as [29]:

$$\omega_x^2(z) = \frac{\iint (x - s_{ix})^2 \langle I(\rho, z)\rangle dxdy}{\iint \langle I(\rho, z)\rangle dxdy} \tag{10.8}$$

$$\omega_y^2(z) = \frac{\iint (y - s_{iy})^2 \langle I(\rho, z)\rangle dxdy}{\iint \langle I(\rho, z)\rangle dxdy} \tag{10.9}$$

Substituting Eq. (10.5) into Eqs. (10.8) and (10.9) for simplification, the root mean square beam width of a partially coherent GSM array beam in the x and y directions can be expressed as:

$$\omega_x(z) = \omega_y(z) = \sqrt{\sum_{i=1}^{4} \left(\Delta^2(z)\sigma_s^2\right) \exp\left[-3c_1 \mathbf{s}_i^2\right]} \tag{10.10}$$

where c_1 appears in Eq. (10.5).

10.2.3 Power in the Bucket (PIB)

Power in the bucket (PIB) [30] is a parameter that characterizes the focusing ability of laser energy in the far-field and represents the percentage of Power in a given bucket in the total Power, which is defined as [31]:

$$PIB = \frac{\int_0^{2\pi} \int_0^{b} \langle I(\boldsymbol{\rho}, z)\rangle \boldsymbol{\rho} d\boldsymbol{\rho} d\varphi}{\int_0^{2\pi} \int_0^{\infty} \langle I(\boldsymbol{\rho}, z)\rangle \boldsymbol{\rho} d\boldsymbol{\rho} d\varphi} \tag{10.11}$$

where b represents the radius of a given bucket. Substituting Eq. (10.5) into Eq. (10.11) for simplification, the PIB expression for a partially coherent GSM array beam is given as follows:

$$PIB = \sum_{i=1}^{4}\left[1 - \frac{\exp(c_1 b^2 - 2c_1 \mathbf{s}_i b)}{1 + 2\pi \mathbf{s}_i \exp(-c_1 \mathbf{s}_i^2)}\right] \tag{10.12}$$

The parameter c_1 is defined in Eq. (10.5).

The range of the PIB is $0 \leq PIB \leq 1$. The closer the value is to 1, the higher the energy concentration of the array beam in the far-field, and the better the beam quality.

10.3 Intensity Distribution and Self-coupling Properties of Partially Coherent GSM Array Beams

The intensity distribution and self-coupling characteristics of partially coherent GSM array beams with different influence factors are analyzed using the parameters $c_0 = 1.7 \times 10^{-14}\text{m}^{-2/3}$, $l_0 = 0.01$ m, $L_0 = 10$ m, $\sigma_g = 0.02$ m, $\sigma_s = 0.05$ m, $\lambda = 650$ nm, $s_x = [0.1\ 0.1\ -0.1\ -0.1]$, $s_y = [-0.1\ 0.1\ -0.1\ 0.1]$ without special instructions.

10.3.1 *Effect of Transmission Distance on Intensity Distribution and Self-coupling Characteristics*

Based on Eq. (10.5), Fig. 10.3 shows the results of a theoretical analysis of the light intensity distribution, self-coupling characteristics, and light intensity envelope of a partially coherent GSM array as a function of the transmission distance. In Fig. 10.3a–e, when a partially coherent GSM array beam is transmitted in atmospheric turbulence, the average light intensity decreases with the increase of the transmission distance. As the transmission distance increases, the four beams superimpose on the axis (Fig. 10.3c), and the intensity value on the axis increases gradually (Fig. 10.3h). When the transmission distance is 4500 m, the four beams can synthesize a flat-topped beam (Fig. 10.3d) and continue to propagate until the intensity value on the axis is a maximum. The flat-topped beam becomes a Gaussian beam (Fig. 10.3e), which has a larger envelope than the Gaussian beam in Fig. 10.3a. This indicates that the partially coherent GSM array beam has self-coupling characteristics during propagation in the atmosphere, i.e., the array beam can naturally synthesize a beam for transmission to a certain distance, and the synthesized beam can be transformed from a flat-top beam into a Gaussian beam. These results are consistent with the findings in reference [29]. During the transmission of the array beam, due to the influence of atmospheric turbulence, the beam spread will occur, and the four beams will be superimposed on the axis. With the increase of the transmission distance, the beam spread will become more and more obvious, and the superimposed parts on the axis will become more and more, finally forming the envelope (e) is shown in Fig. 10.3.

Figure 10.4 shows the variation of the normalized intensity of a partially coherent GSM array beam and a partially coherent single GSM beam as a function of the transmission distance. By comparing Fig. 10.4a and d, it can be shown that the normalized intensity values of an array beam and a single beam are the same in the case of short-distance transmission. However, when the transmission distance reaches 3000 m, as shown in Fig. 10.4b, c and e, f, the array beam is superimposed on the axis, and the intensity of the array beam at the same distance is greater than that of the single beam. Moreover, the intensity difference between the two increases with the increase of the transmission distance. This shows that under the same transmission conditions, a partially coherent GSM array beam is less affected by turbulence than a partially coherent single GSM beam, and has superior turbulence suppression characteristics, which is conducive to the realization of long-distance communication.

10.3.2 Influence of Beam Center Position on Intensity Distribution and Self-Coupling Characteristics

Figure 10.5 shows the changes in the normalized light intensity of a partially coherent GSM array beam in atmospheric media and free space for different beam center positions at $z = 5000$ m. By comparing Fig. 10.5a and b, in the atmospheric medium, when the transmission distance is the same, the closer the distance between the center positions of the array beam, the better the self-coupling characteristics and the greater the light intensity value. This is because under the same transmission conditions, the closer the distance between the center positions of the array beam, the greater the extent of superimposing on the axis, the greater the intensity value on the axis, and the faster the synthesis of the beam. Therefore, the self-coupling characteristics improve. By comparing Fig. 10.5a and c, it can be determined that when the center position of the array beam is the same, it is combined into a single beam, whereas the four beams in the array beam are tangential in free space. This indicates that the self-coupling characteristics of an array beam in an atmospheric medium are better compared to free space. This also indicates that the self-coupling characteristics of partially coherent GSM array beams are primarily caused by atmospheric turbulence, which is a different conclusion from that based on the results for partially coherent Gaussian vortex array beams in the literature [21].

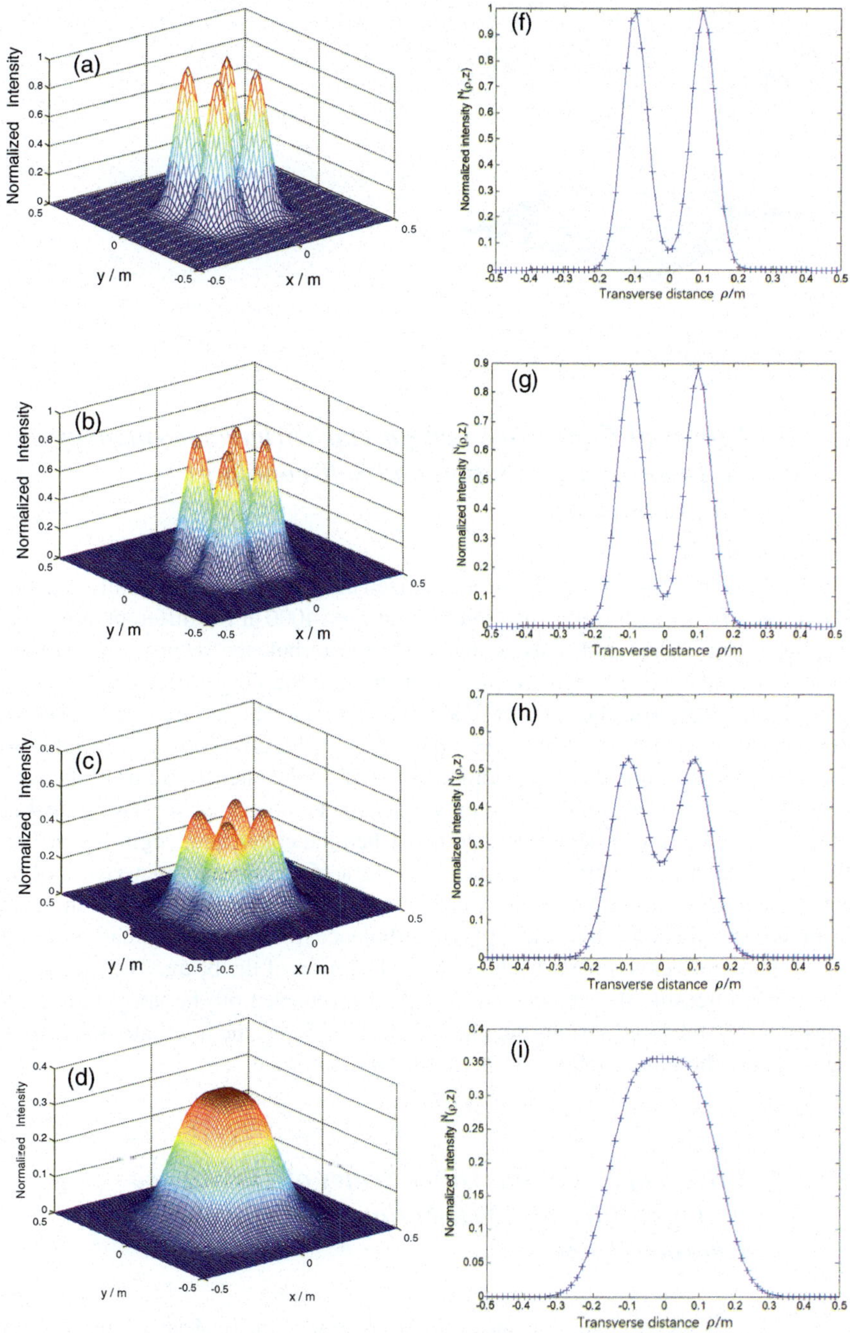

Fig. 10.3 Variation of normalized beam intensity distribution of GSM array with transmission distance **a**, **f** $z = 0$ m; **b**, **g** $z = 1500$ m; **c**, **h** $z = 3000$ m; **d**, **i** $z = 4500$ m; **e**, **j** $z = 5000$ m

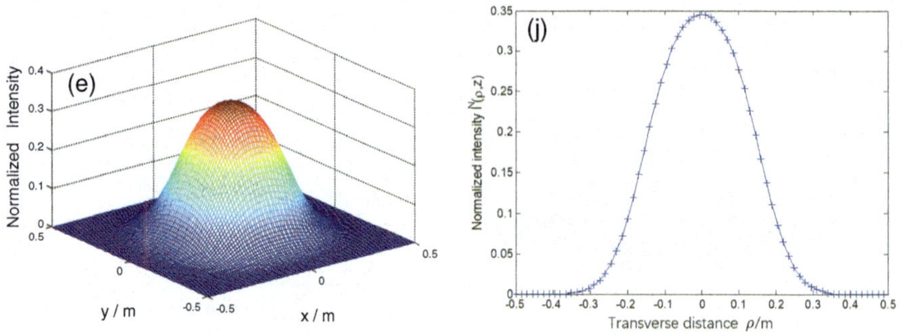

Fig. 10.3 (continued)

10.3.3 Effects of Coherence Length and Turbulence Intensity on Intensity Distribution and Self-coupling Characteristics

Figure 10.6 shows the changes in the normalized light intensity of a partially coherent GSM array beam with turbulence intensity at $z = 4000$ m for different coherence lengths. Figure 10.6a and b show that when the turbulence intensity is constant, the shorter the coherence length, the better the self-coupling characteristics of the array beam. However, the intensity value is also less. This is because the shorter the coherence length, the worse the coherence of the array beam. The intensity distribution at the same distance is more scattered, so the intensity value of the array beam is reduced, although the self-coupling characteristics improve. By comparing Fig. 10.6a and c, it can be determined that when the coherence length is constant, the larger the turbulence intensity, the better the self-coupling characteristics of the array beam. However, the intensity value is reduced. This is because when the array beam is transmitted through atmospheric turbulence, the random fluctuations of the atmospheric refractive index result in phase distortion of the beam. The greater the turbulence intensity, the greater the impact of turbulence on the array beam, and consequently, the more serious the phase distortion. Thus, the light intensity will be reduced, but the self-coupling characteristics will improve.

10.3.4 Effects of Inner and Outer Scales of Turbulence on Light Intensity Distribution and Self-coupling Characteristics

Figure 10.7 shows the changes in the normalized light intensity of a partially coherent GSM array beam for different internal scales with external scales at $z = 3000$ m. Figure 10.7a and b show that under the same outer scale, the intensity of an array

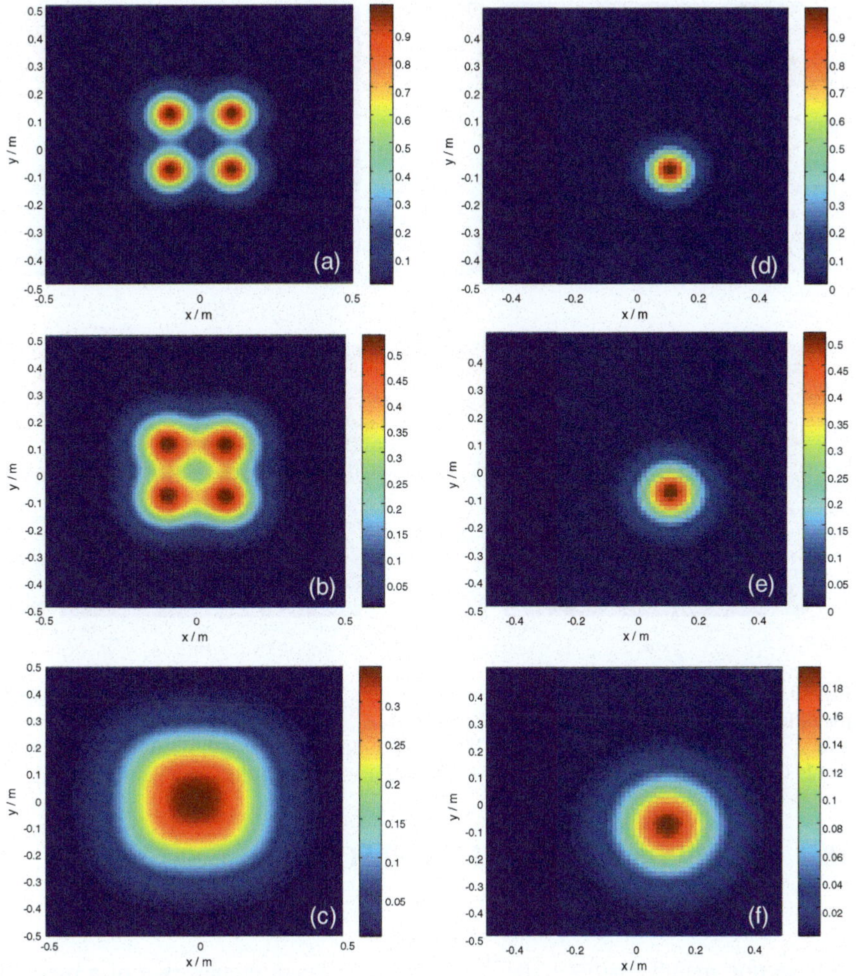

Fig. 10.4 Changes in the normalized intensity of an array beam and a single beam with transmission distance **a–c** is an array beam; **d–f** is a single beam; **a, d** $z = 100$ m; **b, e** $z = 3000$ m; **c, f** $z = 5000$ m

beam increases with the increase of the inner scale, i.e., the larger the inner scale of turbulence, the less is the effect of turbulence on the array beam and the worse is its self-coupling characteristics. This is because the smaller the internal scale of turbulence, the smaller the turbulent vortices in the beam cross-section, and the more significant the diffraction of the beam irradiated on these vortices. Thus, the random distribution of the intensity of the beam in time and space is more pronounced, and the light intensity will is more dispersed. By comparing Fig. 10.7a and c, it can be determined that the outer scale of turbulence has little influence on the intensity distribution and self-coupling characteristics of the array beam, and can be ignored.

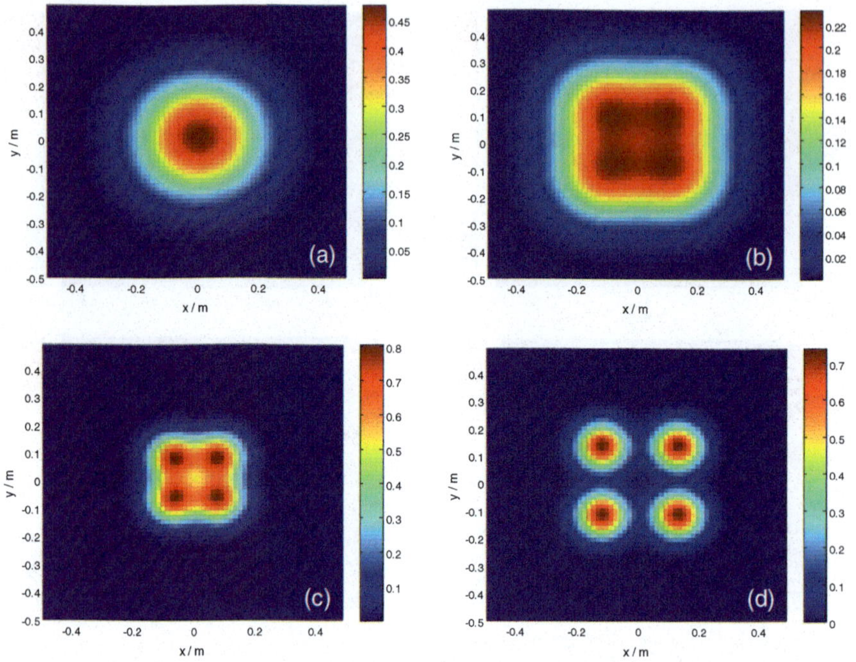

Fig. 10.5 Normalized intensity changes of array beams in atmospheric medium and free space for different center positions at $z = 5000$ m **a, b** $c_0 = 1.7 \times 10^{-14}$ m$^{-2/3}$, **c, d** $c_0 = 0$, **a, c** $s_x = [0.075\ 0.075\ -0.075\ -0.075]$, $s_y = [-0.075\ 0.075\ -0.075\ 0.075]$, **b, d** $s_x = [0.125\ 0.125\ -0.125\ -0.125]$, $s_y = [-0.125\ 0.125\ -0.125\ 0.125]$

10.4 Beam Propagation of Partially Coherent GSM Array Beams

This section mainly analyzes the variation of the beam propagation of a partially coherent GSM array with different influence factors. The parameters were chosen to be $c_0 = 1.7 \times 10^{-14}m^{-2/3}$, $l_0 = 0.01$ m, $L_0 = 10$ m, $\sigma_g = 0.02$ m, $\sigma_s = 0.2$ m, $\lambda = 650$ nm, $s_x = [0.1\ 0.1\ -0.1 -0.1]$, $s_y = [-0.1\ 0.1\ -0.1\ 0.1]$ without special instructions.

10.4.1 Influence of Beam Center Position on Root Mean Square Beam Width

Based on Eq. (10.10), Fig. 10.8 shows the changes in the root mean square beam width of a partially coherent GSM array beam in the x-direction in the atmosphere and free space as a function of transmission distance for different center positions.

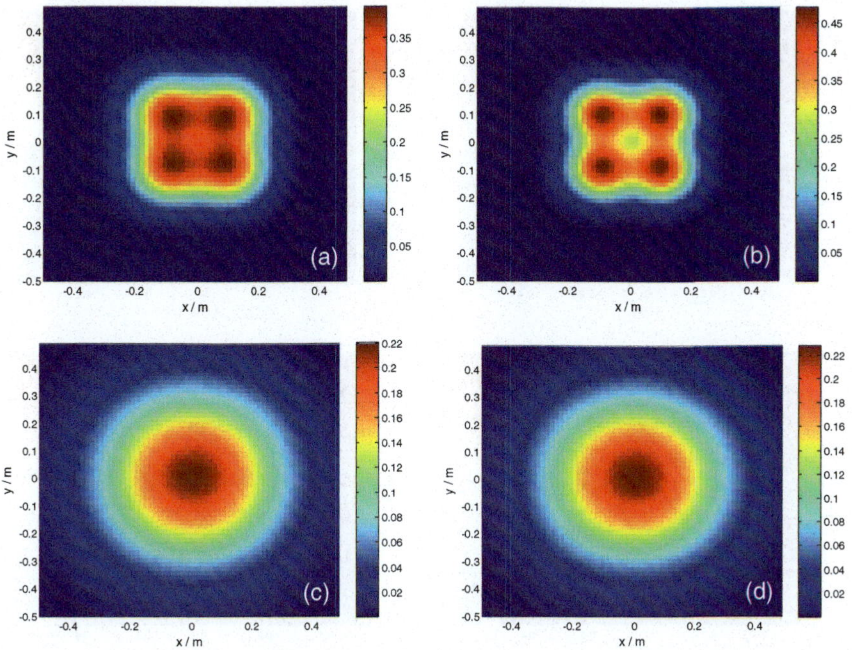

Fig. 10.6 The normalized light intensity of a GSM array beam for different coherence lengths that vary with turbulence intensity at $z = 4000\,\text{m}$ **a, c** $\sigma_g = 0.01$ m, **b, d** $\sigma_g = 0.2$ m, **a, b** $c_0 = 1.0 \times 10^{-14}\,\text{m}^{-2/3}$, **c, d** $c_0 = 1.0 \times 10^{-13}\,\text{m}^{-2/3}$

Figure 10.8 shows that the root mean square beam width of the array beam remains unchanged in both an atmospheric medium and free space during short-distance transmission. For long-distance transmission, when the center position of the array beam is constant, the root mean square beam width increases with the increase of the transmission distance in an atmospheric medium, whereas the change of the root mean square beam width in free space is very small and can be ignored. This is because the propagation of the array beam in atmospheric turbulence is determined by two factors: the diffraction of light in free space and the turbulent atmosphere. It can be seen from Fig. 10.8 that compared to diffraction propagation in free space, atmospheric turbulence has a greater impact on beam propagation. By comparing the solid line (or dashed line), it can be determined that in the same transmission medium, the farther apart the center coordinate positions, the larger the root mean square beam width of the array beam. This is because the closer the centers of the beams are, the more the array beam is superimposed on the axis for transmission over the same distance. Moreover, the array beam expansion is less pronounced, so the root mean square beam width is smaller.

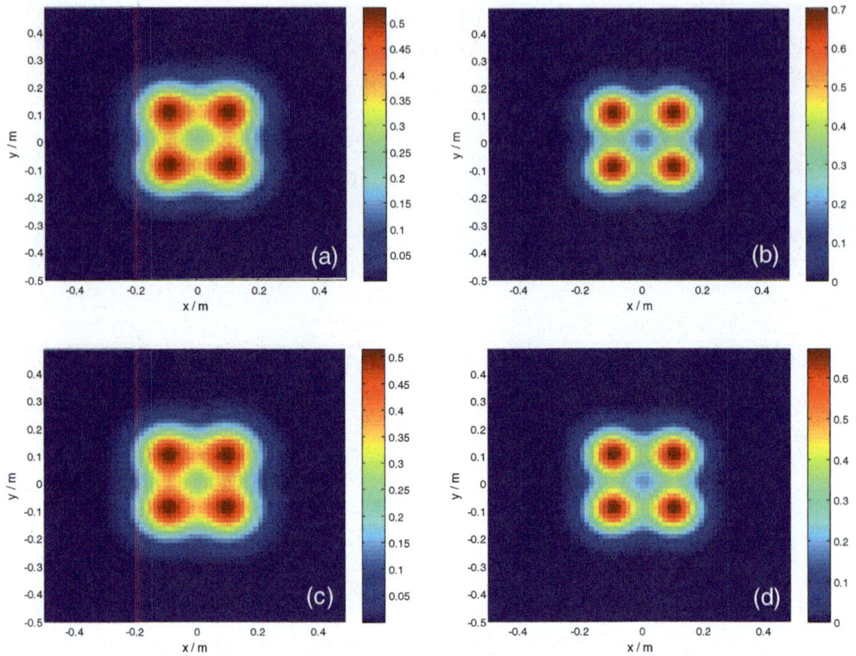

Fig. 10.7 Changes of the normalized light intensity of a GSM array beam at different internal scales with external scales at $z = 3000$ m, **a, c** $l_0 = 0.01$ m, **b, d** $l_0 = 0.1$ m, **a, b** $L_0 = 10$ m, **c, d** $L_0 = 100$ m

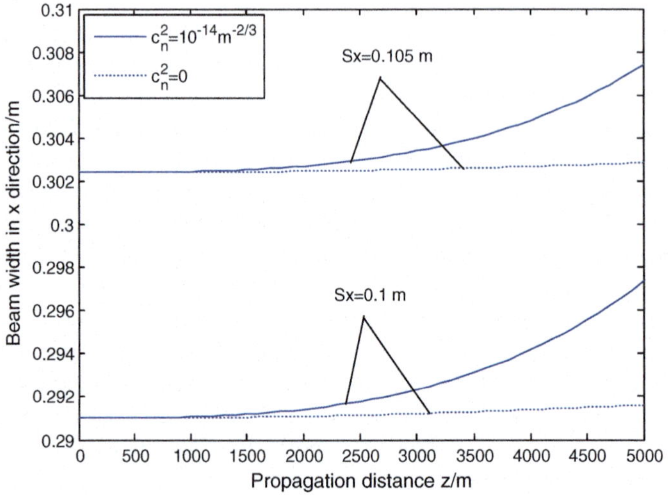

Fig. 10.8 Changes of beam width in x-direction with the transmission distance of GSM array beam at different center positions

10.4.2 Influence of Coherence Length and Turbulence Intensity on Root Mean Square Beam Width

Figure 10.8 shows the variation of the root mean square beam width of partially coherent GSM array beams in the x-direction with the propagation distance for different turbulence intensities and coherence lengths. The figure shows that atmospheric turbulence and coherence length do not affect the propagation of the array beam at short distances. When the propagation distance is long, the shorter the coherence length, the larger the root-mean-square beam width. This indicates that the shorter the coherence length, the greater the effect of turbulence on the array beam, and the more pronounced the beam expansion, which is consistent with published results [29]. When the coherence length is the same, the larger the turbulence intensity, the larger the root-mean-square beam width. This is because when the transmission conditions are constant, the greater the turbulence intensity, the more significant the wavefront phase distortion of the array beam during the transmission process, and the more significant the beam expansion. Therefore, the root mean square beam width of the array beam is larger (Fig. 10.9)

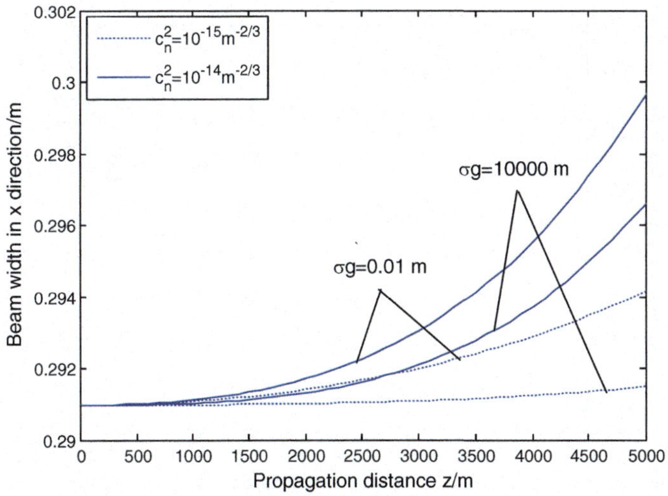

Fig. 10.9 Changes of the beam width in the x-direction with propagation distance for different coherence lengths and turbulence intensities of a GSM array beam

10.4.3 The Influence of Internal and External Scales on the Root Mean Square Beam Width

Figure 10.10 shows the root mean square beam width of a partially coherent GSM array beam propagating in the x-direction as a function of the propagation distance at different internal and external scales. The figure shows that the internal and external scales of turbulence have no effect on the array beam propagation at short distances. A comparison between the solid line and the dashed line shows that with the increase of the transmission distance, the smaller the inner scale of the turbulence, the larger the root mean square beam width of the array beam. In terms of physical properties, a decrease of the inner scale results in an increase in the turbulence intensity, i.e., the smaller the inner scale, the greater the influence of turbulence on the beam, and the more pronounced the beam expansion, so the root mean square beam width is larger. By comparing the solid lines (or dashed lines) in Fig. 10.10, it can be determined that when the inner scale of turbulence is constant, the root mean square beam width of the array beam varies very little at different outer scales of turbulence, i.e., propagation of the array is not affected by the outer scale of turbulence and can be ignored.

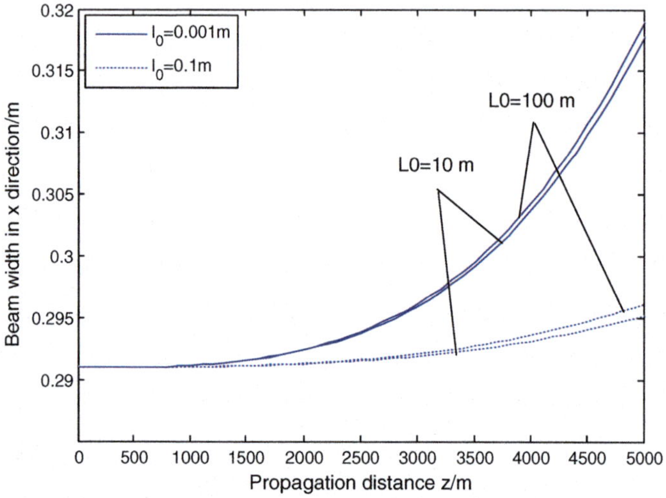

Fig. 10.10 Changes of beam width in the x-direction with transmission distance at different internal and external scales of GSM array beam

10.5 PIB of Partially Coherent GSM Array Beam

This section mainly analyzes the PIB variation of a partially coherent GSM array beam in atmospheric turbulence with transmission distance under the influence of different factors. The investigated parameters were chosen to be $c_0 = 1.7 \times 10^{-14}\ \mathrm{m}^{-2/3}$, $l_0 = 0.01$ m, $L_0 = 10$ m, $\sigma_g = 0.02$ m, $\sigma_s = 0.05$ m, $\lambda = 650$ nm, $b = 0.2$ m, $s_x = [0.1\ 0.1\ \ -0.1 - 0.1]$, $s_y = [-0.1\ 0.1\ \ -0.1\ 0.1]$ without special instructions.

10.5.1 Effects of Array Beam Center Position and Turbulence Intensity on PIB

Based on Eq. (10.12), Fig. 10.11 shows the influence of the beam center position of a partially coherent GSM array on the PIB for different turbulence intensities. According to the figure, turbulence intensity has no effect on the PIB during short-distance transmission. Under the same transmission conditions, the PIB decreases with the increase of the transmission distance. When the center of the array beam is fixed, the larger the turbulence intensity, the smaller the PIB. This indicates that the larger the turbulence intensity, the more the array beam is affected by turbulence, and the worse the beam quality. By comparing the solid lines (or curves) in Fig. 10.11, it can be determined that under the same turbulence intensity, the closer the distance between the center positions of the array beams, the larger the PIB. This is because the closer the distance between the center positions of the array beams, the more they are superimposed on the axis during transmission over the same distance, and the greater the light intensity value near the optical axis. Thus, the PIB is larger, which is consistent with the result in Fig. 10.5.

10.5.2 Influence of Coherence Length and Barrel Radius on PIB

Figure 10.12 shows the influence curve of the coherence length on the PIB of a partially coherent GSM array beam for different barrel radii. The figure shows that the coherence length has no effect on the PIB during short-distance transmission. When the barrel radius is constant, the longer the coherence length, the larger the PIB. This is because under the same transmission conditions, the longer the coherence length, the better the coherence of the array beam, and the less the beam is affected by turbulence. By comparing the solid lines (or curves) in the figure, it can be established that when the coherence length is constant, the larger the radius of the barrel, the larger the PIB of the array beam. Considering the physical properties, the numerical simulation results are consistent with reality.

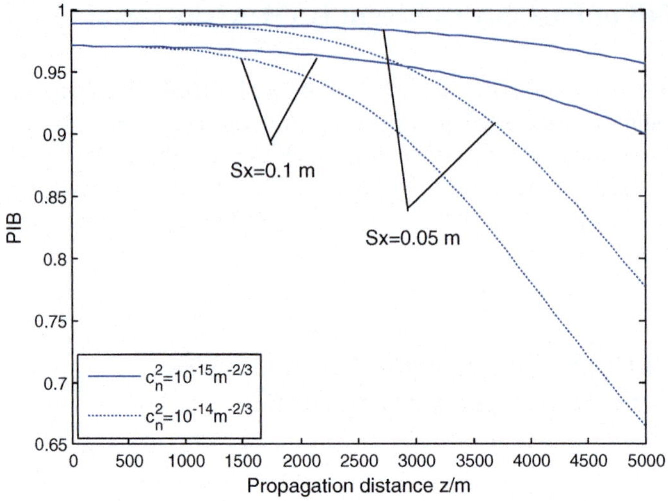

Fig. 10.11 PIB changes with propagation distance of GSM array beam at different center positions and turbulence intensities

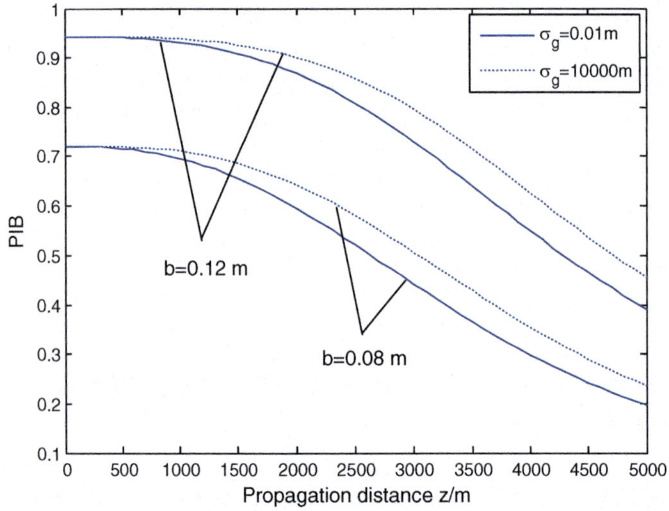

Fig. 10.12 PIB variation for a GSM array beam with transmission distance at different coherence lengths and barrel radii

10.6 Summary

Based on the Huygens-Fresnel principle and the modified Von Karmon spectral model, the intensity distribution, root mean square beam width, and PIB of a partially coherent GSM array beam were numerically derived and analyzed for propagation through atmospheric turbulence. The main conclusions are as follows:

(1) A partially coherent GSM array beam has self-coupling characteristics, i.e., when it travels a certain distance in atmospheric turbulence ($z = 4500$ m), the four beams of the array beam can naturally combine to produce a flat-top beam, which then evolves into a Gaussian beam. The self-coupling characteristics of partially coherent GSM array beams in atmospheric turbulence are superior to free space propagation. Compared to a single GSM beam, the partially coherent GSM array beam has improved turbulence suppression properties.

(2) In the case of short-distance transmission, only the center position of the array beam affects the self-coupling characteristics, whereas, in the case of long-distance transmission, the self-coupling characteristics of a partially coherent GSM array beam improve with the decrease of the coherence length, the center position of the beam, the inner scale, and an increase of the turbulence intensity. However, the outer scale does not affect the self-coupling characteristics.

(3) The average light intensity of a partially coherent GSM array beam decreases with the increase of the propagation distance, but increases with the decrease of the array beam center position, turbulence intensity, coherence length, and internal scale, whereas the external scale has little effect on the light intensity distribution and can be ignored.

(4) The beam propagation of a partially coherent GSM array beam increases with the increase of the beam center position, turbulence intensity, and the decrease of the coherence length and internal scale, whereas the external scale has little effect on the root mean square beam width and can be ignored.

(5) The PIB of a partially coherent GSM array beam increases with the decrease of the turbulence intensity, beam center position, and the increase in the coherence length and barrel radius.

References

1. Guo K, Zhu YD, Zhou P et al (2014) Numerical study on limit output power of single-mode erbium-doped fiber laser. Infrared Laser Eng 43(8):2407–2412
2. Ji XQ, Li XQ (2009) Far-field divergence angle and far-field radiation intensity of gauss-Schell model array beams. J Phys 58(7):4624–4629
3. Ji G, Ji X (2010) Effective radius of curvature of Hermite-Gaussian array beams. Optics Laser Technol 42(6):1054–1058
4. Pan LL, Zhang B, Pan PP (2010) Spectral shift of a partially coherent rectangular array beam in free space. Intense Laser Particle Beams 22(7):1473–1478

5. Li CJ, Luo YM (2013) Propagation characteristics of multiple radial arrays of non-paraxial Hermite-Gaussian beams. Infrared Laser Eng 42(S01):111–117
6. Ji XL (2010) Effects of atmospheric turbulence on beam propagation and directivity of radially distributed Gaussian array. J Phys 59(1):692–698
7. Ji XL, Li XQ (2008) Effect of turbulence on coherent and incoherent Synthesis of Gaussian beams in off-axis array. J Phys 57(12):7674–7679
8. Lu L, Ji XL, Deng JP et al (2014) Effects of non-Kolmogorov atmospheric turbulence on Gaussian array beam propagation. Acta Phys Sin 63(1):14206–14207
9. Zhou P, Ma Y, Wang X et al (2010) Average spreading of a Gaussian beam array in non-Kolmogorov turbulence. Opt Lett 35(7):1043–1045
10. Wu J (1990) Propagation of a Gaussian-Schell beam through turbulent media. J Mod Opt 37(4):671–684
11. Pingping P, Bin Z, Na Q et al (2011) Characteristics of scintillations and bit error rate of partially coherent rectangular array beams in turbulence[J]. Optics Communications 284(4):1019–1025
12. Ji X, Shao X (2010) Influence of turbulence on the beam propagation factor of Gaussian Schell-model array beams. Opt Commun 283(6):869–873
13. Ai Y, Dan Y (2011) Range of turbulence-negligible propagation of Gaussian Schell-model array beams. Opt Commun 284(13):3216–3220
14. Lu F, Han XE (2015) Spatial coherence of Gauss-Schell model array beams in turbulent atmosphere. Infrared Laser Eng 44(1):305–309
15. Yangsheng Y, Xianlong L, Fei W et al (2013) Scintillation index of a multi-Gaussian Schell-model beam in turbulent atmosphere. Opt Commun 305:57–65
16. Yuan Y, Cai Y, Eyyuboğlu HT et al (2012) Propagation factor of partially coherent flat-topped beam array in free space and turbulent atmosphere. Optics Lasers Eng 50(5):752–759
17. Öztan MA, Baykal Y (2015) Scintillations of partially coherent annular and flat-topped array beams in extremely strong turbulent medium. Opt Commun 354:419–427
18. Yang A, Zhang E, Ji X et al (2008) Angular spread of partially coherent Hermite-cosh-Gaussian beams propagating through atmospheric turbulence. Opt Exp 16(12):8366–8380
19. Baykal Y, Eyyuboğlu HT, Cai Y (2009) Scintillations of partially coherent multiple Gaussian beams in turbulence. Appl Opt 48(10):1943–1954
20. Zhou P, Ma Y, Wang X et al (2009) Average intensity of a partially coherent rectangular flat-topped laser array propagating in a turbulent atmosphere. Opt Soc Am 48(28):5251–5258
21. Dajun L, Yaochuan W, Hongming Y (2016) Propagation properties of partially coherent four-petal Gaussian vortex beams in turbulent atmosphere. Opt Laser Technol 78:95–100
22. Liu F, Ji XL (2011) Turbulent distance of partially coherent hyperbolic cosine-Gaussian array beams. China laser 38(7):1–6
23. Wang H, Li X (2010) Propagation properties of radial partially coherent flat-topped array beams in a turbulent atmosphere. Opt Commun 283(21):4178–4189
24. Li CQ, Zhang HY, Wang TF et al (2013) Coherent properties of Gaussian Scherer mode beams propagating in atmospheric turbulence. Acta Phys Sin 62(22):3–7
25. Ke XZ, Wang WT (2015) Intensity and propagation of partially coherent light in oblique and horizontal atmospheric turbulence. J Appl Sci 33(2):142–154
26. Lu W, Liu L, Sun J et al (2007) Change in degree of coherence of partially coherent electromagnetic beams propagating through atmospheric turbulence. Opt Commun 271(1):1–8
27. Andrews LC, Phillips RL (2005) Laser beam propagation through random media. SPIE optical Engineering Press, Bellingham, p 195
28. Tomohiro S, Aristide D, Emil W (2003) Mode analysis of spreading of partially coherent beams propagating through atmospheric turbulence. J Opt Soc Am A Opt Image Sci Vis 20(6):1094–1102
29. Kashani FD, Hedayatirad MR, Mahzoun MR et al (2012) Beam propagation analysis of a multi beam FSO system with partially flat-topped laser beams in turbulent atmosphere. Optik 123(10):879–886
30. Xiang NJ, Wang MJ, Wang TR (2012) Average intensity and broadening of partially coherent Gaussian Schell beams in atmospheric turbulence. Laser J 33(5):4–6

31. Yan H, Li S, Zhang D et al (2000) Numerical simulation of adaptive optics system with laser propagation in the atmosphere. Appl Opt 39(18):3023–3031

Chapter 11
Effect of Non-Kolmogorov Atmospheric Turbulence on Radially Partially Coherent Arrays

In this section, analytical expressions for the Rayleigh interval (z_R) and the turbulent distance (z_T) of a radially distributed partially coherent Gaussian-Schell model (GSM) array beam in non-Kolmogorov atmospheric turbulence are derived. In addition, the variation of these two parameters with turbulent and the beam properties are numerically analyzed.

11.1 Introduction

Beam synthesis technology has attracted significant attention in recent years as an effective means to obtain high-power laser output and maintain good beam quality [1–3]. Atmospheric turbulence effects such as beam expansion, spot drift, and intensity scintillation occur during the transmission of array beams, which severely limit their applications in fields such as laser communications [4]. The transmission characteristics of typical array beams in atmospheric turbulence have been extensively investigated, and several of these studies have addressed the effect of atmospheric turbulence on the expansion of an array beam [5, 6]. To quantitatively describe the expansion of array beams in atmospheric turbulence, the concepts of Rayleigh interval and turbulence distance were introduced. The Rayleigh interval or collimation range is used to describe the transmission distance over which a laser beam does not expand significantly, whereas the turbulence distance is the transmission distance at which expansion starts to be significantly affected by atmospheric turbulence. These two physical quantities are both distinct and related, and they characterize the effect of atmospheric turbulence effect on the beam from different perspectives [7].

In recent years, the Rayleigh interval and turbulence distance have been used by researchers to quantitatively analyze the spread of array beams in von Karman, Tatarskii, or ocean turbulence spectrum models [8–13]. It has been determined that there are significant differences between the theoretical results obtained using

X. Ke, *Partially Coherent Optical Transmission Theory in Optical Wireless Communication*, Optical Wireless Communication Theory and Technology,
https://doi.org/10.1007/978-981-97-7158-5_11

common turbulence models and the experimental data for actual propagation through atmospheric turbulence [14]. Toselli et al. proposed the non-k turbulence power spectrum by introducing the turbulence generalized index α and the generalized amplitude factor [15]. There have been numerous studies on the turbulence distance and Rayleigh interval of a single laser beam in non-k turbulence [16–23], however, there are few studies on array beam expansion. In 2013, Huang et al. [24, 25] derived an analytical expression for the turbulent distance and the Rayleigh interval of partially coherent Hermite-Gaussian array beams in non-k turbulence based on the definition of the beam propagation factor (M^2 factor). It was determined that these two parameters increased with the increase of the number of constituent beams, but the authors of the study did not consider the influence of wavelength, waist radius, turbulence intensity, and other factors on the turbulent distance. In 2014, Liu et al. [7] investigated the turbulence distance and Rayleigh interval of Gaussian array beams in non-k turbulence. The results revealed that these quantities varied non-monotonically with the increase of α for both coherent and incoherent Gaussian array beams, and that there was a minimum value at $\alpha = 3.108$. However, the emphasis of the study was on fully coherent array beams and partially coherent beams were not considered. The aforementioned studies on the turbulent distance and Rayleigh interval of array beams were based on one-dimensional linear arrays, and the turbulent distance and Rayleigh interval of radially distributed partially coherent array beams in non-k atmospheric turbulence were not investigated.

In this chapter, analytical expressions are derived for the Rayleigh interval and the turbulence distance for the propagation of radially distributed partially coherent GSM array beams through non-k atmospheric turbulence. Moreover, their variation with turbulence parameters (turbulence generalized index α, inner scale l_0, outer scale L_0, turbulence intensity $\tilde{c}_n^2$) and beam parameters (coherence parameter β, radial distribution radius r_0, wavelength λ, waist radius ω_0) is analyzed, and the main results are discussed.

11.2 Theoretical Analysis

11.2.1 Rayleigh Interval

In a rectangular coordinate system, it is assumed that n partially coherent GSM beams with a waist radius ω_0 are uniformly distributed on a circle with a radius of r_0, and the angular spacing of adjacent constituent beams is $\alpha_0 = 2\pi / N$. These n sub-beams form a radial distribution array beam, as shown in Fig. 11.1.

In the case of phase locking, i.e., when the radially distributed partially coherent GSM array beam undergoes coherent synthesis, the cross-spectral density function in the z = 0 plane can be expressed as [4]:

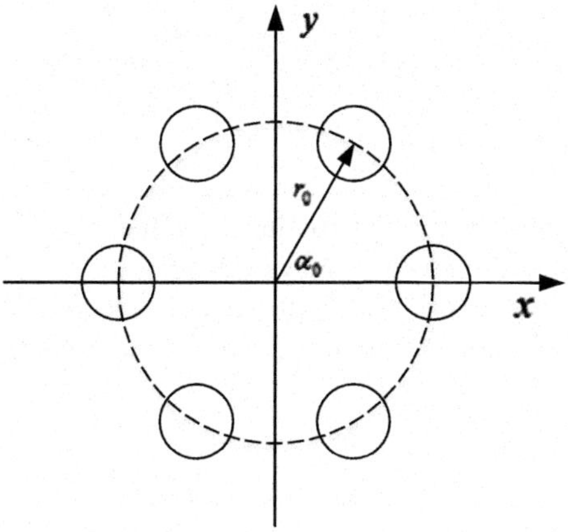

Fig. 11.1 Schematic diagram of radial distribution array beam

$$W^0(\boldsymbol{\rho}_1', \boldsymbol{\rho}_2'; 0) = \sum_{n=1}^{N}\sum_{m=1}^{N} \exp\left[-\frac{(\boldsymbol{\rho}_1' - \mathbf{r_n})^2 + (\boldsymbol{\rho}_2' - \mathbf{r_m})^2}{\omega_0^2}\right] \times \exp\left[-\frac{\left[(\boldsymbol{\rho}_1' - \mathbf{r_n}) - (\boldsymbol{\rho}_2' - \mathbf{r_m})\right]^2}{2\sigma_g^2}\right] \quad (11.1)$$

where $\boldsymbol{\rho}_1'$ and $\boldsymbol{\rho}_2'$ are the coordinate vectors of two points in the source plane, $\mathbf{r_n}$ and $\mathbf{r_m}$ are the vectors that represent the center positions of the n-th and m-th constituent beams respectively. $\mathbf{r_n} = (r_0 \cos\alpha_n, r_0 \sin\alpha_n)$, $\alpha_n = n\alpha_0 (n = 1, 2, ..., N)$, and the parameters ω_0 and σ_g represent the waist width and coherent length of the partially coherent GSM beams, respectively. If $\sigma_g \to \infty$, Eq. (11.1) represents the cross-spectral density function of the fully coherent array beam at $z = 0$.

Based on the Huygens-Fresnel principle, the cross-spectral density function that arrives at the surface $z = 0$ after propagation through atmospheric turbulent can be expressed as [22]:

$$I(\boldsymbol{\rho}, z) = \frac{1}{(\lambda z)^2} \iint d^2\boldsymbol{\rho}_1' \iint d^2\boldsymbol{\rho}_2' W^0(\boldsymbol{\rho}_1', \boldsymbol{\rho}_2'; 0) \times \left\langle \exp\left[\psi(\boldsymbol{\rho}_1', \boldsymbol{\rho}, z) + \psi^*(\boldsymbol{\rho}_2', \boldsymbol{\rho}, z)\right]\right\rangle \times \exp\left\{-\frac{i\kappa}{2z}\left[(\boldsymbol{\rho} - \boldsymbol{\rho}_1')^2 - (\boldsymbol{\rho} - \boldsymbol{\rho}_2')^2\right]\right\} \quad (11.2)$$

where $\boldsymbol{\rho}$ represents a point in the coordinate system of the receiving plane; $\psi(\boldsymbol{\rho}', \boldsymbol{\rho}, z)$ is the complex random phase fluctuation introduced by a spherical wave propagating from point $(\boldsymbol{\rho}', 0)$ to point $(\boldsymbol{\rho}, z)$ in turbulent atmosphere,* represents complex conjugation, and $\langle\cdot\rangle$ is the ensemble average for the turbulent medium [16].

$$\left\langle \exp\left[\psi(\boldsymbol{\rho}_1', \boldsymbol{\rho}_1, z) + \psi^*(\boldsymbol{\rho}_2', \boldsymbol{\rho}_2, z)\right]\right\rangle$$

$$= \exp\left[-4\pi^2k^2z\int_0^1\int_0^\infty k\Phi_n(k,\alpha)\times\left[1-J_0(k\xi\left|\boldsymbol{\rho}_2'-\boldsymbol{\rho}_1'\right|)\right]dkd\xi\right] \quad (11.3)$$

where $J_0(\cdot)$ is the zero-order Bessel function, $\Phi_n(k)$ is the spatial spectral density function of refractive index fluctuation of the turbulent medium, and the wave number $k = 2\pi/\lambda$ (λ is the wavelength).

A non-k statistical model is adopted for atmospheric turbulence, and the power spectrum function has a von Karman form. Thus, $\Phi_n(k,\alpha)$ can be expressed as [12, 17]:

$$\Phi_n(k,\alpha) = H(\alpha)\tilde{c}_n^2\frac{\exp\left[-(\kappa^2/\kappa_m^2)\right]}{\left(\kappa^2+\kappa_0^2\right)^{\alpha/2}} \quad 0 \le \kappa < \infty,\ 3 < \alpha < 4 \quad (11.4)$$

where $\kappa_0 = 2\pi/L_0$ (L_0 is the outer scale of turbulence), $\kappa_m = c(\alpha)/l_0$ (l_0 is the inner scale of turbulence), $\tilde{c}_n^2$ is the generalized refractive index structure constant with a unit of $\mathrm{m}^{3-\alpha}$. In addition, $H(\alpha) = \Gamma(\alpha-1)\cos(\alpha\pi/2)/4\pi^2$ and $c(\alpha) = \left[\Gamma(5-\alpha/2)\cdot H(\alpha)\cdot 2/(3\pi)\right]^{1/(\alpha-5)}$, where $\Gamma(\cdot)$ is the gamma function and α is the generalized exponent of turbulence. If $\alpha = 11/3$, $H(11/3) = 0.033$, $\tilde{c}_n^2 = c_n^2$, Eq. (11.4) can be simplified to the conventional Kolmogorov power spectrum. According to the definition for a second-order moment beam width, we have [11]:

$$\omega^2(z) = \frac{4\iint \rho^2 I(\boldsymbol{\rho},z)d^2\boldsymbol{\rho}}{\iint I(\boldsymbol{\rho},z)d^2\boldsymbol{\rho}} \quad (11.5)$$

By substituting Eq. (11.2) into Eq. (11.5), an expression for the second-order pitch beam width of a radially distributed partially coherent GSM synthetic array beam propagating through non-k atmospheric turbulence can be obtained as follows:

$$\omega^2(z)\big|_{co} = P + \frac{4}{\kappa^2}Bz^2 + Mz^3 \quad (11.6)$$

where the beam coherence parameter $\beta = \sigma_g/\omega_0$.

$$P = \sum_{n=1}^{N}\sum_{m=1}^{N}\left\{r_0^2[1+\cos(\alpha(n)-\alpha(m))]+\omega_0^2\right\}D/Q,$$

$$B = \frac{1}{\omega_0^2}\left(1+\frac{1}{\beta^2}\right)\sum_{n=1}^{N}\sum_{m=1}^{N}\left\{1-\frac{r_0^2[1-\cos(\alpha(n)-\alpha(m))]}{\omega_0^2}\left(1+\frac{1}{\beta^2}\right)\right\}D/Q,$$

$$D = \exp\left[-\frac{r_0^2(1-\cos(\alpha(n)-\alpha(m)))}{\omega_0^2}\left(1+\frac{1}{\beta^2}\right)\right],\ Q = \sum_{n=1}^{N}\sum_{m=1}^{N}D$$

$$M = \frac{8}{3}\pi^2 \int_0^\infty \kappa^3 \Phi_n(\kappa, \alpha) d\kappa$$

When N = 1, Eq. (11.6) represents the second-order moment beam width of a partially coherent single GSM beam in non-k atmospheric turbulence, which is consistent with Eq. 5 in Ref. [17]. When $\beta \to \infty$, Eq. (11.6) can be reduced to represent the second-order moment beam width of a radial fully coherent GSM array beam in non-k atmospheric turbulence, which is similar to Eq. 8 in Ref. [11]. If the array beam propagates in free space ($M = 0$), Eq. (11.6) can be simplified as:

$$\omega^2(z)\Big|_{free} = P + \frac{4}{\kappa^2} B z^2 \tag{11.7}$$

where P and B have the same definition as in Eq. (11.6).

The Rayleigh interval, z_R, of a beam is defined as the propagation distance when the cross-sectional area is twice that in the $z = 0$ plane, which is also the collimation interval of the beam [19]. From Eq. (11.7), it can be shown that the Rayleigh interval $z_R\big|_{free}^{co}$ of a radially distributed partially coherent GSM synthetic array in free space satisfies the following equation:

$$\omega^2\left(z_R\big|_{free}^{co}\right) = P + \frac{4}{\kappa^2} B z_R^2\big|_{free}^{co} = 2P \tag{11.8}$$

By solving Eq. (11.8), the Rayleigh interval for free space propagation can be obtained as follows:

$$z_R\big|_{free}^{co} = \frac{k}{2}\sqrt{\frac{P}{B}} \tag{11.9}$$

Similarly, it can be concluded that the Rayleigh interval of the radially distributed partially coherent GSM array beam propagating through non-k atmospheric turbulence is given as:

$$z_R\big|_{co} = \frac{1}{3M}\left(C + \frac{16B^2}{\kappa^4 C} - \frac{4}{\kappa^2} B\right) \tag{11.10}$$

where, $C = \left[\frac{27}{2} M^2 P - \frac{64}{\kappa^6} B^3 + \frac{3}{2} M \left(81 M^2 P^2 - \frac{768 B^3 P}{\kappa^6}\right)^{1/2}\right]^{1/3}$.

When the phase is not locked, the radially distributed partially coherent GSM array beams are incoherent composites, and the average intensity of the composite beams can be expressed as [23]:

$$I(\boldsymbol{\rho}, z) = \sum_{i=1}^{N} I_i(\boldsymbol{\rho}, z) \tag{11.11}$$

where $I_i(\boldsymbol{\rho}, z)$ represents the intensity distribution of the *i-th* GSM beam in the radially distributed partially coherent array beam at the receiving plane $z = 0$.

Substituting Eq. (11.11) into Eq. (11.5), it can be shown that the second-order moment beam width of the incoherent synthetic radially distributed partially coherent GSM array beam propagating in non-k atmospheric turbulence is given by:

$$\omega^2(z)\big|_{inco} = P' + \frac{4}{\kappa^2}B'z^2 + Mz^3 \tag{11.12}$$

where

$$P' = \omega_0^2 \tag{11.12a}$$

$$B' = \frac{1}{\omega_0^2}\left(1 + \frac{1}{\beta^2}\right) \tag{11.12b}$$

According to Eq. (11.12) and the definition of the Rayleigh interval, it is shown that this quantity for an incoherent synthetic radially distributed partially coherent GSM array beam in non-k atmospheric turbulence is given by:

$$z_R\big|_{inco} = \frac{1}{3M}\left(C' + \frac{16B'^2}{\kappa^4 C'} - \frac{4}{\kappa^2}B'\right) \tag{11.13}$$

where C' can be obtained by substituting P' and B' for P and B in Eq. (11.10). Similarly, the Rayleigh interval for free space propagation of the GSM array beam is given by:

$$z_R\big|_{free}^{inco} = \frac{k}{2}\sqrt{\frac{P'}{B'}} \tag{11.14}$$

11.2.2 Turbulence Distance

The turbulence distance z_T is a physical quantity used to quantitatively describe the influence of turbulence on beam propagation. It is defined as the beam propagation distance at which turbulence causes the beam cross-sectional area to expand by 10%, which is defined as [18]:

$$\frac{\omega^2(z_T) - \omega^2(z_T)\big|_{free}}{\omega^2(z_T)} = 10\% \tag{11.15}$$

When $z < z_T$, it can be considered that the beam spread is not affected by turbulence, and when $z > z_T$, it is considered that the beam spread is affected by turbulence, i.e., the greater the value z_T, the smaller the impact of atmospheric turbulence on beam spread.

By substituting Eqs. (11.6) and (11.7) into Eq. (11.15), the cubic equation about z_T can be simplified as follows:

$$9Mz_T^3 - \frac{4}{\kappa^2}Bz_T^2 - P = 0 \tag{11.16}$$

The unique real solution of Eq. (11.16) is given by:

$$z_T\big|_{co} = \frac{1}{27M}\left[\frac{1}{\sqrt[3]{2}}\left(Y^{1/3} + \frac{\sqrt[3]{4} \times 16B^2}{\kappa^4 Y^{1/3}}\right) + \frac{4}{\kappa^2}B\right] \tag{11.17}$$

where $Y = \left[27M\left(6561M^2P^2 + \frac{768B^3P}{\kappa^6}\right)^{1/2} + \frac{128B^3}{\kappa^6} + 2187M^2P\right]$.

Equation (11.17) is an analytical expression for the turbulent distance of a coherent synthetic radially distributed partially coherent GSM array beam propagating in non-k atmospheric turbulence.

Similarly, an analytical expression for the turbulent distance of an incoherent synthetic radially distributed partially coherent GSM array beam propagating in non-k atmospheric turbulence is given by:

$$z_T\big|_{inco} = \frac{1}{27M}\left[\frac{1}{\sqrt[3]{2}}\left(Y'^{1/3} + \frac{\sqrt[3]{4} \times 16B'^2}{\kappa^4 Y'^{1/3}}\right) + \frac{4}{\kappa^2}B'\right] \tag{11.18}$$

where Y' can be obtained by substituting P' and B' for P and B in Eq. (11.17).

Combining Eqs. (11.6) and (11.4), it is determined that the turbulence distance of a radially distributed partially coherent GSM array beam in non-k atmospheric turbulence is related to the turbulence generalized index α, the beam coherence parameter β, the turbulence inner and outer scales, the radial distribution radius, the beam waist radius, the turbulence intensity, and the wavelength.

11.3 Numerical Calculation and Analysis

In this section, the variation of the Rayleigh range and the turbulence distance with various parameters of radially distributed partially coherent GSM array beams propagating in non-k atmospheric turbulence is numerically evaluated. The parameter settings are, $\omega_0 = 0.03$ m, $\beta = 0.5$, $l_0 = 0.01$ m, $\mathrm{L}_0 = 10$ m, $N = 16$, $r_0 = 0.02$ m, and $\tilde{c}_n^2 = 1.0 \times 10^{-14}\ \mathrm{m}^{3-\alpha}$.

11.3.1 Comparison of Rayleigh Intervals

Figures 11.2 and 11.3 show the variation curves of the Rayleigh interval z_R and the relative Rayleigh interval z_R / z_{Rfree}, respectively, of radially distributed partially coherent GSM array beams with turbulence generalized index α, where z_R and z_R / z_{Rfree} represent the corresponding Rayleigh interval in atmospheric turbulence and free space, respectively. Figures 11.2 and 11.3 show that both z_R and z_R / z_{Rfree} of the synthetic beam change non-monotonically with the increase of α; there is an initial decrease followed by an increase. Moreover, there is a minimum value at $\alpha = 3.11$, i.e., the collimation range of the optical beam is the smallest, the array beam expands the fastest, and the beam is also most affected by the non-k atmospheric turbulence. Figure 11.2 shows that when $\alpha > 3.11$, with the increase in α, the difference between the Rayleigh range of coherently and incoherently combined partially coherent GSM array beams increases and the coherently combined beam is larger than the incoherently combined Rayleigh range z_R. However, it is also significantly affected by atmospheric turbulence (see Fig. 11.3). Comparing the two curves in Fig. 11.3, the relative Rayleigh interval z_R / z_{Rfree} of the two synthesis modes is < 1, and the closer z_R / z_{Rfree} is to 1, the less the effect of turbulence. Moreover, the value for the non-coherent synthesis of z_R / z_{Rfree} is larger than that obtained for coherent synthesis. This is because the Rayleigh interval z_{Rfree} in free space for coherent synthesis and non-coherent synthesis of partially coherent GSM array beams are 1639.5 m and 1192.9 m respectively. Beam propagation in atmospheric turbulence is determined by two factors: one is the diffraction of light in free space, and the other is the turbulent atmosphere. Physically, the more pronounced the spread of light in free space, the smaller the influence of atmospheric turbulence on beam spread [11]. Therefore, compared to incoherent synthesis, coherent synthesis is more affected by non-k atmospheric turbulence.

Figure 11.4 shows the variation curve of the Rayleigh interval of a radially distributed partially coherent GSM array beam with turbulence generalized exponent α when the radial distribution radius r_0 and the beam coherence parameter β are different. The red line in Fig. 11.4a shows that for coherent synthesis, the larger the radial distribution radius r_0, the larger the Rayleigh interval of the synthetic beam. This is because P in Eq. (11.6) increases with the increase of the radial distribution radius r_0, therefore, the Rayleigh interval also increases. However, for incoherent synthesis, the radial distribution radius r_0 does not affect the Rayleigh interval of the synthetic beam. The Rayleigh interval expression of Eq. (11.12) shows that there is no relationship between z_R and r_0 in the incoherent synthesis mode, i.e., z_R does not change with the change of r_0. Thus, the numerical analysis results are consistent with the theoretical conclusions. By comparing the red line (or blue line) in Fig. 11.4b, it can be determined that the larger the beam coherence parameter β, the larger the

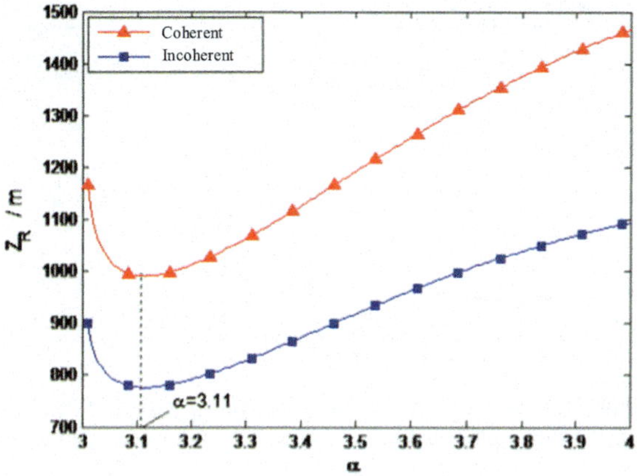

Fig. 11.2 Variation of Rayleigh interval z_R with turbulence generalized index α

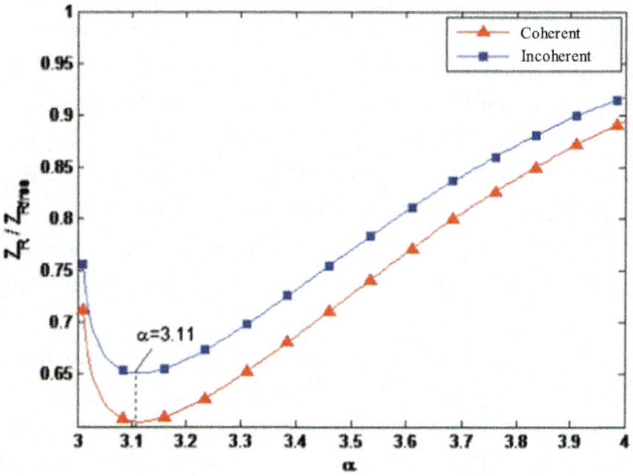

Fig. 11.3 Variation of z_R / z_{Rfree} in relative Rayleigh interval with turbulence generalized index α

Rayleigh range of the composite beam, whether it is a case of coherent synthesis or noncoherent synthesis. When $\beta = 0.2$, the Rayleigh interval of the array beam changes little with the increase of α. Further analysis shows that when β is sufficiently small, the influence of turbulence on the Rayleigh interval is very small, and may be ignored. From the expression for β, the larger the value of this parameter, the better the coherence of most coherent GSM array beams, and the larger the collimation range, which is consistent with the results obtained in Ref. [24].

Figure 11.5 shows the plot of the Rayleigh interval z_R of radially distributed partially coherent GSM array beams versus the turbulence generalized exponent α for different turbulence intensity and beam waist radii. The red line (or blue line) in Fig. 11.5a indicates that for the same synthesis mode, the smaller the turbulence intensity, the larger the Rayleigh interval of the synthetic beam, which indicates that the enhancement of turbulence reduces the Rayleigh interval. The stronger the turbulence, the greater the influence of the non-k atmospheric turbulence on the propagation of the array beam. Moreover, for the same synthesis mode, when the

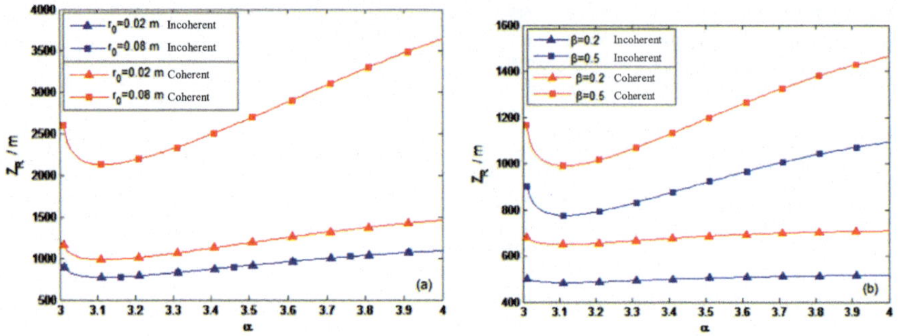

Fig. 11.4 Variation of Rayleigh interval z_R with turbulence generalized exponent α when radial distribution radius and beam coherence parameters are different

turbulence intensity is different, and $\alpha > 3.11$, the difference in the Rayleigh range of partially coherent GSM array beams gradually increases with the increase in α. Compared to the red line (or blue line) in Fig. 11.5b, it can be determined that for the same synthesis mode, the smaller the waist radius of the array beam, the smaller the Rayleigh interval of the composite beam. This implies that an array beam with a small waist radius is more affected by the non-k atmospheric turbulence. However, when $\omega_0 = 0.02$ m, the change in z_R in the Rayleigh region with the increase in α is very small, whether it is coherent or incoherent. When ω_0 is sufficiently small, the influence of non-k atmospheric turbulence on the Rayleigh region is very small and may be ignored.

Figure 11.6 shows the variation curve of the Rayleigh interval z_R with turbulence generalized index α for different internal and external turbulence scales. Comparing Fig. 11.6a with b, it is determined that compared with the outer scale of turbulence, the inner scale of turbulence has a very significant impact on the Rayleigh interval. The red line (or blue line) in Fig. 11.6a shows that for the same synthesis mode, the larger the turbulence internal scale, the larger the Rayleigh interval of the synthetic

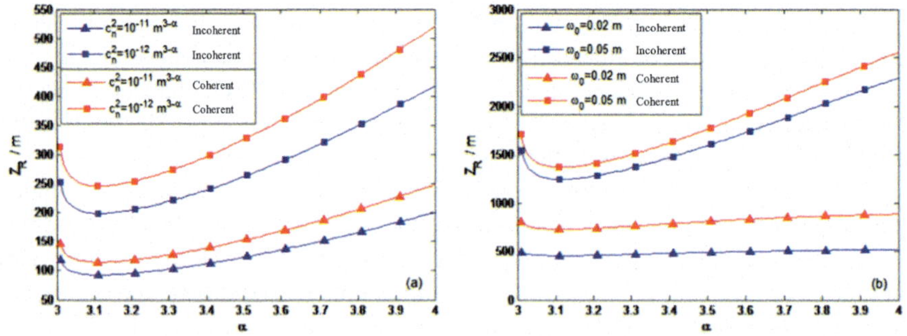

Fig. 11.5 Variation of Rayleigh interval z_R with turbulence generalized index α when turbulence intensity and beam waist radius are different

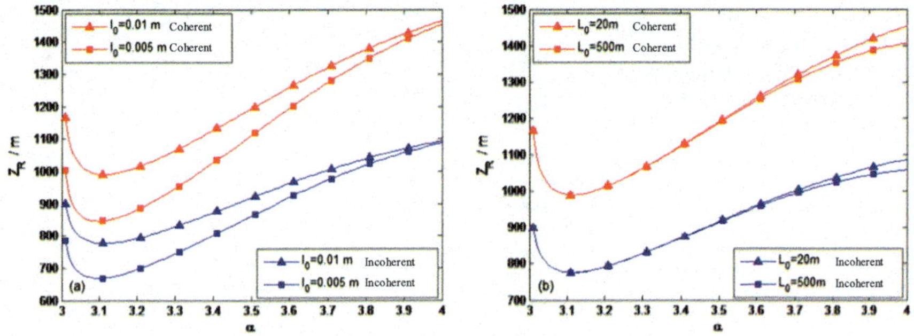

Fig. 11.6 Variation of Rayleigh interval z_R with turbulence generalized index α for different internal and external scales

beam. When $\alpha > 3.11$, the difference in the Rayleigh interval for different turbulence internal scales gradually decreases with the increase in α, which indicates that the larger the turbulence generalized index α, the smaller the impact of the non-k atmospheric turbulence internal scale on the Rayleigh interval of the synthetic beam. The red line (or blue line) in Fig. 11.6b shows that when $0 < \alpha < 3.6$, the external scale does not affect the Rayleigh range of the synthetic beam, whether it is a case of coherent synthesis or incoherent synthesis. Moreover, when $3.6 < \alpha < 4$, the smaller the outer scale of the non-k atmospheric turbulence, the larger the Rayleigh interval of the synthetic beam. As α continues to increase, the difference in the Rayleigh interval for different outer scales increases. Under the same conditions, non-k atmospheric turbulence with larger internal scales and smaller external scales is more conducive to beam propagation. A possible physical mechanism is that the M-Term for non-k atmospheric turbulence in Eq. (11.6) becomes smaller when the internal scale of turbulence increases or the external scale decreases. The weaker the atmospheric turbulence, the smaller the impact on the propagation of the array beam, and the larger the Rayleigh range of the composite beam.

Figure 11.7 shows the variation curve of the Rayleigh interval z_R with turbulence generalized index α for different wavelengths. By comparing the red line (or blue line) in the figure, it can be determined that for the same synthesis mode, the smaller the wavelength of the partially coherent GSM array beam, the larger the Rayleigh range of the synthetic beam. This indicates that the alignment range of an array beam is longer for short wavelengths. Moreover, when the array beam wavelength $\lambda = 1.55 \times 10^{-6}$ m, compared to $\lambda = 0.48 \times 10^{-6}$ m, the Rayleigh range z_R of the synthetic beam changes little with the increase in α, whether it is coherent or incoherent synthesis. Further analysis shows that when the array beam wavelength λ is sufficiently large, the influence of atmospheric turbulence on the Rayleigh range of the synthetic beam is small and can be ignored.

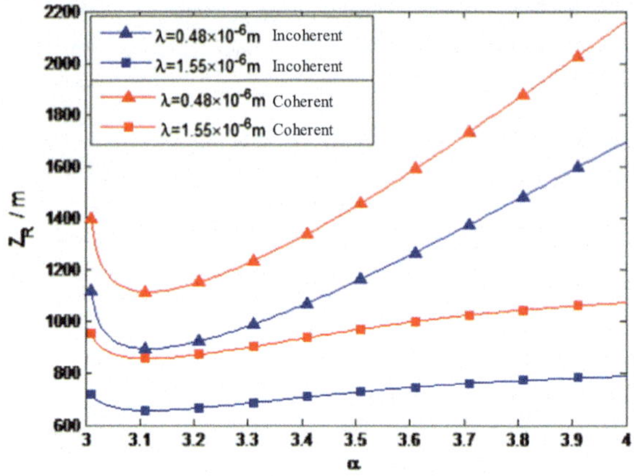

Fig. 11.7 Variation of Rayleigh interval z_R with turbulence generalized index α for different wavelengths

11.3.2 Comparison of Turbulence Distance

Turbulence distance can be used to quantitatively describe the effect of atmospheric turbulence on beam propagation. The shorter the turbulence distance, the greater the influence of turbulence on beam spread. The variation of turbulence distance with the inner and outer scales of atmospheric turbulence, turbulence intensity, radial radius, and beam waist radius is similar to that in the Rayleigh interval, thus, it is not repeated. In the following section, the variation of the turbulence distance of a partially coherent radial GSM array beam is analyzed for non-k turbulence propagation for different wavelengths and beam coherence parameters. Unless otherwise specified, $\omega_0 = 0.04$ m, and other parameters are consistent with Sect. 11.3.1.

Figure 11.8 shows the variation curves for turbulence distance z_T and Rayleigh interval z_R as a function of the turbulence generalized index α. The figure shows that the change of z_T with α is similar to that of z_R, and exhibits a non-monotonic change. A minimum value is obtained at $\alpha = 3.11$, and z_R is larger than z_T under the same transmission conditions, whether it is coherent or incoherent synthesis, which indicates that the0 non-k atmospheric turbulence has an impact on beam expansion in the Rayleigh range. By comparing the blue lines in Fig. 11.8, it can be determined that under the same transmission conditions, the turbulence distance for coherent synthesis is larger than that of incoherent synthesis, and with the increase in α, the difference in the turbulence distance obtained using the two methods increases.

The physical reason for this result is that the turbulence term in Eq. (11.6) is a function of α, m first increases then decreases with the increase of α, and a maximum value is obtained at $\alpha = 3.11$. A maximum turbulence value results in maximum beam spread, i.e., a very small turbulence distance.

Figure 11.9 shows the variation curve of turbulence distance z_T as a function of the turbulence generalized exponent α for different wavelengths and coherent parameters. By comparing the red line (or blue line) in Fig. 11.9a, it can be determined that the longer the wavelength of the array beam, the greater the turbulence distance of

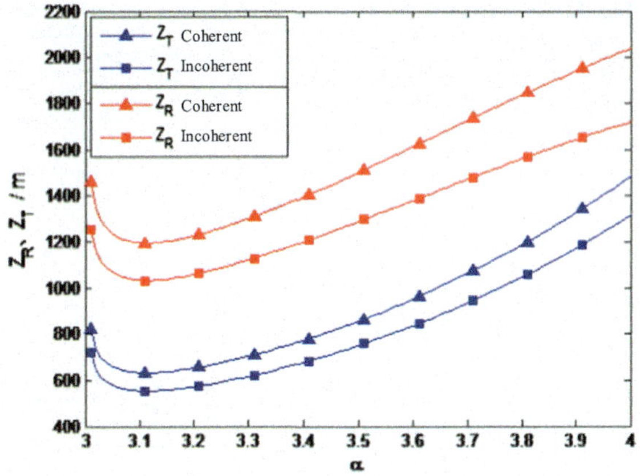

Fig. 11.8 Turbulence distance z_T and Rayleigh range z_R versus generalized parameter α

the composite beam, whether it is coherent or incoherent synthesis. This implies that a long wavelength array beam is less affected by non-k atmospheric turbulence, which is conducive to propagation in atmospheric turbulence. Figure 11.9b shows that for the same combination mode, the larger the beam coherence parameter, the smaller the turbulence distance of the combined beam, indicating that an array beam with low coherence has a long turbulence distance. When $\alpha > 3.11$, the difference in the turbulence distance of the combined beam with different beam coherence parameters increases with the increase of α, which is different from the change in the Rayleigh interval with β. Further analysis of Fig. 11.9b shows that the turbulence distance of partially coherent GSM array beams in non-k atmospheric turbulence is larger compared to that of fully coherent GSM array beams. This indicates that partially coherent array beams are better able to suppress non-k atmospheric turbulence than fully coherent array beams, which is consistent with the main conclusion in reference [16].

Figure 11.10 shows the variation of z_T / z_R with α for different beam coherence parameters β. The figure shows that the larger the beam coherence parameter, the

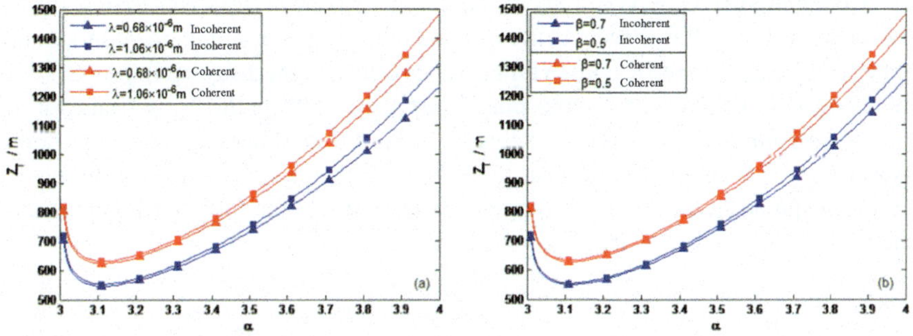

Fig. 11.9 variation of turbulence distance z_T with turbulence generalized index α for different beam wavelengths and coherence parameters

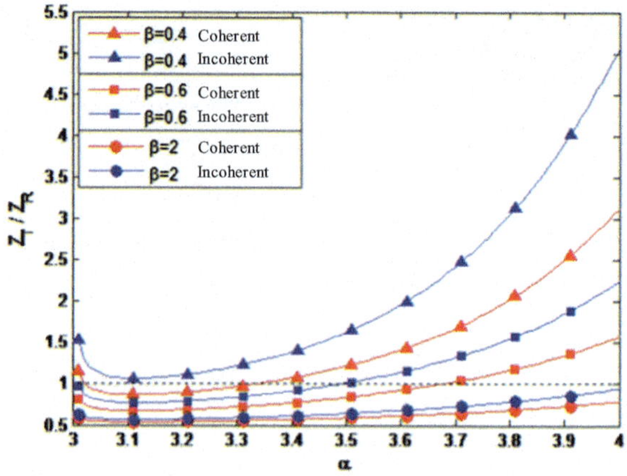

Fig. 11.10 Variation curve of $z_T \big/ z_R$ with α when the beam coherence parameter β is different $\omega_0 = 0.02$ m

smaller the $z_T \big/ z_R$ of the array beam, whether it is a case of coherent synthesis or incoherent synthesis. When $\beta = 2$, $z_T \big/ z_R$ is always less than 1 regardless of the value in α. This indicates that in the Rayleigh range, non-k atmospheric turbulence has an impact on beam propagation. When $\beta = 0.6$, for coherent synthesis, if $3 < \alpha < 3.71$, then $z_T \big/ z_R < 1$, i.e., $z_T < z_R$. Thus, non-k atmospheric turbulence has an impact on beam propagation in the Rayleigh range. If $3.71 < \alpha < 4$, $z_T > z_R$, the non-k atmospheric turbulence has no effect on beam propagation in the Rayleigh range. The trend of incoherent synthesis is similar. When $\beta = 0.4$, for coherent synthesis, the curve $z_T \big/ z_R$ is divided into three parts: $3 < \alpha < 3.02$, $z_T > z_R$; $3.02 < \alpha < 3.33$, $z_T < z_R$, and $3.33 < \alpha < 4$, $z_T > z_R$. This shows that different turbulence generalized index α determines whether non-k atmospheric turbulence can affect the array beam spread in the Rayleigh range. For incoherent synthesis, when $\beta = 0.4$, $z_T \big/ z_R$ is always > 1 regardless of the value of α. This indicates that in the Rayleigh range, non-k atmospheric turbulence has little effect on the spread of an array beam.

Figure 11.11 shows the variation of $z_T \big/ z_R$ as a function of the turbulence generalized index α for different beam waist radii ω_0. The figure shows that the larger the waist radius ω_0, the smaller the $z_T \big/ z_R$ of the array beam, whether it is coherent or incoherent synthesis. When ω_0 is small, $z_T \big/ z_R$ is always greater than 1 for the two array beam synthesis modes, irrespective of the value of α, i.e., the non-k atmospheric turbulence in the Rayleigh range has no effect on the array beam propagation. When ω_0 is large, the non-k atmospheric turbulence affects the array beam for the two beam combination modes irrespective of the value of α in the Rayleigh range. When ω_0 is within a certain range, the influence of non-k atmospheric turbulence on beam spread in the Rayleigh range is determined by the turbulence generalized index α.

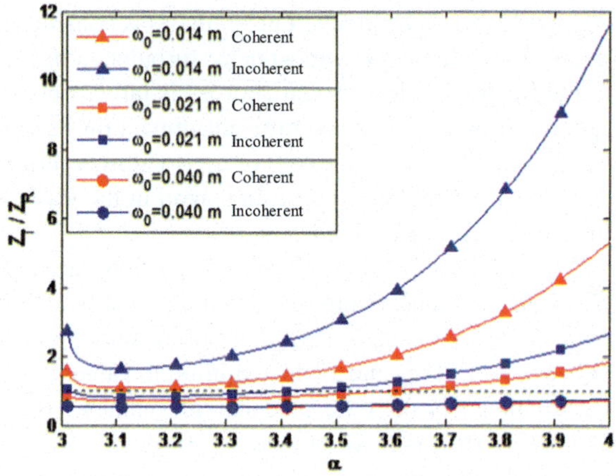

Fig. 11.11 Variation curve of z_T / z_R with α for different waist radii ω_0

11.4 Summary

Analytical expressions for the Rayleigh range z_R and turbulence distance z_T of radially distributed partially coherent GSM array beams propagating in non-k atmospheric turbulence were derived. A numerical analysis of the changes in the Rayleigh range z_R and the turbulence distance z_T with turbulence and beam parameters was performed. The main conclusions are as follows:

(1) In the cases of both coherent and incoherent synthesis, z_R and z_T of the radially distributed partially coherent GSM array beams change non-monotonically with the increase of the turbulence generalized exponent α, first decreasing then increasing. When $\alpha = 3.11$, z_R and z_T have minimum values, i.e., the beam spread is maximum. Compared to incoherent synthesis, coherent synthesis is more affected by atmospheric turbulence.

(2) For coherent synthesis, the larger the radial distribution radius r_0, the larger the z_R of the synthetic beam, whereas r_0 has no effect on z_R for incoherent synthesis. When β, ω_0 and l_0 are large and $\tilde{c}_n^2$, λ and L_0 (only when $3.6 < \alpha < 4$) are small, the Rayleigh range z_R of the synthetic beam for the same synthetic mode is large. When β and ω_0 are sufficiently small or λ is sufficiently large, the influence of non-k atmospheric turbulence on the Rayleigh range is small and can be ignored.

(3) Different from the Rayleigh range of a radially distributed partially coherent GSM array beam, when the synthesis mode is the same, the longer the wavelength λ of the array beam or the smaller the beam coherence parameter β, the larger the turbulence distance of the combined beam. Moreover, the partially coherent array beam is better able to suppress non-k atmospheric turbulence than the fully coherent array beam.

The size of z_R and z_T are compared under different beam coherence parameters β, beam waist radius ω_0, and turbulence generalized index α. The results show that whether coherent or incoherent synthesis is considered, if β or ω_0 is small, irrespective of the value of α, atmospheric turbulence has no effect on beam propagation in the Rayleigh range. When β or ω_0 is large, irrespective of the value of α, atmospheric turbulence has a significant impact on beam spread in the Rayleigh range. Moreover, when the value of β or ω_0 is within a certain range, the influence of non-k atmospheric turbulence on beam spread in the Rayleigh range is determined by the turbulence generalized index α.

The results show that the z_R and z_T of the radially distributed partially coherent GSM array beams change non-monotonically with the increase of the turbulence generalized exponent α, regardless of the cases of coherent or incoherent synthesis. When $\alpha = 3.11$, z_R and z_T have minimum values, and the beam spread is maximum. Beam spread in the case of coherent synthesis is smaller than that of incoherent synthesis, but it is more affected by turbulence. When the beam coherence parameter β and the beam waist radius ω_0 are sufficiently small or the wavelength λ is sufficiently large, the influence of atmospheric turbulence on the array beam z_R can be ignored. It is worth noting that, regardless of coherent or incoherent synthesis, if β or ω_0 is small, irrespective of the value of α, atmospheric turbulence does not affect array beam spread in the Rayleigh range. When β or ω_0 is large, regardless of the value of α, atmospheric turbulence has a significant impact on the array beam spread in the Rayleigh range. Finally, when β or ω_0 is within a certain range, the influence of non-k atmospheric turbulence on beam spread in the Rayleigh range is determined by α.

References

1. Ji X, Li X (2008) Influence of turbulence on coherent and incoherent combination of off-axis Gaussian beams. Acta Phys Sin 57(12):7674–7679
2. Zhou P, Liu Z, Xu X (2009) Influence of atmospheric turbulence on far-field beam quality of coherent and incoherent combined laser. Chin J Lasers 36(5):1042–1046
3. Geng C, Zhao B, Zhang E et al (2013) 15 kW incoherent beam combining of four fiber lasers using adaptive fiber-optics collimators. IEEE Photon Technol Lett 25(13):1286–1289
4. Lu F, Han X, Spatial coherence properties of GSM array beams in turbulent atmosphere. Infrared Laser Eng
5. Li X, Ji X, Eyyuboğlu HT et al (2010) Turbulence distance of radial Gaussian Schell-model array beams. Appl Phys B 98(2):557–565
6. Ai Y, Dan Y (2011) Range of turbulence-negligible propagation of Gaussian Schell-model array beams. Opt Commun 284(13):3216–3220
7. Liu F, Ji X (2011) Turbulence distance of partially coherent Cosh-Gaussian array beams. Chin J lasers 38(7):225–230
8. Lu L, Ji X, Li X et al (2014) Influence of oceanic turbulence on propagation characteristics of Gaussian array beams. Optik 125(24):7154–7161
9. Miaomiao T, Daomu Z (2015) Regions of spreading of Gaussian array beams propagating through oceanic turbulence. Appl Opt 54(11):3407–3411
10. Wu X, Huang Y, Mei H et al (2014) Measurement of Non-Kolmogorov turbulence characteristic parameter in atmospheric surface layer. Acta Opt Sin 34(6):0601001
11. Lu L, Ji X, Deng J et al (2014) Influence of non-Kolmogorov turbulence on the spreading of Gaussian array beams. Acta Phys Sin 63(1):014207

12. Toselli I, Andrews LC, Phillips RL et al (2007) Angle of arrival fluctuations for free space laser beam propagation through non-Kolmogorov turbulence. SPIE 6551:65510E
13. Deng J, Ji X, Pu Z et al (2013) Influence of polychroism and decentrationon spreading of laser beams propagating in non-Kolmogorov turbulence. Opt Commun 301:19–28
14. Yongping H, Zenghui G, Bin Z (2013) The Rayleigh range of elegant Hermite-Gaussian beams in non-Kolmogorov turbulence. Opt Laser Technol 50(2):125–129
15. Ji X, Pu Z, Ji G et al (2014) Regions of spreading of partially coherent beams propagating through non-Kolmogorov turbulence. Opt Commun 322(322):109–113
16. Huang Y, Duan Z, He Y et al (2014) Change of turbulence distance for partially coherent flat-topped beams propagating through non-Kolmogorov turbulence. Chin J Lasers 41(9):245–250
17. Tang M, Li B, Wang Y et al (2014) Influence of Non-Kolmogorov turbulence on the Rayleigh range and turbulence distance of Gaussian-Schell beams. Chin J Lasers 41(10):1013002
18. Chen X, Li X, Li B (2015) Turbulence distance of partially coherent flat-topped beams in non-Kolmogorov turbulence. High Power Laser Particle Beams 27(1):33–38
19. Yang T, Ji X (2015) Regions of spreading of laser beams propagating through non-Kolmogorov atmospheric turbulence. Acta Opt Sin 35(5):0501001
20. Tang M, Li B (2015) Turbulence distance of cosh-Gaussian beams in non-Kolmogorov turbulence. Laser Technol 39(4):581–584
21. Yongping H, Bin Z (2013) Turbulence distance for laser beams propagating through non-Kolmogorov turbulence. J Opt Soc Am A 30(11):2339–2346
22. Ke X, Wang J (2015) Intensity of reflected wave from corner reflector illuminated by partially coherent beam in the atmospheric turbulence. Acta Opt Sin 35(10):1001001
23. Ji X-L (2010) Influence of atmospheric turbulence on the spreading and directionality of radial Gaussian array beams. Acta Phys Sin 59(1):692–698
24. Ji X-L (2011) Influence of turbulence on the Rayleigh range of partially coherent cosh-Gaussian beams. Acta Phys Sin 60(6):064207

Chapter 12
Speckle Characteristics of Partially Coherent Beams

When a beam propagates in atmospheric turbulence, its intensity and phase are randomly modulated, which results in the formation of speckles in the far-field. In this chapter, the Huygens-Fresnel principle and the modified von Karmon spectral model are used to derive expressions for the effective radius and the average speckle radius at the receiving end for a partially coherent Gaussian Schell model (GSM) beam as it propagates through atmospheric turbulence. The effects of the light source parameters and atmospheric turbulence on the effective radius and average speckle radius are analyzed based on numerical calculations.

12.1 Introduction

When a beam propagates in atmospheric turbulence, it is affected by the refractive index fluctuations caused by atmospheric turbulence, and other related effects such as beam drift, light intensity fluctuation, and spot jitter [1–4]. The effects and impact of atmospheric turbulence on free-space optical communication, lidar, and laser imaging, are of practical interest to engineers [5–7]. In particular, in wireless laser communication systems, an analysis of the characteristics of light spots in an atmospheric turbulence environment is important in the design of APT (acquisition, pointing, tracing) systems [8]. Therefore, it is necessary to investigate the changes in light beam characteristics under the influence of atmospheric turbulence.

Atmospheric turbulence randomly modulates the intensity and phase of a propagating beam, and speckle is eventually formed in the far-field. The speckle field is divided into two types according to the light path [9]: one is formed by propagation in free space, which is called objective speckle, and the other is formed during lens imaging, which is called subjective speckle. This chapter focuses on objective speckle. In recent years, research on objective speckle has mainly focused on the scattering characteristics of a target. For instance, Wu et al. investigated the statistical

X. Ke, *Partially Coherent Optical Transmission Theory in Optical Wireless Communication*, Optical Wireless Communication Theory and Technology,
https://doi.org/10.1007/978-981-97-7158-5_12

characteristics of speckle at the receiving plane when a laser was used to irradiate a far-field target [10]. Xing et al. theoretically investigated the scattering statistical characteristics of a Gaussian Schell beam in atmospheric turbulence after transmission through a diffused target [11–13]. However, in space optical communication systems, the effects of atmospheric turbulence on speckle and the light intensity flicker associated with beam propagation directly affect the spot-centroid detection accuracy of an APT system. Therefore, it is necessary to conduct a detailed theoretical analysis of the characteristics of beam speckle caused by atmospheric turbulence as a theoretical reference for the design of space optical communication systems.

In this section, using the partially coherent GSM (Gaussian-Schell model) beam as the model, expressions are derived for the effective radius and the average speckle radius of a partially coherent beam in atmospheric turbulence, where the speckle radius refers to the statistical mean value. In addition, the effects of the light source and the atmospheric turbulence parameters on the effective radius and average speckle radius are systematically analyzed.

12.2 Average Speckle Radius

The cross-spectral density function (CSDF) of partially coherent GSM collimated beams at L = 0 can be expressed as [14]:

$$\begin{aligned} W_0(\mathbf{r}_1, \mathbf{r}_2; 0) &= \langle u_0(\mathbf{r}_1) u_0^*(\mathbf{r}_2) \rangle \\ &= \exp\left(-\frac{\mathbf{r}_1^2 + \mathbf{r}_2^2}{w_0^2} \right) \exp\left(-\frac{|\mathbf{r}_1 - \mathbf{r}_2|^2}{2\sigma_g^2} \right) \end{aligned} \tag{12.1}$$

where $\mathbf{r} = (x, y)$ is a vector that is perpendicular to the beam propagation direction at the transmitting end, w_0 is the beam waist radius of the source beam, σ_g denotes the coherent length of the source field, and $\langle \cdot \rangle$ represents the statistical average. When σ_g increases, the degree of coherence of the beam increases. When σ_g decreases, the degree of coherence of the beam decreases. When $\sigma_g \to \infty$, the source field is a completely coherent beam. The transmission diagram for a partially coherent GSM beam from the z = 0 plane to the z > 0 plane under atmospheric turbulence is shown in Fig. 12.1.

Assuming that the transmitting source and target are small relative to the transmission path, according to the Huygens-Fresnel principle [11], the CSDF of the GSM beam uplink of the field at the target can be expressed as:

$$\begin{aligned} W(\rho_1, \rho_2; L) &= \langle u_i(\rho_1) u_i^*(\rho_2) \rangle \\ &= \left(\frac{k}{2\pi L} \right)^2 \int d\mathbf{r}_1 \int d\mathbf{r}_2 \langle u_0(\mathbf{r}_1) u_0^*(\mathbf{r}_2) \rangle \end{aligned}$$

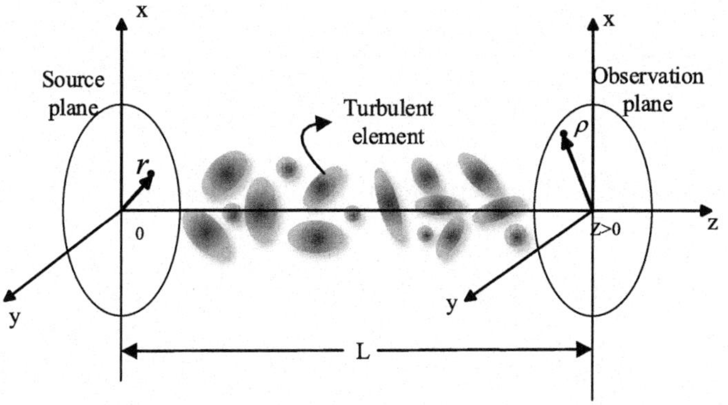

Fig. 12.1 Schematic diagram of the transmission path in atmospheric turbulence

$$\exp\left[\frac{ik}{2L}\left(|\rho_1 - \mathbf{r}_1|^2 - |\rho_2 - \mathbf{r}_2|^2\right)\right]$$
$$\times \left\langle \exp\left[\psi(\rho_1, \mathbf{r}_1) + \psi^*(\rho_2, \mathbf{r}_2)\right]\right\rangle \quad (12.2)$$

where ρ is a two-dimensional vector on the z > 0 plane, L is the beam propagation distance and $\psi(\rho, \mathbf{r})$ is the random disturbance of the complex phase of a spherical wave that propagates from the transmitting aperture ($\mathbf{r}$; 0) to the point (ρ; L) on the target through the turbulent medium. The expression for $\langle \exp[\psi(\rho_1, \mathbf{r}_1) + \psi^*(\rho_2, \mathbf{r}_2)]\rangle$ is:

$$\left\langle \exp\left[\psi(\rho_1, \mathbf{r}_1) + \psi^*(\rho_2, \mathbf{r}_2)\right]\right\rangle = \exp\left[-\frac{1}{2}D_\psi(\rho_1 - \rho_2, \mathbf{r}_1 - \mathbf{r}_2)\right] \quad (12.3)$$

where $D_\psi(\rho_1 - \rho_2, \mathbf{r}_1 - \mathbf{r}_2)$ is the wave structure function, which can be expressed as:

$$D_\psi(\rho_1 - \rho_2, \mathbf{r}_1 - \mathbf{r}_2) = 8\pi^2 k^2 L \int_0^\infty \kappa \phi_n(\kappa)[1 - J_0(\mathbf{r}_1 - \mathbf{r}_2)\kappa]d\kappa \quad (12.4)$$

where J_0 is the first type of the zero-order Bessel function and $\phi_n(\kappa)$ represents the refractive index spectral function of the turbulent medium. The modified von Karmon spectrum [14] was adopted in this study, and can be expressed as:

$$\phi_n(\kappa) = 0.033 C_n^2 \exp(-\kappa^2/\kappa_m^2)(\kappa^2 + \kappa_0^2)^{-11/6} \quad (12.5)$$

where $\kappa_m = 5.92/l_0$, $\kappa_0 \approx 2\pi/L_0$, and l_0 and L_0 represent the internal and external scales of turbulence, respectively. C_n^2 is atmospheric refractive index structure constant. By simplifying Eq. 12.4, we obtain:

$$D_\psi(\rho_1 - \rho_2, \mathbf{r}_1 - \mathbf{r}_2) = M \cdot (\mathbf{r}_1 - \mathbf{r}_2)^2 \quad (12.6)$$

where,

$$M = 0.0055\pi^2 k^2 L C_n^2 \kappa_0^{1/3} \Gamma(2) U\left(2; \frac{7}{6}; \frac{\kappa_0^2}{\kappa_m^2}\right) \quad (12.7)$$

By substituting Eqs. 12.1 and 12.6 into Eq. 12.2, we obtain:

$$W(\rho_1, \rho_2; L) = \left(\frac{k}{2\pi L}\right)^2 \int d\mathbf{r}_1 \int d\mathbf{r}_2 \exp\left(-\frac{\mathbf{r}_1^2 + \mathbf{r}_2^2}{w_0^2}\right) \exp\left(-\frac{|\mathbf{r}_1 - \mathbf{r}_2|^2}{2\sigma_g^2}\right)$$
$$\times \exp\left[\frac{ik}{2L}\left(|\rho_1 - \mathbf{r}_1|^2 - |\rho_2 - \mathbf{r}_2|^2\right)\right] \cdot \exp\left[-\frac{1}{2}M(\mathbf{r}_1 - \mathbf{r}_2)^2\right] \quad (12.8)$$

If we let $\mathbf{r}_c = \frac{1}{2}(\mathbf{r}_1 + \mathbf{r}_2)$, $\mathbf{r}_d = \mathbf{r}_1 - \mathbf{r}_2$, $\rho_c = \frac{1}{2}(\rho_1 + \rho_2)$, $\rho_d = \rho_1 - \rho_2$ and simplify (12.8), we obtain:

$$W(\rho_1, \rho_2; L) = \frac{k^2}{4\pi^2 L^2} \int d\mathbf{r}_c \int d\mathbf{r}_d \exp\left(-\frac{4\mathbf{r}_c^2 + \mathbf{r}_d^2}{2w_0^2} - \frac{\mathbf{r}_d^2}{2\sigma_g^2}\right) \times$$
$$\exp\left[\frac{ik}{L}\rho_c \cdot \rho_d - \frac{ik}{L}\mathbf{r}_c \cdot \rho_d - \frac{ik}{L}\mathbf{r}_d \cdot \rho_c + \frac{ik}{L}\mathbf{r}_c \cdot \mathbf{r}_d\right] \cdot \exp\left[-\frac{M\mathbf{r}_d^2}{2}\right] \quad (12.9)$$

The two-dimensional Fourier transform of the two-dimensional Gaussian function $g(\boldsymbol{\rho}) = e^{-a\rho^2}$ is given as:

$$Ft\{g(\rho)\} = \int_{-\infty}^{+\infty} \left[e^{-ax^2}\right] e^{-i2\pi w_x x} dx \cdot \int_{-\infty}^{+\infty} \left[e^{-ay^2}\right] e^{-i2\pi w_y y} dy$$
$$= \frac{\pi}{a} \cdot e^{-\frac{\pi^2 w^2}{a}} \quad (12.10)$$

Using Eq. 12.10, the integral of the variables $\mathbf{r}_c$ and $\mathbf{r}_d$ in Eq. 12.9 can be performed to yield the following:

$$W(\rho_1, \rho_2; L) = \frac{w_0^2}{8L^2 A/k^2} \exp\left[\frac{ik}{L}\left(1 - \frac{w_0^2 k^2}{16L^2 A}\right)\rho_c \cdot \rho_d\right]$$
$$\times \exp\left(-\frac{k^2 \cdot 2\rho_c^2}{8L^2 A}\right) \exp\left[-\frac{w_0^2 k^2}{8L^2}\left(1 - \frac{w_0^2 k^2}{32L^2 A}\rho_d^2\right)\right] \quad (12.11)$$

where $A = \frac{1}{2w_0^2} + \frac{1}{2\sigma_g^2} + \frac{M}{2} + \frac{w_0^2 k^2}{8L^2}$.

Setting $\rho_1 = \rho_2 = \rho$, the beam intensity at $z > 0$ is given by:

$$I(\rho; L) = \frac{w_0^2}{8L^2A/k^2}\exp\left(-\frac{2\rho^2}{8L^2A/k^2}\right) \tag{12.12}$$

Therefore, from (12.11), it can be determined that the spot radius at z > 0 is given by:

$$\begin{aligned} w_i &= \sqrt{\frac{8L^2A}{k^2}} \\ &= \sqrt{w_0^2 + \frac{4L^2}{w_0^2k^2} + \frac{4L^2}{\sigma_g^2k^2} + \frac{4L^2M}{k^2}} \end{aligned} \tag{12.13}$$

Considering the CSDF of the beam at z > 0, the complex coherence coefficient [9] can be obtained as follows:

$$\begin{aligned} \mathrm{DOC}(\rho_1, \rho_2; L) &= \frac{|W(\rho_1, \rho_2; L)|}{\sqrt{W(\rho_1, \rho_1; L)W(\rho_2, \rho_2; L)}} \\ &= \exp\left\{\left[-\frac{w_0^2k^2}{8L^2}\left(1 - \frac{w_0^2k^2}{32L^2A}\right) + \frac{k^2}{16L^2A}\right]\cdot\rho_d^2\right\} \end{aligned} \tag{12.14}$$

When the complex coherence coefficient of the beam reaches $1/e$, the corresponding beam radius is called the average speckle radius and is given by:

$$\begin{aligned} \rho_{speckle} &= \frac{1}{\sqrt{\frac{w_0^2k^2}{8L^2}\left(1 - \frac{w_0^2k^2}{32L^2A}\right) - \frac{k^2}{16L^2A}}} \\ &= \frac{4\sqrt{2}L}{\sqrt{32L^2w_0^2A - w_0^4k^2 - 16L^2}}\cdot w_i \end{aligned} \tag{12.15}$$

12.3 Numerical Analysis

Using Eqs. 12.14 and 12.15, the effective radius of a beam and the average speckle radius can be theoretically calculated and analyzed. The parameters used were beam waist radius $w_0 = 0.01$ m, wavelength $\lambda = 650$ nm, beam coherence length $\sigma_g = 0.01$ m, turbulence external scale $L_0 = 10$ m, turbulence internal scale $l_0 = 0.01$ m, and atmospheric refractive index structure constant $C_0 = 1.7 \times 10^{-14}\,\mathrm{m}^{-2/3}$.

12.3.1 Influence of Light Source Parameters

The influence of the waist radius, wavelength, and coherent length of the light source on the effective radius and average speckle radius was analyzed, and the influence of the waist radius is shown in Fig. 12.2. The influence of the source wavelength on the effective beam radius and the average speckle radius is shown in Fig. 12.3. Finally, the influence of the coherent length of the light source on the effective radius and average speckle radius of the beam is shown in Fig. 12.4.

Figure 12.2 shows the changes in the effective radius w, the average speckle radius $\rho_{speckle}$, and the ratio $\rho_{speckle}/w$ of the two at the receiving end as a function of the transmission distance for different waist radii w_0 of the light source. Figure 12.2a shows that the smaller w_0, the faster w changes with the increase of the transmission distance, i.e., the smaller w_0, the faster the beam expands. In Fig. 12.2b, the smaller the value of w_0, the faster $\rho_{speckle}$ changes at the receiving end with the increase of the transmission distance, which is consistent with the change in the effective radius of the beam. Figure. 12.2c shows that when w_0 is large, for a transmission distance

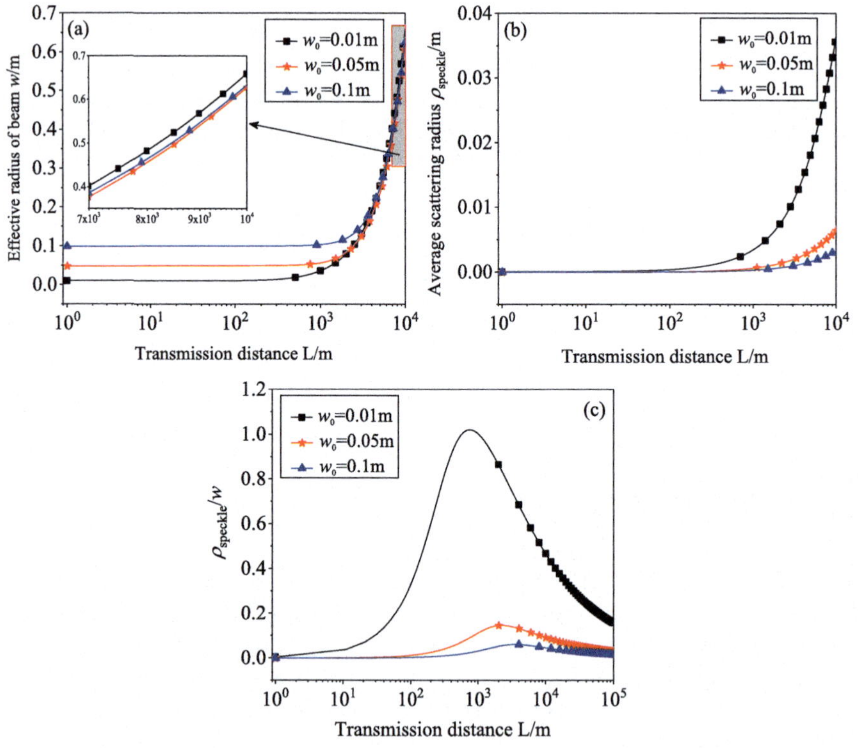

Fig. 12.2 Beam characteristics **a** effective radius at the receiving end with different beam waist radius; **b** average speckle radius; **c** the ratio of the average speckle radius to the effective radius of the beam

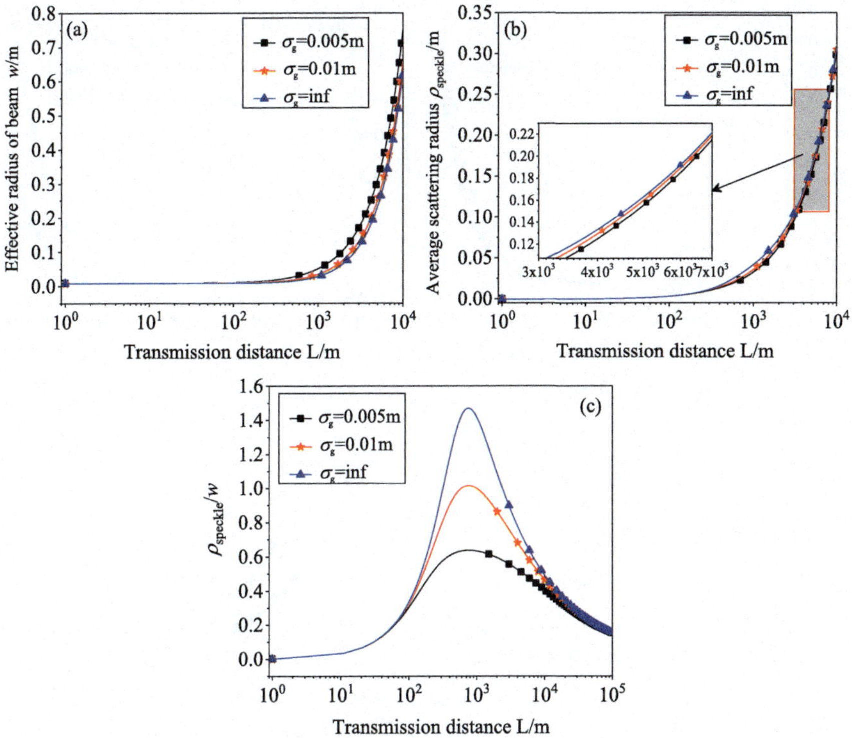

Fig. 12.3 Beam characteristics at the receiving end for different coherent lengths **a** effective radius of the beam; **b** average speckle radius; **c** the ratio of the average speckle radius to the effective radius of the beam

$L < 1$ km, the beam experiences minimal broadening, and the average speckle half diameter is zero, i.e., the larger the value of w_0, the smaller the impact of turbulence on the beam.

Figure. 12.3 shows the changes in the effective radius w, the average speckle radius $\rho_{speckle}$, and the ratio $\rho_{speckle}/w$ of the two at the receiving end as a function of the transmission distance for different light source coherence lengths σ_g. In this figure, when the coherence length $\sigma_g \to \infty$, the light source is completely coherent. When the coherence length G is small, the coherence of the light source is low. Based on an analysis of Fig. 12.3a–c, when $\sigma_g \to \infty$, the maximum value of $\rho_{speckle}/w$ is the smallest, i.e., when $\sigma_g \to \infty$, the average speckle radius $\rho_{speckle}$ is very close to that of the effective radius w of the beam. Thus, for the same transmission distance, the better the coherence of the beam, the greater the average speckle radius of the beam at the receiving end, the greater the beam diffusion, and the greater the influence of turbulence.

Figure. 12.4 shows the changes in the effective radius w, the average speckle radius $\rho_{speckle}$, and the ratio $\rho_{speckle}/w$ of the two at the receiving end as a function of the transmission distance for different wavelengths λ. According to Fig. 12.4,

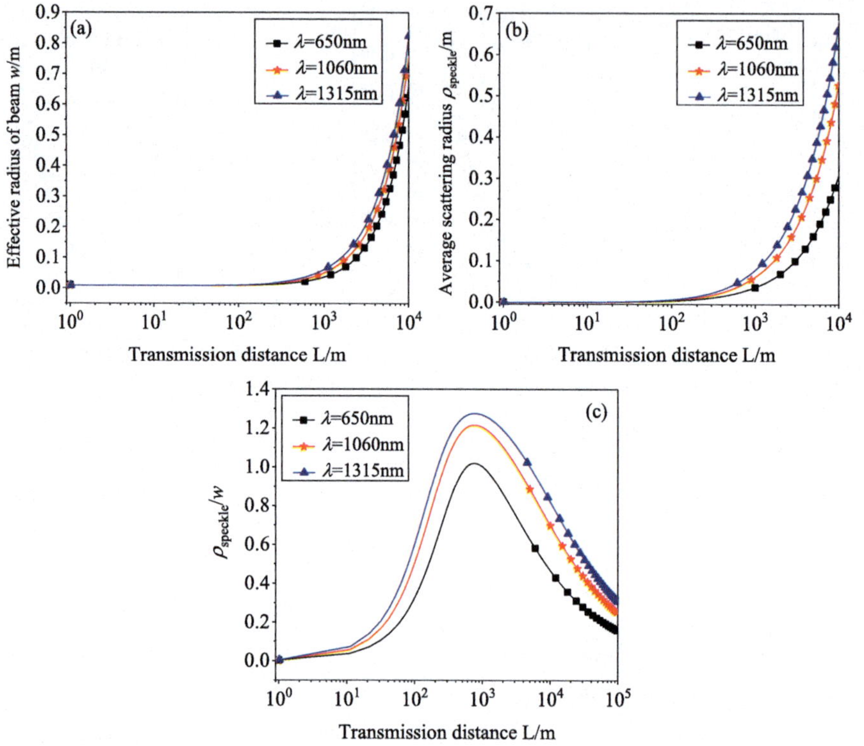

Fig. 12.4 Beam characteristics at the receiving end for different wavelengths **a** effective radius of the beam; **b** average speckle radius; **c** ratio of average speckle radius to effective beam radius

the smaller the wavelength, the smaller the change in the effective radius w with the increase in the transmission distance, and the smaller the change in the average speckle radius $\rho_{speckle}$ with the increase of the transmission distance. This shows that the smaller the wavelength of the beam, the less the influence of turbulence on the propagation through atmospheric turbulence.

12.3.2 Impact of Atmospheric Turbulence

The effects of the atmospheric refractive index structure constant and the internal and external scales of turbulence on the effective radius and average speckle radius are analyzed. The influence of atmospheric refractive index structure constant the effective radius of the beam and the average speckle radius is shown in Fig. 12.5, and the influence of the internal and external scales of turbulence on the effective radius of the beam and the average speckle radius is shown in Fig. 12.6.

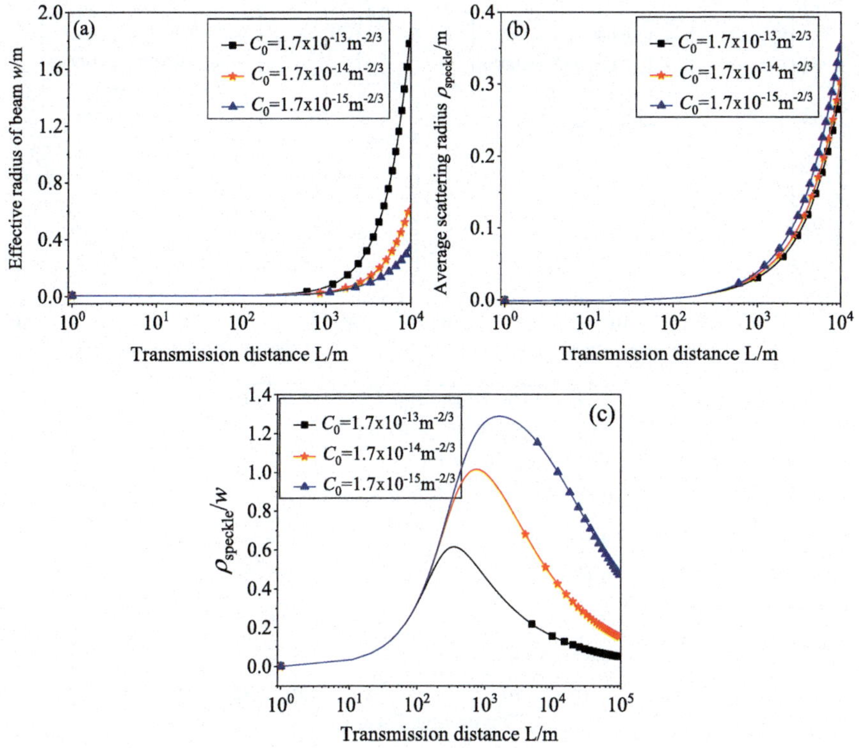

Fig. 12.5 Beam characteristics at the receiving end for different atmospheric refractive index structure constants **a** effective radius of the beam; **b** average speckle radius; **c** ratio of average speckle radius to effective beam radius

Figure. 12.5 shows the changes in the effective radius w, the average speckle radius $\rho_{speckle}$, and the ratio $\rho_{speckle}/w$ of the two at the receiving end as a function of the propagation distance for different atmospheric refractive index structure constants C_0. This figure shows that when C_0 is large, i.e., the turbulence intensity is large, w changes rapidly with the increase of the transmission distance, whereas $\rho_{speckle}$ changes slowly. This is because when the beam is transmitted in atmospheric turbulence, the random fluctuation of the atmospheric refractive index causes phase distortion of the beam, resulting in beam expansion after a certain distance, and more scattered light spots at the receiving end. Thus, the speckle radius is relatively small.

Figure. 12.6 shows the variation of the effective radius w, the average speckle radius $\rho_{speckle}$, and the ratio $\rho_{speckle}/w$ of the two at the receiving end as a function of the propagation distance for different turbulent inner scales l_0 and outer scales L_0. This figure shows that for the same external scale L_0, the larger the internal scale, the slower the change of the effective radius w of the receiving end beam with the transmission distance, and the change of the average speckle radius $\rho_{speckle}$ with the transmission distance is relatively fast. In the case of different external scales L_0, the curves of both w and $\rho_{speckle}$ with transmission distance are similar.

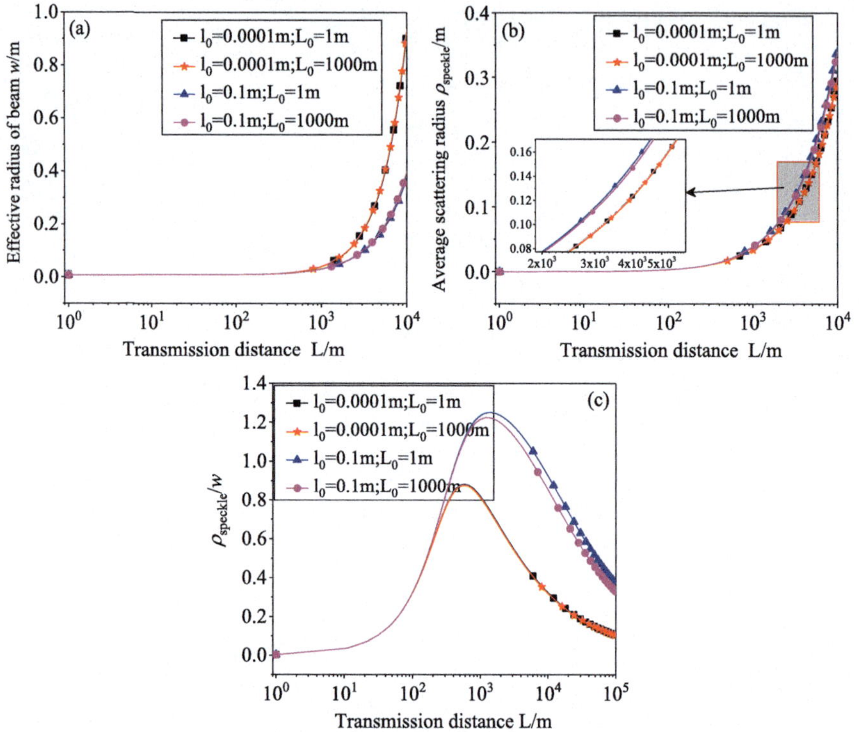

Fig. 12.6 Beam characteristics at the receiving end for different internal and external scales of turbulence **a** effective radius of the beam; **b** average speckle radius; **c** ratio of average speckle radius to effective beam radius

The results indicated that the effective radius and the average speckle radius of the beam are significantly affected by the inner scale of turbulence, and the influence of the outer scale of turbulence is negligible. This is because as the turbulent internal scale becomes smaller, more small turbulent vortices are contained in the beam cross-section. Thus, beams that are incident on these vortices experience significant diffraction, which increases the random distribution of the beam intensity in space and time, resulting in more scattered light spots.

12.4 Summary

The effective radius and average speckle radius of partially coherent beams in atmospheric turbulence have been theoretically investigated. The effects of the light source parameters and atmospheric turbulence on the effective radius and average speckle radius were also numerically analyzed, and the main conclusions are as follows:

(1) When the waist radius of the light source is large, the coherent length is small, and the wavelength is short. Consequently, the effective radius and average speckle radius of the receiving beam are less affected by turbulence.
(2) When the atmospheric refractive index structure constant is large, i.e., the turbulence intensity is stronger, the effective radius of the beam is larger, the propagation is more serious, and the average speckle radius is small.
(3) When the beam propagates in atmospheric turbulence, the effective radius and average speckle radius at the receiving end are significantly affected by the inner scale of turbulence, whereas the outer scale of turbulence has little effect.

References

1. Shirai T, Dogariu A, Wolf E (2003) Mode analysis of spreading of partially coherent beams propagation through atmospheric turbulence. J Opt Soc Am A 20(6):1094–1102
2. Wei HY, Wu ZS (2008) Study on the effect of laser beam propagation on the slant path through atmospheric turbulence. J Electromagn Waves Appl 22(5–6):787–802
3. Wei L, Liu LR, Sun JF et al (2007) Change in degree of coherence of partially coherent electromagnetic beams propagating through atmospheric turbulence. Opt Commun 271:1–8
4. Wang SCH, Plonus MA (1979) Optical beam propagation for a partially coherent source in the turbulent atmosphere. J Opt Soc Am 69(9):1297–1304
5. Andrews LC, Phillips RL (2001) Monostatic Lidar in weak-to-strong turbulence. Waves Random Media 11:233–245
6. Belmonte A (2005) Coherent return turbulent fluctuations in ground Lidar systems profiling along slant paths. Opt Express 13(23):9598–9604.
7. Wu DS, Bai YZ, Liu BQ et al (2013) Dynamic imaging detection and target recognition for cat-eye effect echo. Acta Optica Sinica 33(8):0811003-1–0811003-6
8. Fenghua SHAN, Shoufeng TONG, Chunlei LV (2013) Beacon detection technology of APT system in free space optical communications. J Changchun Univ Sci Technol 36(3–4):53–56
9. Xiong L, Zhang G, Ge J (2008) Measurement of laser speckle statistical radius based on self-correlation function. Opt Optoelectron Technol 6(1):59–62
10. Wu Y, Wu Z (2011) Study on statistical characteristics of rough surfaces scattering in double transmission. Laser Technol 35(2):234–239
11. Ningjing X, Zhensen W, Xuexia H, Mingjun W et al (2014) Statistical properties of Gaussian-Schell beam from diffuse target in turbulent atmosphere. High Power Laser Particle Beams26(2):021003-1–021003-7
12. Xiang J, Wu Z, Hua X et al (2014) Statistical properties of a Gaussian-Schell model beam from a semi-rough target in the atmospheric turbulence. Chin Sci Phys Mech Astron 44(5):457–464
13. Wei H-Y, Wu Z-S, Peng H (2008) Scattering from a diffuse target in the slant atmospheric turbulence. Acta Phys Sin 57(10):6666–6672
14. Andrews LC, Phillips RL (2005) Laser beam propagation through random media. SPIE Press, Bellingham, pp 671–680

100 Important Papers on Partially Coherent Light100

1. Xiao X, Voelz DG (2012) Beam wander analysis for focused partially coherent beams propagating in turbulence. Opt Eng 51(2):026001
2. Tofsted H (1992) Outer-scale effects in beam-wander and angle-of-arrival variances. Appl Opt 31(27):5865–5870
3. Xiang J, Wu Z, Wang M (2013) Spreading and wander of Gaussian-Schell model beam propagation through atmospheric turbulence. Infrared Laser Eng 42(3):658–662
4. Recolons J, Andrews LC, Phillips RL (2007) Analysis of beam wander effects for a horizontal-path propagating Gaussian-beam wave: focused beam case. Opt Eng 46(8):086002
5. Fante RL (1980) Electromagnetic beam propagation in turbulent media: an update. Proc IEEE 68(11):1424–1443
6. Mironov VL, Nosov V (1977) On the theory of spatially limited light beam displacements in a randomly inhomogeneous medium. J Opt Soc Am 67(8):1073–1080
7. Hill RJ (1981) Theory of saturation of optical scintillation by strong turbulence-plane-wave variance and covariance and spherical-wave covariance. J Opt Soc Am A 72(2):212–222
8. Hill RJ, Clifford SF (1981) Theory of saturation of optical scintillation by strong turbulence for arbitrary refractive-index spectra. J Soc Am A 71(6):675–686
9. Wu Z (1986) Intensity dependence and scintillation in saturated state of wave propagation in random media. J Radio Propag 1(1):72–83
10. Flatte SM, Gerber JS (2000) Irradiance-variance behavior by numerical simulation for plane-wave and spherical-wave optical propagation through strong turbulence. J Opt Soc Am A 17(6):1092–1097
11. Wang SCH, Plonus MA (1979) Optical beam propagation for a partially coherent source in the turbulent atmosphere. J Opt Soc Am A 69(9):1297–1304
12. Yixin Z, Zhengfang S (1986) Fluctuation of laser beam propagation direction in turbulent atmosphere. Acta Optica Sinica 6(12):1111–1117
13. Whitman AM, Beran MJ (1970) Beam spread of laser light propagating in a random medium. J Opt Soc Am A 60(12):1595–1602
14. Poirier JL (1972) Beam spreading in a turbulent medium. J Opt Soc Am A 62(7):893–898
15. Fante RL (1975) Electromagnetic beam propagation in turbulent media. Proc IEEE 63(12):1669–1688
16. Fan CY (1999) The effect of Gaussian-beam wave in slant propagation on atmospheric turbulence. Chin J Quantum Electron 16(6):519–525
17. Andrews LC (1995) Single-pass and double-pass propagation through complex paraxial optical systems. J Opt Soc Am A 12(1):137–150
18. Andrews LC, Miller WB (1995) The mutual coherence function and the backscatter amplification effect for a reflected Gaussian-beam wave in atmospheric turbulence. Wave Random Media 5(2):167–182

X. Ke, *Partially Coherent Optical Transmission Theory in Optical Wireless Communication*, Optical Wireless Communication Theory and Technology,
https://doi.org/10.1007/978-981-97-7158-5

19. Andrews LC, Philips RL, Al-Halash MA (1999) Theory of optical scintillation. J Opt Soc Am A 16(6):1417–1429
20. Andrews LC, Phillips RL, Hopen CY et al (2001) Theory of optical scintillation: Gaussian-beam wave model. Wave Random Media 11(2):271–291
21. Andrews LC, Phillips RL, Miller WB (1997) Mutual coherence function for a double-passage retroreflected atmospheric turbulence. App Opt 36(3):698–708
22. Xiaoxin Z, Youquan D, Bin Z (2012) Spreading of partially coherent flat-topped beams propagating along a slant path in turbulent atmosphere. Acta Optica Sinica 32(12):8–14
23. Duan M, Li J, Wei J (2013) Spreading of partially coherent Hermite-Gaussian beams in slant atmospheric turbulence. High Power Laser Particle Beams 25:2252–2256
24. Leader JC (1978) Atmospheric propagation of partially irradiation. Opt Soc Am 68(2):175–185
25. Ouyang CF. Irradiance scintillation of a partially coherent source in extremely strongly turbulence. Appl Opt 18:1133–1135
26. Wu J (1990) Propagation of a Gaussian-Schell beam through turbulent media. J Mod Opt 37(4):671–684
27. Wu J, Boardman AD (1991) Coherence length of a Gaussian-Schell model beam and atmospheric turbulence. J Mod Opt 38(7):1355–1363
28. Ponomarenko AS, Greffet J, Wolf E (2002) The diffusion of partially coherent beams in turbulent media. Opt Commun 208(1):1–8
29. Gbur G, Wolf E (2002) Spreading of partially coherent bean in random media. J Opt Soc Am A 19(8):1592–1598
30. Ricklin JC, Davidson FM (2002) Atmospheric turbulence effects on a partially coherent Gaussian beam: implications for free-space laser communication. J Opt Soc Am A 19(8):1794–1802
31. Salem M, Shirai T, Sogariu A et al (2003) Long-distance propagation of partially coherent beams through atmospheric turbulence. Opt Commun 216(2):216–265
32. Dogariu A, Arnarande S (2003) Propagation of partially coherent beams: turbulence induced degradation. Opt Lett 28(1):10–12
33. Shirai T, Dogariu A, Wolf E (2003) Mode analysis of spreading of partially coherent beams propagation through atmospheric turbulence optics. J Opt Soc Am A 20(6):1094–1102
34. Borghi R, Guattari G (2003) Evaluation of the spatial coherence of a light beam through transverse intensity measurements. J Opt Soc Am A 20(9):1763–1770
35. Bechmann P (1965) Signal degeneration in laser beams propagated through a turbulent atmosphere. Radio Sci J Res 69D(4):629–640
36. Berman GP, Chumak A, Gorshkov VN (2007) Beam wandering in the atmosphere: the effect of partial coherence. Phys Rev E 76(5):056606
37. Xianmei Q, Wenyue Z, Ruizhong R (2008) Numerical simulation of turbulent effects of laser propagation along a ground-space slant atmospheric path. Infrared Laser Eng 37(5):787–792
38. Yixin Z, Zhengfang S, Zhiben G (1986) Discussion on some problems of beam drift in turbulent atmosphere. Chin J Quant Electron 3(1):57–61
39. Yixin Z, Zhengfang S, Zhiben G (1986) Theory of laser beam displacement in turbulent atmosphere. Acta Optica Sinica 4:373–381
40. Ruizhong R, Shipeng W, Xiaochun L, Zhiben G (2000) Experimental study of spot dancing of laser beam in a turbulent atmosphere. Chin J Lasers 27(11):1011–1015
41. Wang C, Li J (2007) Detection of laser spot drift. J Appl Opt 28(2):205–208
42. Xianmei Q (2010) Numerical simulation for coherent and partially coherent beam propagation through atmospheric turbulence. Chin Phys Lett 27(4):044214
43. Wu GH, Luo B, Yu S et al (2011) The propagation of electromagnetic Gaussian–Schell model beams through atmospheric turbulence in a slanted path. J Opt A Pure Appl Opt 13(3):0357061
44. Li Y, Wu Z (2012) Characteristics of a partially coherent Gaussian Schell-model beam propagating in slanted atmospheric turbulence. Chin Phys B 21(5):054203
45. Gao M, Liu Y, Wang F, Lv H (2014) Spread and wander characteristics of polarized and partially coherent laser beam propagated in turbulent atmosphere. Acta Photonica Sinica 43(10):1001002

46. Belen'kii MS (1994) Effect of residual turbulent scintillation and a remote sensing technique for simultaneous determination of turbulence and scattering parameters of the atmosphere. J Opt Soc Am A 11(3):1150–1158
47. Lutomirski RF, Yura HT (1971) Propagation of a finite optical beam in an inhomogeneous medium. Appl Opt 1(10):1652
48. Xiao XF, Voelz D (2006) Wave optics simulation approach for partial spatially coherent beams. Opt Express 14(16):6986–6992
49. Yura HT (1972) Mutual coherence function of a finite cross section optical beam propagating in a turbulent medium. Appl Opt 11(6):1399–1406
50. Korotkova O, Andrews LC, Phillips RL. Model for a partially coherent Gaussian beam in atmospheric turbulence with application in LaserCom. Opt Eng 43(2):330–341
51. Ji X, Chen X, Lü B (2008) Spreading and directionality of partially coherent Hermite-Gaussian beams propagating through atmospheric turbulence. J Opt Soc Am A 25(1):21–28
52. Ke X, Han M, Wang M (2014) Spreading and wander of partially coherent beam propagating along a horizontal-path in the atmospheric turbulence. Acta Optica Sinica 34(11):78–82
53. Ke X, Han M, Wang M (2015) Spreading and wander of partially coherent beam through atmospheric turbulence in a slanted path. Acta Optica Sinica 44(3):0306001-1–0306001-5
54. Xianmei Q, Wenyue Z, Ruizhong R (2008) Research progress on partially coherent beam propagation in turbulent atmosphere. J Atmosph Environ Opt 3(2):81–90
55. Feizulin ZI, Kravtsov YA (1967) Broadening of a laser beam in a turbulent medium. Radio Phys Quant Electron 10:33–35
56. Leader JC (1976) An analysis of the spatial coherence of laser light scattered from a surface with two scales of roughness. J Opt Soc Am 66(6):536–546
57. Foley JT, Zubairy MS (1978) The directionality of Gaussian Schell-model beams. Opt Commun 26(3):297–300
58. Santarsiero M, Gori F, Borghi R et al (1999) Spreading properties of beams radiated by partially coherent Schell-model sources. J Opt Soc Am A 16(1):106–112
59. Motlagh AC, Ahmadi V, Ghassemlooy Z et al (2008) The effect of atmospheric turbulence on the performance of the free space optical communications. In: International symposium on communication systems, networks and digital signal processing, 540–543
60. Borah DK, Voelz DG (2010) Spatially partially coherent beam parameter optimization for free space optical communications. Opt Express 20(18):20746–20758
61. Chu X, Liu Z, Wu Y (2008) Propagation of a general multi-Gaussian beam in turbulent atmosphere in a slant path. J Opt Soc Am A 25(1):74–79
62. Chu XX, Liu ZJ, Wu Y (2010) Comparison between relay propagation and direct propagation of Gaussian-Schell-model beam in turbulent atmosphere along a slant path. Chin Phys B 19(9):094201-1-6
63. Duan ML, Li JH, Wei JL (2013) Influence of different propagation paths on the propagation of laser in atmospheric turbulence. Opt Elec Lett 9(6):0477–0480
64. Zhang Y, Liu L, Wang F et al (2014) Average intensity and spectral shifts of a partially coherent standard or elegant Laguerre-Gaussian beam beyond paraxial approximation. Opt Quant Electron 46(2):365–379
65. Dan YQ, Zhang B (2008) Beam propagation factor of partially coherent flat-topped beams in a turbulent atmosphere. Opt Express 16(20):15563–15575
66. Wei L, Liu LR, Sun JF et al (2007) Change in degree of coherence of partially coherent electromagnetic beams propagating through atmosphere turbulence. Opt Commun 27(1):1–8
67. Hua W, Xiangzhao W, Aijun Z et al (2007) Effect of atmospheric turbulence on temporal coherence of Gaussian Schell-model beams propagating in slant path. Acta Optica Sinica 27(9):1548–1552
68. Xizheng K, Wanting W (2015) Intensity and expansion of partially coherent beam propagating in slant and horizontal atmospheric turbulence. J Appl Sci 33(2):142–154
69. Yang AL, Zhang ET, Ji XL et al (2008) Angular spread of partially coherent Hermite-Cosh-Gaussian beams propagating through atmospheric turbulence. Opt Express 16(12):8366–8380

70. Xizheng K, Wanting W (2015) Expansion and angular spread of partially coherent beam propagating in atmospheric turbulence. Infrared Laser Eng 44(9):2726–2733
71. Fried DL (1967) Optical heterodyne detection of an atmospherically distorted signal wave front. Proc IEEE 55(1):57–67
72. Fried DL (1967) Atmospheric modulation noise in an optical heterodyne receiver. IEEE Quantum Electron QE-3:213–221
73. Churnside JH, McIntyre CM (1978) Signal current probability distribution for optical heterodyne receivers in the turbulent atmosphere. Theory Appl Opt 17(14):2141–2147
74. Churnside JH, McIntyre CM (1980) Heterodyne receivers for atmospheric optical communications. Appl Opt 19(4):582–590
75. Winick KA (1986) Atmospheric turbulence-induced signal fades on optical heterodyne communication links. Appl Opt 25(11):1817–1825
76. Belmonte A, Kahn JM (2008) Performance of synchronous optical receivers using atmospheric compensation techniques. Opt Express 16(18):14151–14162
77. Okayama H, Li ZW (1999) Measurement of the spatial coherence of light influenced by turbulence. Appl Opt 38(12):2342–2345
78. Belmonte A (2000) Feasibility study for the simulation of beam propagation: consideration of coherent lidar performance. Appl Opt 39(30):5426–5445
79. Morozov AV (2002) Two-frequency mutual coherence function of electromagnetic waves in random media: a path-integral variational solution. J Opt Soc Am A 19(10):2074–2084
80. Cliford F, Yura HT (1974) Equivalence of two theories of Strong optical scintillation. J Opt Soc Am 64:1641–1644
81. Wolf E (1978) Coherence and radiometry. J Opt Soc Am 68:6–17
82. Yura HT (1974) Physical model for strong optical-amplitude fluctuations in a turbulent medium. J Opt Soc Am 64:59–67
83. Wu J, Boardman AD (1991) Coherence length of a Gaussian-Schell model beam and atmospheric turbulence. Mod Opt 38:1355–1368
84. Ponomarenko AS, Greffet JJ, Wolf E (2002) The diffusion of partially coherent beams in turbulent media. Opt Commun 208:1–8
85. Gagliardi R, Robbins J, Taylor H (1987) Acquisition sequences in PPM communication. IEEE Trans Inf Theory IT-33(5):739–745
86. Boroson DM (1993) LITE engineering model-I: operation and performance of the communications and beam-control subsystems. Proc SPIE 1866:73–82
87. Pillsbury AD (1993) LITE engineering model-II: the flight-qualified subsystems. Proc SPIE 1866:58–72
88. Weeks AR, Xu J, Philips RR et al (1998) Experiment verification and theory for an eight-element multiple-aperture equal-gain coherent laser receiver for laser communications. Appl Opt 37(21):4782–4788
89. Xu J, Philips RR, Andrews LC (1999) A new scheme of coherent array detection to cancel out phase fluctuations and Doppler frequency shift due to atmospheric turbulence and target movement for laser communications. Proc SPIE 3615:325–330
90. Arnon S (1999) Power versus stabilization for laser satellite communication. Appl Opt 38(15):3229–3233
91. Vilnrotter VA, Lau CW (2002) Quantum detection and channel capacity for communications applications. Proc SPIE 4635:103–115
92. Biswas A, Wright MW (2002) Mountain-top-to-mountain-top optical link demonstration: part I[R]. IPN progress report, 1–27
93. Biswas A, Wright MW (2002) Mountain-top-to-mountain-top optical link demonstration: part II[R]. IPN progress report, 1–16
94. Yamakawa S, Aroki T, Hisada Y (1999) Trade-off between IM/DD and coherent system in high-data-rate optical interorbit links. Proc SPIE 3615:80–89
95. Yamakawa S, Araki T, Morikawa H (2000) Analysis of noises generated in high-power-operated erbium-doped fiber amplifiers and their influences on IM-DD/Coheren OIC system performance. Proc SPIE 3932:196–204

96. Yamakawa S, Takata N (2003) Coherent lightwave receivers with a laser diode local oscillator for interorbit optical communication. Proc SPIE 4975:69–79
97. Weihui L, Jian W (2004) Impact of partial coherent beams transmitting through strong turbulence on log-intensity variance. Opt Optoelectron Technol 31(1):30–33
98. Yixin Z, Chunkan T (2005) Angle-of-arrival of Gaussian Schell beam propagation in atmospheric turbulence. Acta Photonica Sinica 34(3):424–427
99. Hill RJ (1978) Models of the scalar spectrum for turbulent advection. J Fluid Mech 88(88):541–562
100. Xizheng K, Yu Z (2015) Scintillation of partially coherent beam in atmospheric turbulence. Acta Optica Sinica 35(1):0106001

SPRINGER NATURE

GPSR Compliance

The European Union's (EU) General Product Safety Regulation (GPSR) is a set of rules that requires consumer products to be safe and our obligations to ensure this.

If you have any concerns about our products, you can contact us on ProductSafety@springernature.com

In case Publisher is established outside the EU, the EU authorized representative is:

Springer Nature Customer Service Center GmbH
Europaplatz 3
69115 Heidelberg, Germany

The manufacturer's authorised representative in the EU is Springer Nature Customer Service Centre GmbH, Europaplatz 3, 69115 Heidelberg, Germany. If you have any concerns regarding our products, please contact ProductSafety@springernature.com

Printed and bound by CPI Group (UK) Ltd, Croydon, CR0 4YY
01/10/2026
02225907-0005